Theory & Calculation
Of
Electric Circuits

By

Charles Proteus Steinmetz

First Edition

Watchmaker Publishing

1918

PREFACE

In the twenty years since the first edition of "Theory and Calculation of Alternating Current Phenomena" appeared, electrical engineering has risen from a small beginning to the world's greatest industry; electricity has found its field, as the means of universal energy transmission, distribution and supply, and our knowledge of electrophysics and electrical engineering has increased many fold, so that subjects, which twenty years ago could be dismissed with a few pages discussion, now have expanded and require an extensive knowledge by every electrical engineer.

In the following volume I have discussed the most important characteristics of the fundamental conception of electrical engineering, such as electric conduction, magnetism, wave shape, the meaning of reactance and similar terms, the problems of stability and instability of electric systems, etc., and also have given a more extended application of the method of complex quantities, which the experience of these twenty years has shown to be the most powerful tool in dealing with alternating current phenomena.

In some respects, the following work, and its companion volume, "Theory and Calculation of Electrical Apparatus," may be considered as continuations, or rather as parts of "Theory and Calculation of Alternating Current Phenomena." With the 4th edition, which appeared nine years ago, "Alternating Current Phenomena" had reached about the largest practical bulk, and when rewriting it for the 5th edition, it became necessary to subdivide it into three volumes, to include at least the most necessary structural elements of our knowledge of electrical engineering. The subject matter thus has been distributed into three volumes: "Alternating Current Phenomena," "Electric Circuits," and "Electrical Apparatus."

CHARLES PROTEUS STEINMETZ.

SCHENECTADY,
January, 1917.

v

CONTENTS

SECTION III

**CHAPTER XIV. CONSTANT POTENTIAL CONSTANT CURRENT TRANS-
FORMATION**

CHAPTER XV. CONSTANT POTENTIAL SERIES OPERATION

CHAPTER XVIII. OSCILLATING CURRENTS

THEORY AND CALCULATION OF ELECTRIC CIRCUITS

SECTION I

CHAPTER I

ELECTRIC CONDUCTION. SOLID AND LIQUID CONDUCTORS

1. When electric power flows through a circuit, we find phenomena taking place outside of the conductor which directs the flow of power, and also inside thereof. The phenomena outside of the conductor are conditions of stress in space which are called the electric field, the two main components of the electric field being the electromagnetic component, characterized by the circuit constant inductance, L, and the electrostatic component, characterized by the electric circuit constant capacity, C. Inside of the conductor we find a conversion of energy into heat; that is, electric power is consumed in the conductor by what may be considered as a kind of resistance of the conductor to the flow of electric power, and so we speak of resistance of the conductor as an electric quantity, representing the power consumption in the conductor.

Electric conductors have been classified and divided into distinct groups. We must realize, however, that there are no distinct classes in nature, but a gradual transition from type to type.

Metallic Conductors

2. The first class of conductors are the metallic conductors. They can best be characterized by a negative statement—that is, metallic conductors are those conductors in which the conduction of the electric current converts energy into no other form but heat. That is, a consumption of power takes place in the metallic con-

1

ductors by conversion into heat, and into heat only. Indirectly, we may get light, if the heat produced raises the temperature high enough to get visible radiation as in the incandescent lamp filament, but this radiation is produced from heat, and directly the conversion of electric energy takes place into heat. Most of the metallic conductors cover, as regards their specific resistance, a rather narrow range, between about 1.6 microhm-cm. (1.6×10^{-6}) for copper, to about 100 microhm-cm. for cast iron, mercury, high-resistance alloys, etc. They, therefore, cover a range of less than 1 to 100.

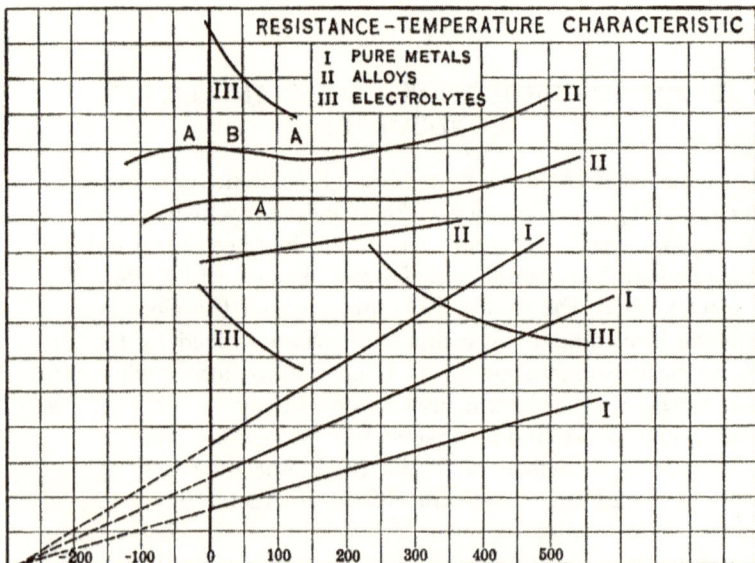

Fig. 1.

A characteristic of metallic conductors is that the resistance is approximately constant, varying only slightly with the temperature, and this variation is a rise of resistance with increase of temperature—that is, they have a positive temperature coefficient. In the pure metals, the resistance apparently is approximately proportional to the absolute temperature—that is, the temperature coefficient of resistance is constant and such that the resistance plotted as function of the temperature is a straight line which points toward the absolute zero of temperature, or, in other words, which, prolonged backward toward falling tem-

perature, would reach zero at $-273°$C., as illustrated by curves
I on Fig. 1. Thus, the resistance may be expressed by

$$r = r_0 T \tag{1}$$

where T is the absolute temperature.

In alloys of metals we generally find a much lower temperature
coefficient, and find that the resistance curve is no longer a straight
line, but curved more or less, as illustrated by curves II, Fig. 1,
so that ranges of zero temperature coefficient, as at A in curve II,
and even ranges of negative temperature coefficient, as at B in
curve II, Fig. 1, may be found in metallic conductors which are
alloys, but the general trend is upward. That is, if we extend the
investigation over a very wide range of temperature, we find that
even in those alloys which have a negative temperature coefficient
for a limited temperature range, the average temperature coefficient is positive for a very wide range of temperature—that is,
the resistance is higher at very high and lower at very low temperature, and the zero or negative coefficient occurs at a local
flexure in the resistance curve.

3. The metallic conductors are the most important ones in
industrial electrical engineering, so much so, that when speaking of a "conductor," practically always a metallic conductor is
understood. The foremost reason is, that the resistivity or
specific resistance of all other classes of conductors is so very
much higher than that of metallic conductors that for directing
the flow of current only metallic conductors can usually come
into consideration.

As, even with pure metals, the change of resistance of metallic
conductors with change of temperature is small—about ⅓ per
cent. per degree centigrade—and the temperature of most apparatus during their use does not vary over a wide range of temperature, the resistance of metallic conductors, r, is usually
assumed as constant, and the value corresponding to the operating temperature chosen. However, for measuring temperature
rise of electric currents, the increase of the conductor resistance
is frequently employed.

Where the temperature range is very large, as between room
temperature and operating temperature of the incandescent lamp
filament, the change of resistance is very considerable; the resistance of the tungsten filament at its operating temperature is about

nine times its cold resistance in the vacuum lamp, twelve times in the gas-filled lamp.

Thus the metallic conductors are the most important. They require little discussion, due to their constancy and absence of secondary energy transformation.

Iron makes an exception among the pure metals, in that it has an abnormally high temperature coefficient, about 30 per cent. higher than other pure metals, and at red heat, when approaching the temperature where the iron ceases to be magnetizable, the temperature coefficient becomes still higher, until the temperature is reached where the iron ceases to be magnetic. At this point its temperature coefficient becomes that of other pure metals. Iron wire—usually mounted in hydrogen to keep it from oxidizing —thus finds a use as series resistance for current limitation in vacuum arc circuits, etc.

Electrolytic Conductors

4. The conductors of the second class are the electrolytic conductors. Their characteristic is that the conduction is accompanied by chemical action. The specific resistance of electrolytic conductors in general is about a million times higher than that of the metallic conductors. They are either fused compounds, or solutions of compounds in solvents, ranging in resistivity from 1.3 ohm-cm., in 30 per cent. nitric acid, and still lower in fused salts, to about 10,000 ohm-cm. in pure river water, and from there up to infinity (distilled water, alcohol, oils, etc.). They are all liquids, and when frozen become insulators.

Characteristic of the electrolytic conductors is the negative temperature coefficient of resistance; the resistance decreases with increasing temperature—not in a straight, but in a curved line, as illustrated by curves III in Fig. 1.

When dealing with electrical resistances, in many cases it is more convenient and gives a better insight into the character of the conductor, by not considering the resistance as a function of the temperature, but the voltage consumed by the conductor as a function of the current under stationary condition. In this case, with increasing current, and so increasing power consumption, the temperature also rises, and the curve of voltage for increasing current so illustrates the electrical effect of increasing temperature. The advantage of this method is that in many cases we get a better view of the action of the conductor in an electric circuit by eliminating the temperature, and relating only electrical quantities with each other. Such volt-ampere characteristics of electric conductors can easily and very accurately be determined, and, if desired, by the radiation law approximate values of the temperature be derived, and therefrom the temperature-resistance curve calculated, while a direct measurement of the resist-

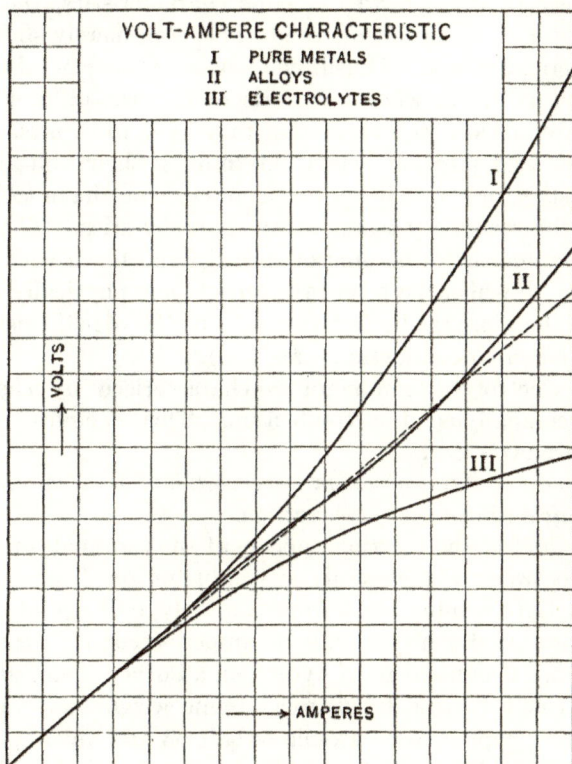

VOLT-AMPERE CHARACTERISTIC
I   PURE METALS
II ALLOYS
III ELECTROLYTES

Fig. 2.

ance over a very wide range of temperature is extremely difficult, and often no more accurate.

In Fig. 2, therefore, are shown such volt-ampere characteristics of conductors. The dotted straight line is the curve of absolutely constant resistance, which does not exist. Curves I and II show characteristics of metallic conductors, curve III of electrolytic conductors. As seen, for higher currents I and II rise faster, and III slower than for low currents.

It must be realized, however, that the volt-ampere character-istic depends not only on the material of the conductor, as the temperature-resistivity curve, but also on the size and shape of the conductor, and its surroundings. For a long and thin con-ductor in horizontal position in air, it would be materially differ-ent numerically from that of a short and thick conductor in dif-ferent position at different surrounding temperature. However, qualitatively it would have the same characteristics, the same characteristic deviation from straight line, etc., merely shifted in their numerical values. Thus it characterizes the general nature of the conductor, but where comparisons between different con-ductor materials are required, either they have to be used in the same shape and position, when determining their volt-ampere characteristics, or the volt-ampere characteristics have to be re-duced to the resistivity-temperature characteristics. The volt-ampere characteristics become of special importance with those conductors, to which the term resistivity is not physically appli-cable, and therefore the "effective resistivity" is of little meaning, as in gas and vapor conduction (arcs, etc.).

5. The electrolytic conductor is characterized by chemical action accompanying the conduction. This chemical action follows Faraday's law:

The amount of chemical action is proportional to the current and to the chemical equivalent of the reaction.

The product of the reaction appears at the terminals or "elec-trodes," between the electrolytic conductor or "electrolyte," and the metallic conductors. Approximately, 0.01 mg. of hydro-gen are produced per coulomb or ampere-second. From this electrochemical equivalent of hydrogen, all other chemical reac-tions can easily be calculated from atomic weight and valency. For instance, copper, with atomic weight 63 and valency 2, has the equivalent $63/2 = 31.5$ and copper therefore is deposited at the negative terminal or "cathode," or dissolved at the positive terminal or "anode," at the rate of 0.315 mg. per ampere-second; aluminum, atomic weight 28 and valency 3, at the rate of 0.093 mg. per ampere-second, etc.

The chemical reaction at the electrodes represents an energy transformation between electrical and chemical energy, and as the rate of electrical energy supply is given by current times vol-tage, it follows that a voltage drop or potential difference occurs at the electrodes in the electrolyte. This is in opposition to the current, or a counter e.m.f., the "counter e.m.f. of electrochemical polarization," and thus consumes energy, if the chemical reaction requires energy—as the deposition of copper from a solution of a copper salt. It is in the same direction as the current, thus producing electric energy, if the chemical reaction produces energy, as the dissolution of copper from the anode.

As the chemical reaction, and therefore the energy required for it, is proportional to the current, the potential drop at the electrodes is independent of the current density, or constant for the same chemical reaction and temperature, except in so far as secondary reactions interfere. It can be calculated from the chemical energy of the reaction, and the amount of chemical reaction as given by Faraday's law. For instance: 1 amp.-sec. deposits 0.315 mg. copper. The voltage drop, e, or polarization voltage, thus must be such that e volts times 1 amp.-sec., or e watt-sec. or joules, equals the chemical reaction energy of 0.315 mg. copper in combining to the compound from which it is deposited in the electrolyte.

If the two electrodes are the same and in the same electrolyte at the same temperature, and no secondary reaction occurs, the reactions are the same but in opposite direction at the two electrodes, as deposition of copper from a copper sulphate solution at the cathode, solution of copper at the anode. In this case, the two potential differences are equal and opposite, their resultant thus zero, and it is said that "no polarization occurs."

If the two reactions at the anode and cathode are different, as the dissolution of zinc at the anode, the deposition of copper at the cathode, or the production of oxygen at the (carbon) anode, and the deposition of zinc at the cathode, then the two potential differences are unequal and a resultant remains. This may be in the same direction as the current, producing electric energy, or in the opposite direction, consuming electric energy. In the first case, copper deposition and zinc dissolution, the chemical energy set free by the dissolution of the zinc and the voltage produced by it, is greater than the chemical energy consumed in the deposition of the copper, and the voltage consumed by it, and the resultant of the two potential differences at the electrodes thus is in the same direction as the current, hence may produce this current. Such a device, then, transforms chemical energy into electrical energy, and is called a *primary cell* and a number of them, a *battery*. In the second case, zinc deposition and oxygen production at the anode, the resultant of the two potential differences at the electrodes is in opposition to the current; that is, the device consumes electric energy and converts it into chemical energy, as *electrolytic cell.*"

Both arrangements are extensively used: the battery for producing electric power, especially in small amounts, as for hand lamps, the operation of house bells, etc. The electrolytic cell is used extensively in the industries for the production of metals as aluminum, magnesium, calcium, etc., for refining of metals as copper, etc., and constitutes one of the most important industrial applications of electric power.

A device which can efficiently be used, alternately as battery and as electrolytic cell, is the *secondary cell* or *storage battery*. Thus in the lead storage battery, when discharging, the chemical reaction at the anode is conversion of lead peroxide into lead oxide, at the cathode the conversion of lead into lead oxide; in charging, the reverse reaction occurs.

6. Specifically, as "polarization cell" is understood a combination of electrolytic conductor with two electrodes, of such character that no permanent change occurs during the passage of the current. Such, for instance, consists of two platinum electrodes in diluted sulphuric acid. During the passage of the current, hydrogen is given off at the cathode and oxygen at the anode, but terminals and electrolyte remain the same (assuming that the small amount of dissociated water is replaced).

In such a polarization cell, if e_0 = counter e.m.f. of polarization (corresponding to the chemical energy of dissociation of water, and approximately 1.6 volts) at constant temperature and thus constant resistance of the electrolyte, the current, i, is proportional to the voltage, e, minus the counter e.m.f. of polarization, e_0:

$$i = \frac{e - e_0}{r} \tag{2}$$

In such a case the curve III of Fig. 2 would with decreasing current not go down to zero volts, but would reach zero amperes at a voltage $e = e_o$, and its lower part would have the shape as shown in Fig. 3. That is, the current begins at voltage, e_0, and below this voltage, only a very small "diffusion" current flows.

When dealing with electrolytic conductors, as when measuring their resistance, the counter e.m.f. of polarization thus must be considered, and with impressed voltages less than the polarization voltage, no permanent current flows through the electrolyte, or rather only a very small "leakage" current or "diffusion" current, as shown in Fig. 3. When closing the circuit, however, a transient current flows. At the moment of circuit closing, no counter e.m.f. exists, and current flows under the full impressed voltage. This current, however, electrolytically produces a hydrogen and an oxygen film at the electrodes, and with their gradual formation, the counter e.m.f. of polarization increases and decreases the current, until it finally stops it. The duration of this transient depends on the resistance of the electrolyte and on the surface of the electrodes, but usually is fairly short.

7. This transient becomes a permanent with alternating impressed voltage. Thus, when an alternating voltage, of a maxi-

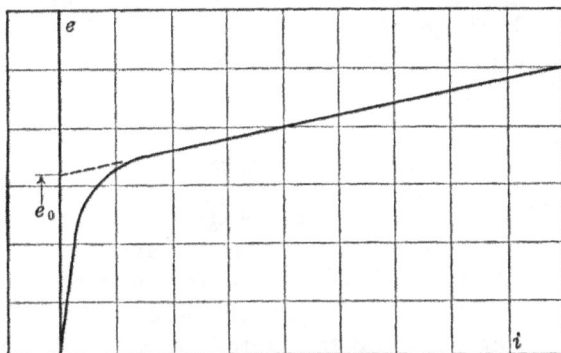

FIG. 3.

mum value lower than the polarization voltage, is impressed upon an electrolytic cell, an alternating current flows through the cell, which produces the hydrogen and oxygen films which hold back the current flow by their counter e.m.f. .The current thus flows ahead of the voltage or counter e.m.f. which it produces, as a leading current, and the polarization cell thus acts like a condenser, and is called an "electrolytic condenser." It has an enormous electrostatic capacity, or "effective capacity," but can stand low voltage only —1 volt or less—and therefore is of limited industrial value. As chemical action requires appreciable time, such electrolytic condensers show at commercial frequencies high losses of power by what may be called "chemical hysteresis," and therefore low efficiences, but they are alleged to become efficient at very low frequencies. For this reason, they have been proposed in the secondaries of induction motors, for power-factor compensation. Iron plates in alkaline solution, as sodium carbonate, are often considered for this purpose.

NOTE.—The aluminum cell, consisting of two aluminum plates with an electrolyte which does not attack aluminum, often is called an electrolytic condenser, as its current is leading; that is, it acts as capacity. It is, however, not an electrolytic condenser, and the counter e.m.f., which gives the capacity effect, is not electrolytic polarization. The aluminum cell is a true electrostatic condenser, in which the film of alumina, formed on the positive aluminum plates, is the dielectric. Its characteristic is, that the condenser is self-healing; that is, a puncture of the alumina film causes a current to flow, which electrolytically produces alumina at the puncture hole, and so closes it. The capacity is very high, due to the great thinness of the film, but the energy losses are considerable, due to the continual puncture and repair of the dielectric film.

Pyroelectric Conductors

8. A third class of conductors are the *pyroelectric conductors* or *pyroelectrolytes*. In some features they are intermediate between the metallic conductors and the electrolytes, but in their essential characteristics they are outside of the range of either. The metallic conductors as well as the electrolytic conductors give a volt-ampere characteristic in which, with increase of current, the voltage rises, faster than the current in the metallic conductors, due to their positive temperature coefficient, slower than the current in the electrolytes, due to their negative temperature coefficient.

The characteristic of the pyroelectric conductors, however, is such a very high negative temperature coefficient of resistance, that is, such rapid decrease of resistance with increase of temperature, that over a wide range of current the voltage decreases with increase of current. Their volt-ampere characteristic thus has a shape as shown diagrammatically in Fig. 4—though not all such conductors may show the complete curve, or parts of the curve may be physically unattainable: for small currents, range (1), the voltage increases approximately proportional to the current, and sometimes slightly faster, showing the positive temperature coefficient of metallic conduction. At *a* the temperature coefficient changes from positive to negative, and the voltage begins to increase slower than the current, similar as in electrolytes, range (2). The negative temperature coefficient rapidly increases, and the voltage rise become slower, until at point *b* the negative temperature coefficient has become so large, that the voltage begins to decrease again with increasing current, range (3). The maximum voltage point *b* thus divides the range of rising characteristic (1) and (2), from that of decreasing characteristic, (3). The negative temperature coefficient reaches a maximum and then decreases again, until at point *c* the negative temperature coefficient has fallen so that beyond this minimum voltage point *c* the voltage again increases with increasing current, range (4),

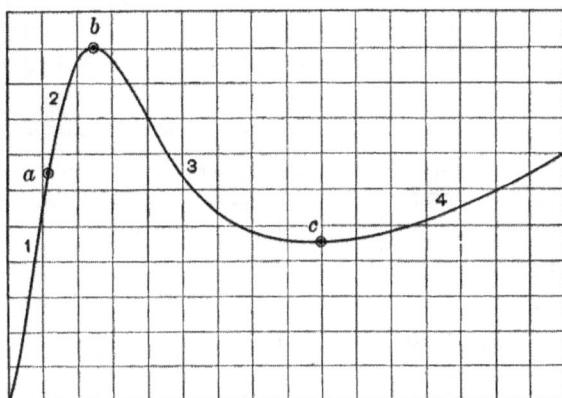

Fig. 4.

though the temperature coefficient remains negative, like in electrolytic conductors.

In range (1) the conduction is purely metallic, in range (4) becomes purely electrolytic, and is usually accompanied by chemical action.

Range (1) and point *a* often are absent and the conduction begins already with a slight negative temperature coefficient.

The complete curve, Fig. 4, can be observed only in few substances, such as magnetite. Minimum voltage point *c* and range (4) often is unattainable by the conductor material melting or being otherwise destroyed by heat before it is reached. Such, for instance, is the case with cast silicon. The maximum voltage point *b* often is unattainable, and the passage from range (2) to range (3) by increasing the current therefore not feasible, because the maximum voltage point b is so high, that disruptive discharge occurs before it is reached. Such for instance is the case in glass, the Nernst lamp conductor, etc.

9. The curve, Fig. 3, is drawn only diagrammatically, and the lower current range exaggerated, to show the characteristics. Usually the current at point b is very small compared with that at point c; rarely more than one-hundredth of it, and the actual proportions more nearly represented by Fig. 5. With pyro-electric conductors of very high value of the voltage b, the currents in the range (1) and (2) may not exceed one-millionth of that at (3). Therefore, such volt-ampere characteristics are

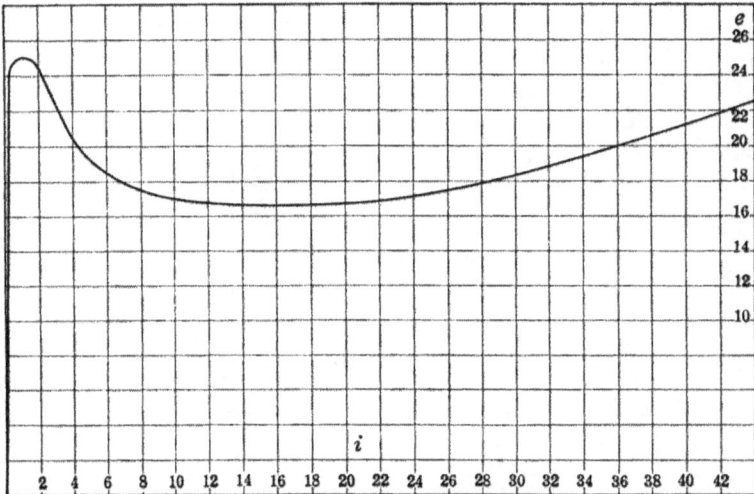

FIG. 5.

often plotted with $\sqrt{i}$ as abscissæ, to show the ranges in better proportions.

Pyroelectric conductors are metallic silicon, boron, some forms of carbon as anthracite, many metallic oxides, especially those of the formula $M^{(2)} M_2^{(3)} O_4$, where $M^{(2)}$ is a bivalent, $M^{(3)}$ a trivalent metal (magnetite, chromite), metallic sulphides, silicates such as glass, many salts, etc.

Intimate mixtures of conductors, as graphite, coke, powdered metal, with non-conductors as clay, carborundum, cement, also have pyroelectric conduction. Such are used, for instance, as "resistance rods" in lightning arresters, in some rheostats, as

cement resistances for high-frequency power dissipation in reactances, etc. Many, if not all so-called "insulators" probably
are in reality pyroelectric conductors, in which the maximum
voltage point *b* is so high, that the range (3) of decreasing characteristic can be reached only by the application of external heat,
as in the Nernst lamp conductor, or can not be reached at all,
because chemical dissociation begins below its temperature, as
in organic insulators.

Fig. 6 shows the volt-ampere characteristics of two rods of
cast silicon, 10 in. long and 0.22 in. in diameter, with $\sqrt{i}$ as ab-

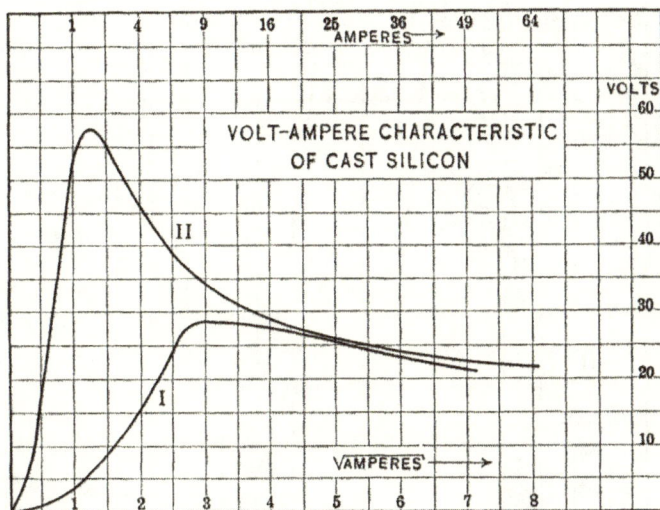

FIG. 6.

scissæ and Fig. 7 their approximate temperature-resistance
characteristics. The curve II of Fig. 7 is replotted in Fig. 8,
with log *r* as ordinates. Where the resistivity varies over a very
wide range, it often is preferable to plot the logarithm of the
resistivity. It is interesting to note that the range (3) of curve
II, between 700° and 1400°, is within the errors of observation
represented by the expression

$$r = 0.01E^{-\frac{9080}{T}}$$

where *T* is the absolute temperature (−273°C. as zero point).
The difference between the two silicon rods is, that the one con-

tains 1.4 per cent., the other only 0.1 per cent. carbon; besides this, the impurities are less than 1 per cent.

As seen, in these silicon rods the range (4) is not yet reached at the melting point.

Fig. 9 shows the volt-ampere characteristic, with $\sqrt{i}$ as abscissæ, and Fig. 10 the approximate resistance temperature char-

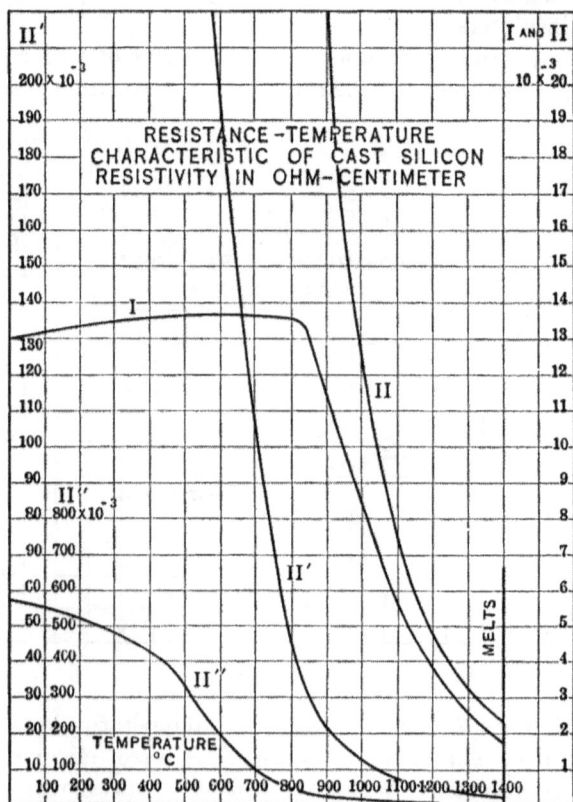

RESISTANCE –TEMPERATURE
CHARACTERISTIC OF CAST SILICON
RESISTIVITY IN OHM–CENTIMETER

Fig. 7.

acteristic derived therefrom, with log r as ordinates, of a magnetic rod 6 in. long and ¾ in. in diameter, consisting of 90 per cent. magnetite (Fe_3O_4), 9 per cent. chromite ($FeCr_2O_4$) and 1 per cent. sodium silicate, sintered together.

10. As result of these volt-ampere characteristics, Figs. 4 to 10, pyroelectric conductors as structural elements of an electric circuit show some very interesting effects, which may be illustrated on the magnetite rod, Fig. 9. The maximum terminal voltage, which can exist across this rod in stationary conditions, is 25 volts at 1 amp. With increasing terminal voltage, the current thus gradually increases, until 25 volts is reached, and then without further increase of the impressed voltage the current rapidly rises to short-circuit values. Thus, such resistances can be used as excess-voltage cutout, or, when connected between circuit and ground, as excess-voltage grounding device: below 24 volts, it

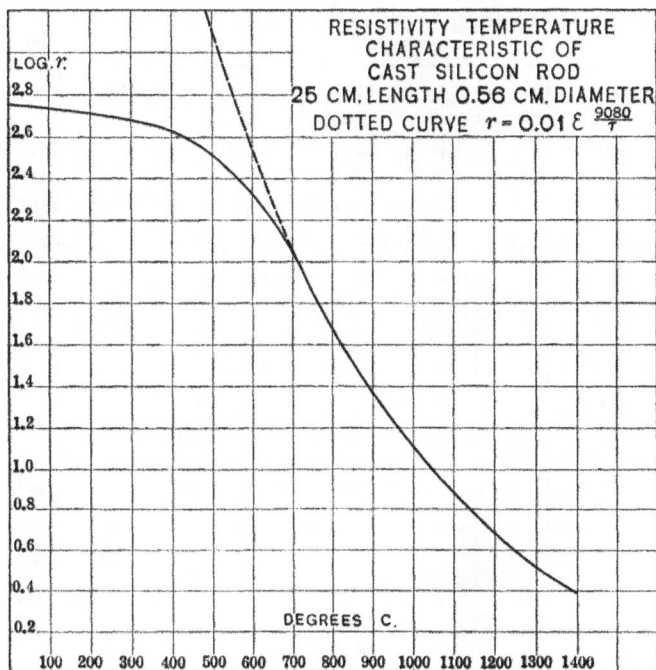

RESISTIVITY TEMPERATURE CHARACTERISTIC OF CAST SILICON ROD 25 CM. LENGTH 0.56 CM. DIAMETER DOTTED CURVE $r = 0.01 \varepsilon^{\frac{9080}{r}}$

Fig. 8.

bypasses a negligible current only, but if the voltage rises above 25 volts, it short-circuits the voltage and so stops a further rise, or operates the circuit-breaker, etc. As the decrease of resistance is the result of temperature rise, it is not instantaneous; thus the rod does not react on transient voltage rises, but only on lasting ones.

Within a considerable voltage range—between 16 and 25 volts —three values of current exist for the same terminal voltage. Thus at 20 volts between the terminals of the rod in Fig. 9, the current may be 0.02 amp., or 4.1 amp., or 36 amp. That is, in series in a constant-current circuit of 4.1 amp. this rod would show the same terminal voltage as in a 0.02-amp. or a 36-amp. constant-current circuit, 20 volts. On constant-potential supply, however, only the range (1) and (2), and the range (4) is stable, but the range (3) is unstable, and here we have a conductor, which is unstable in a certain range of currents, from point b at 1 amp. to point c at 20 amp. At 20 volts impressed upon the rod, 0.02 amp. may pass through it, and the conditions are stable. That is, a tendency to increase of current would check itself by requiring an increase of voltage beyond that supplied, and a decrease of

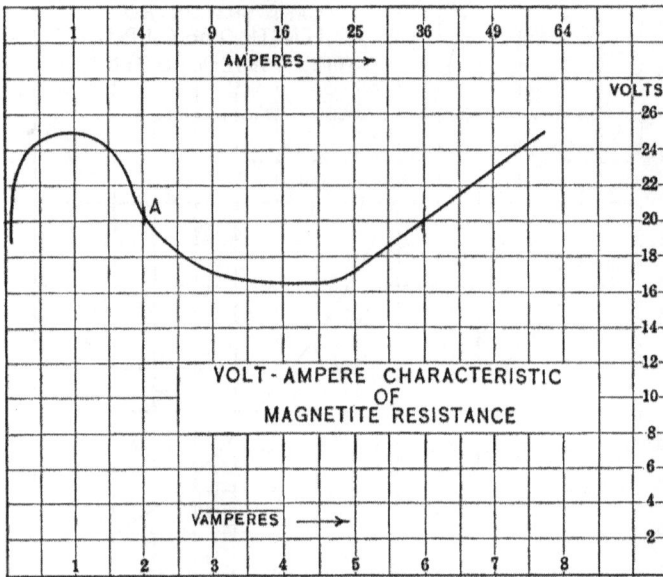

FIG. 9.

current would reduce the voltage consumption below that employed, and thus be checked. At the same impressed 20 volts, 36 amp. may pass through the rod—or 1800 times as much as before—and the conditions again are stable. A current of 4.1 amp. also would consume a terminal voltage of 20, but the condition now is unstable; if the current increases ever so little, by a momentary voltage rise, then the voltage consumed by the rod decreases, becomes less than the terminal voltage of 20, and the current thus increases by the supply voltage exceeding the consumed voltage. This, however, still further decreases the consumed voltage and thereby increases the current, and the current rapidly rises, until conditions become stable at 36 amp. Inversely, a momentary decrease of the current below 4.1 amp. increases the voltage required by the rod, and this higher voltage not being available at constant supply voltage, the current decreases.

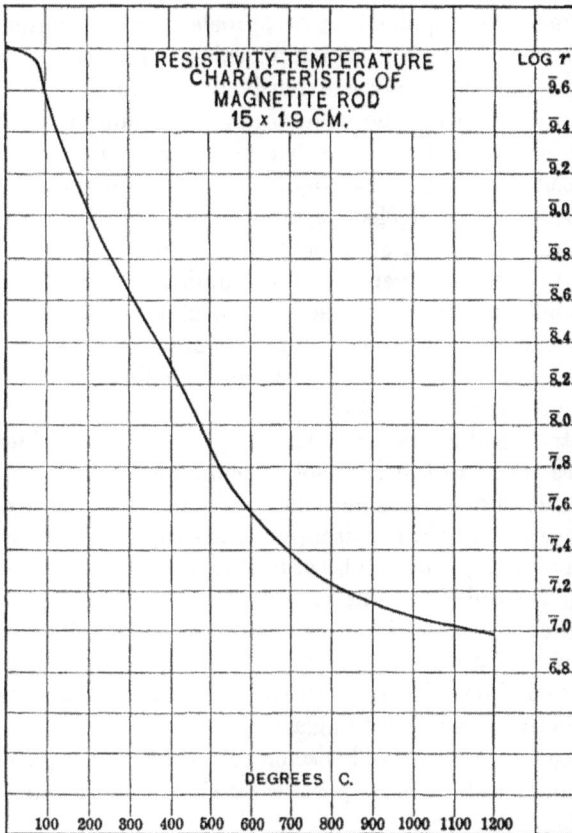

RESISTIVITY-TEMPERATURE CHARACTERISTIC OF MAGNETITE ROD 15 x 1.9 CM.

FIG. 10.

This, however, still further increases the required voltage and decreases the current, until conditions become stable at 0.02 amp.

With the silicon rod II of Fig. 6, on constant-potential supply, with increasing voltage the current and the temperature increases gradually, until 57.5 volts are reached at about 450°C.; then, without further voltage increase, current and temperature rapidly increase until the rod melts. Thus:

2

Condition of stability of a conductor on constant-voltage supply is, that the volt-ampere characteristic is rising, that is, an increase of current requires an increase of terminal voltage.

A conductor with falling volt-ampere characteristic, that is, a conductor in which with increase of current the terminal voltage decreases, is unstable on constant-potential supply.

11. An important application of pyroelectric conduction has been the glower of the Nernst lamp, which before the development of the tungsten lamp was extensively used for illumination.

Pyroelectrolytes cover the widest range of conductivities; the alloys of silicon with iron and other metals give, depending on their composition, resistivities from those of the pure metals up to the lower resistivities of electrolytes: 1 ohm per cm.^{3}; borides, carbides, nitrides, oxides, etc., gave values from 1 ohm per cm.^{3} or less, up to megohms per cm.^{3}, and gradually merge into the materials which usually are classed as "insulators."

The pyroelectric conductors thus are almost the only ones available in the resistivity range between the metals, 0.0001 ohm-cm. and the electrolytes, 1 ohm-cm.

Pyroelectric conductors are industrially used to a considerable extent, since they are the only solid conductors, which have resistivities much higher than metallic conductors. In most of the industrial uses, however, the dropping volt-ampere characteristic is not of advantage, is often objectionable, and the use is limited to the range (1) and (2) of Fig. 3. It, therefore, is of importance to realize their pyroelectric characteristics and the effect which they have when overlooked beyond the maximum voltage point. Thus so-called "graphite resistances" or "carborundum resistances," used in series to lightning arresters to limit the discharge, when exposed to a continual discharge for a sufficient time to reach high temperature, may practically short-circuit and thereby fail to limit the current.

12. From the dropping volt-ampere characteristic in some pyroelectric conductors, especially those of high resistance, of very high negative temperature coefficient and of considerable cross-section, results the tendency to unequal current distribution and the formation of a "luminous streak," at a sudden application of high voltage. Thus, if the current passing through a graphite-clay rod of a few hundred ohms resistance is gradually increased, the temperature rises, the voltage first increases and then decreases, while the rod passes from range (2) into the range (3) of the volt-ampere characteristic, but the temperature and thus the current density throughout the section of the rod is fairly uniform. If, however, the full voltage is suddenly applied, such as by a lightning discharge throwing line voltage on the series resistances of a lightning arrester, the rod heats up very rapidly, too rapidly for the temperature to equalize throughout the rod section, and a part of the section passes the maximum voltage point *b* of Fig. 4 into the range (3) and (4) of low resistance, high current and high temperature, while most of the section is still in the high-resistance range (2) and never passes beyond this range, as it is practically short-circuited. Thus, practically all the current passes by an irregular luminous streak through a small section of the rod, while most of the section is relatively cold and practically does not participate in the conduction. Gradually, by heat conduction the temperature and the current density may become more uniform, if before this the rod has not been destroyed by temperature stresses. Thus, tests made on such conductors by gradual application of voltage give no information on their behavior under sudden voltage application. The liability to the formation of such luminous streaks naturally increases with decreasing heat conductivity of the material, and with increasing resistance and temperature coefficient of resistance, and with conductors of extremely high temperature coefficient, such as silicates, oxides of high resistivity, etc., it is practically impossible to get current to flow through any appreciable conductor section, but the conduction is always streak conduction.

Some pyroelectric conductors have the characteristic that their resistance increases permanently, often by many hundred per cent. when the conductor is for some time exposed to high-frequency electrostatic discharges.

Coherer action, that is, an abrupt change of conductivity by an electrostatic spark, a wireless wave, etc., also is exhibited by some pyroelectric conductors.

13. Operation of pyroelectric conductors on a constant-voltage circuit, and in the unstable branch (3), is possible by the insertion of a series resistance (or reactance, in alternating-current circuits) of such value, that the resultant volt-ampere characteristic is stable, that is, rises with increase of current. Thus, the conductor in Fig. 4, shown as *I* in Fig. 11, in series with the metallic resistance giving characteristic *A*, gives the resultant characteristic *II* in Fig. 11, which is stable over the entire range. *I* in series with a smaller resistance, of characteristic *B*, gives the resultant characteristic *III*.   In this, the unstable range has contracted to from *b'* to *c'*.   Further discussion of the instability of such conductors, the effect of resistance in stablizing them, and the result-

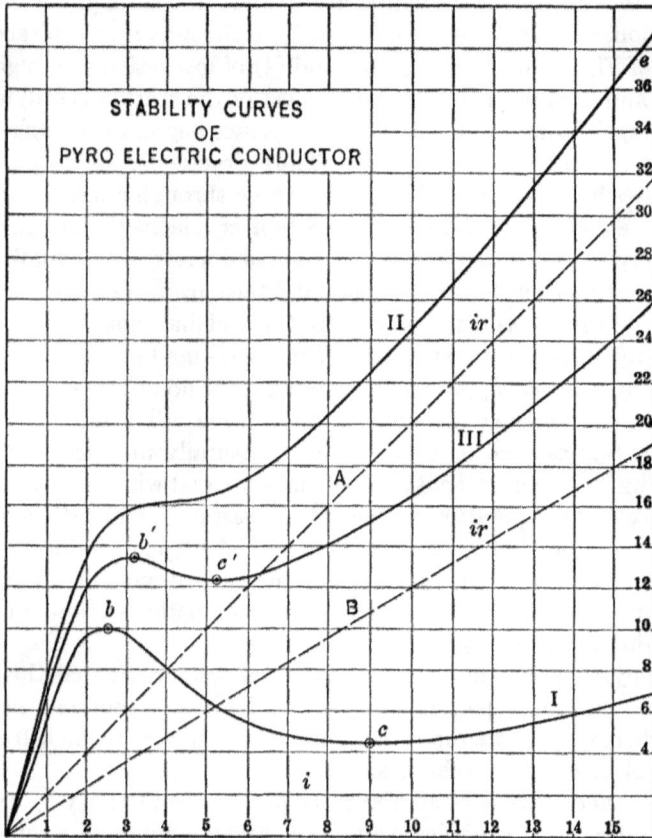

STABILITY CURVES
OF
PYRO ELECTRIC CONDUCTOR

Fig. 11.

ant "stability curve" are found in the chapter on "Instability of Electric Circuits," under "Arcs and Similar Conductors."

14. It is doubtful whether the pyroelectric conductors really form one class, or whether, by the physical nature of their conduction, they should not be divided into at least two classes:

1. True pyroelectric conductors, in which the very high negative temperature coefficient is a characteristic of the material.

In this class probably belong silicon and its alloys, boron, magnetite and other metallic oxides, sulphides, carbides, etc.

2. Conductors which are mixtures of materials of high conductivity, and of non-conductors, and derive their resistance from the contact resistance between the conducting particles which are separated by non-conductors. As contact resistance shares with arc conduction the dropping volt-ampere characteristic, such mixtures thereby imitate pyroelectric conduction. In this class probably belong the graphite-clay rods industrially used. Powders of metals, graphite and other good conductors also belong in this class.

The very great increase of resistance of some conductors under electrostatic discharges probably is limited to this class, and is the result of the high current density of the condenser discharge burning off the contact points.

Coherer action probably is limited also to those conductors, and is the result of the minute spark at the contact points initiating conduction.

Carbon

15. In some respects outside of the three classes of conductors thus far discussed, in others intermediate between them, is one of

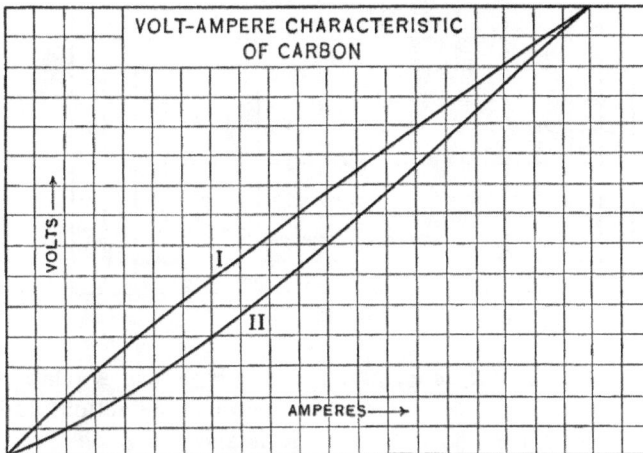

Fig. 12.

the industrially most important conductors, *carbon*. It exists in a large variety of modifications of different resistance characteristics, which all are more or less intermediate between three typical forms:

1. Metallic Carbon.—It is produced from carbon deposited on an incandescent filament, from hydrocarbon vapors at a partial vacuum, by exposure to the highest temperatures of the electric furnace. Physically, it has metallic characteristics: high elas-

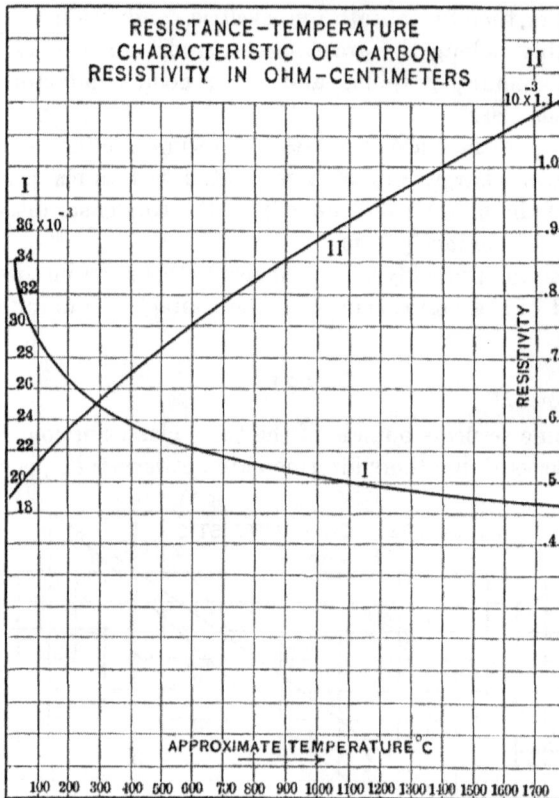

Fig. 13.

ticity, metallic luster, etc., and electrically it has a relatively low resistance approaching that of metallic conduction, and a positive temperature coefficient of resistance, of about 0.1 per cent. per degree C.—that is, of the same magnitude as mercury or cast iron.

The coating of the "Gem" filament incandescent lamp consists of this modification of carbon.

2. **Amorphous carbon,** as produced by the carbonization of cellulose. In its purest form, as produced by exposure to the highest temperatures of the electric furnace, it is characterized by a relatively high resistance, and a negative temperature coefficient of resistance, its conductivity increasing by about 0.1 per cent. per degree C.

3. **Anthracite.**—It has an extremely high resistance, is practically an insulator, but has a very high negative temperature coefficient of resistance, and thus becomes a fairly good conductor at high temperature, but its heat conductivity is so low, and the negative temperature coefficient of resistance so high, that the conduction is practically always streak conduction, and at the high temperature of the conducting luminous streak, conversion to graphite occurs, with a permanent decrease of resistance.

(1) thus shows the characteristics of metallic conduction, (2) those of electrolytic conduction, and (3) those of pyroelectric conduction.

Fig. 12 shows the volt-ampere characteristics, and Fig. 13 the resistance-temperature characteristics of amorphous carbon— curve I—and metallic carbon— curve II.

Insulators

16. As a fourth class of conductors may be considered the so-called "insulators," that is, conductors which have such a high specific resistance, that they can not industrially be used for conveying electric power, but on the contrary are used for restraining the flow of electric power to the conductor, or path, by separating the conductor from the surrounding space by such an insulator. The insulators also have a conductivity, but their specific resistance is extremely high. For instance, the specific resistance of fiber is about 10^{12}, of mica 10^{14}, of rubber 10^{16} ohm-cm., etc.

As, therefore, the distinction between conductor and insulator is only qualitative, depending on the application, and more particularly on the ratio of voltage to current given by the source of power, sometimes a material may be considered either as insulator or as conductor. Thus, when dealing with electrostatic machines, which give high voltages, but extremely small currents, wood, paper, etc., are usually considered as conductors, while for the low-voltage high-current electric lighting circuits they are insulators, and for the high-power very high-voltage transmission circuits they are on the border line, are poor conductors and poor insulators.

Insulators usually, if not always, have a high negative temperature coefficient of resistance, and the resistivity often follows approximately the exponential law,

$$r = r_0 E^{-aT} \tag{3}$$

where T = temperature. That is, the resistance decreases by the same percentage of its value, for every degree C. For instance, it decreases to one-tenth for every 25°C. rise of temperature, so that at 100°C. it is 10,000 times lower than at 0°C. Some temperature-resistance curves, with log r as ordinates, of insulating materials are given in Fig. 14.

As the result of the high negative temperature coefficient, for a sufficiently high temperature, the insulating material, if not destroyed by the temperature, as is the case with organic materials, becomes appreciably conducting, and finally becomes a fairly good conductor, usually an electrolytic conductor.

Thus the material of the Nernst lamp (rare oxides, similar to the Welsbach mantle of the gas industry), is a practically perfect insulator at ordinary temperatures, but becomes conducting at high temperature, and is then used as light-giving conductor.

Fig. 15 shows for a number of high-resistance insulating materials the temperature-resistance curve at the range where the resistivity becomes comparable with that of other conductors.

17. Many insulators, however, more particularly the organic materials, are chemically or physically changed or destroyed, before the temperature of appreciable conduction is reached, though even these show the high negative temperature coefficient. With some, as varnishes, etc., the conductivity becomes sufficient, at high temperatures, though still below carbonization temperature, that under high electrostatic stress, as in the insulation of high-voltage apparatus, appreciable energy is represented by the leakage current through the insulation, and in this case rapid i^2r heating and final destruction of the material may result. That is, such materials, while excellent insulators at ordinary temperature, are unreliable at higher temperature.

It is quite probable that there is no essential difference between the true pyroelectric conductors, and the insulators, but the latter are merely pyroelectric conductors in which the initial resistivity and the voltage at the maximum point *b* are so high, that the change from the range (2) of the pyroelectrolyte, Fig. 4, to the range (3) can not be produced by increase of voltage. That is, the distinction between pyroelectric conductor and insulator would be the quantitative one, that in the former the maximum

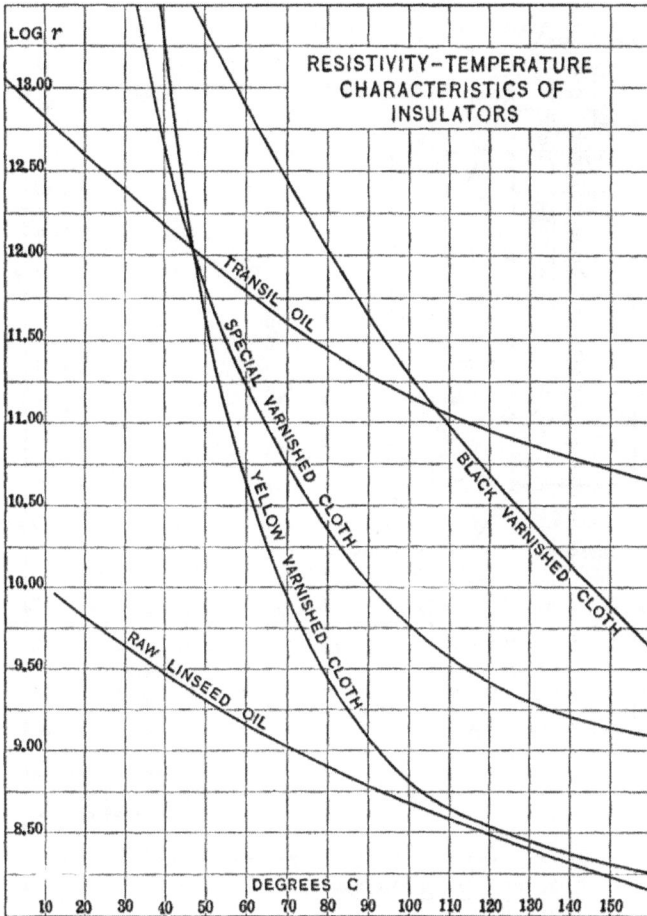

Fig. 14.

voltage point of the volt-ampere characteristic is within experimental reach, while with the latter it is beyond reach.

Whether this applies to all insulators, or whether among organic compounds as oils, there are true insulators, which are not pyroelectric conductors, is uncertain.

Positive temperature coefficient of resistivity is very often met in insulating materials such as oils, fibrous materials, etc. In this case, however, the rise of resistance at increase of temperature usually remains permanent after the temperature is again lowered,

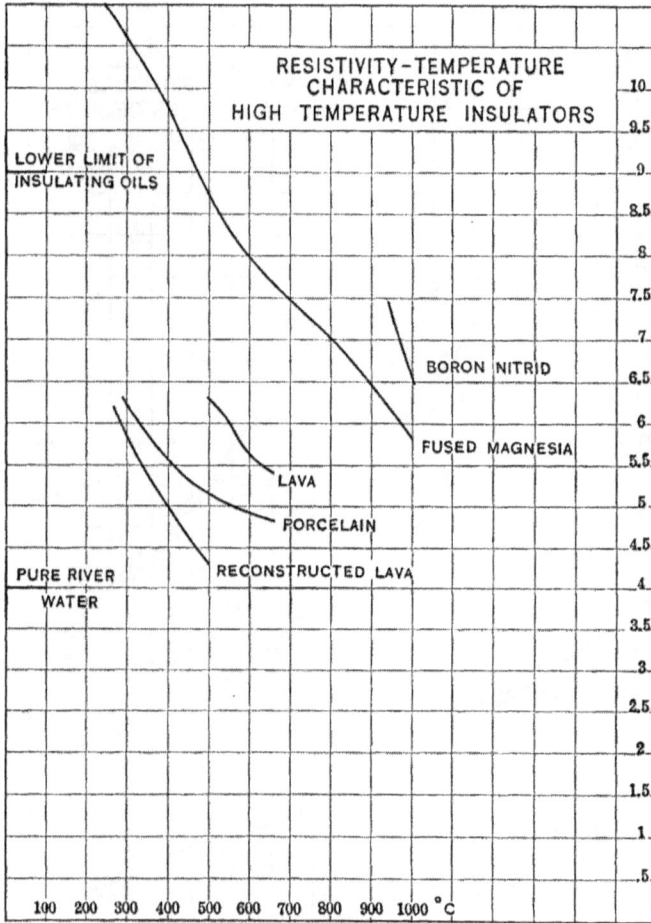

Fig. 15.

and the apparent positive temperature coefficient was due to the expulsion of moisture absorbed by the material. With insulators of very high resistivity, extremely small traces of moisture may decrease the resistivity many thousandfold, and the conductivity of insulating materials very often is almost entirely moisture conduction, that is, not due to the material proper, but due to the moisture absorbed by it. In such a case, prolonged drying may increase the resistivity enormously, and when dry, the material then shows the negative temperature coefficient of resistance, incident to pyroelectric conduction.

CHAPTER II

ELECTRIC CONDUCTION. GAS AND VAPOR CONDUCTORS

Gas, Vapor and Vacuum Conduction

18. As further, and last class may be considered vapor, gas and vacuum conduction. Typical of this is, that the volt-ampere characteristic is dropping, that is, the voltage decreases with increase of current, and that luminescence accompanies the conduction, that is, conversion of electric energy into light.

Thus, gas and vapor conductors are unstable on constant-potential supply, but stable on constant current. On constant potential they require a series resistance or reactance, to produce stability.

Such conduction may be divided into three distinct types: spark conduction, arc conduction, and true electronic conduction.

In spark conduction, the gas or vapor which fills the space between the electrodes is the conductor. The light given by the gaseous conductor thus shows the spectrum of the gas or vapor which fills the space, but the material of the electrodes is immaterial, that is, affects neither the light nor the electric behavior of the gaseous conductor, except indirectly, in so far as the section of the conductor at the terminals depends upon the terminal surface.

In arc conduction, the conductor is a vapor stream issuing from the negative terminal or cathode, and moving toward the anode at high velocity. The light of the arc thus shows the spectrum of the negative terminal material, but not that of the gas in the surrounding space, nor that of the positive terminal, except indirectly, by heat luminescence of material entering the arc conductor from the anode or from surrounding space.

In true electronic conduction, electrons existing in the space, or produced at the terminals (hot cathode), are the conductors. Such conduction thus exists also in a perfect vacuum, and may be accompanied by practically no luminescence.

Disruptive Conduction

19. Spark conduction at atmospheric pressure is the disruptive spark, streamers, and corona. In a partial vacuum, it is the Geissler discharge or glow discharge. Spark conduction is discontinuous, that is, up to a certain voltage, the "disruptive voltage," no conduction exists, except perhaps the extremely small true electronic conduction. At this voltage conduction begins and continues as long as the voltage persists, or, if the source of power is capable of maintaining considerable current, the spark conduction changes to arc conduction, by the heat developed at the negative terminal supplying the conducting arc vapor stream. The current usually is small and the voltage high. Especially at atmospheric pressure, the drop of the 'volt-ampere characteristic is extremely steep, so that it is practically impossible to secure stability by series resistance, but the conduction changes to arc conduction, if sufficient current is available, as from power generators, or the conduction ceases by the voltage drop of the supply source, and then starts again by the recovery of voltage, as with an electrostatic machine. Thus spark conduction also is called *disruptive conduction* and *discontinuous conduction.*

Apparently continuous—though still intermittent—spark conduction is produced at atmospheric pressure by capacity in series to the gaseous conductor, on an alternating-voltage supply, as corona, and as Geissler tube conduction at a partial vacuum, by an alternating-supply voltage with considerable reactance or resistance in series, or from a direct-current source of very high voltage and very limited current, as an electrostatic machine.

In the Geissler tube or vacuum tube, on alternating-voltage supply, the effective voltage consumed by the tube, at constant temperature and constant gas pressure, is approximately constant and independent of the effective current, that is, the volt-ampere characteristic a straight horizontal line. The Geissler tube thus requires constant current or a steadying resistance or reactance for its operation. The voltage consumed by the Geissler tube consists of a potential drop at the terminals, the "terminal drop," and a voltage consumed in the luminous stream, the "stream voltage." Both greatly depend on the gas pressure, and vary, with changing gas pressure, in opposite directions: the terminal drop decreases and the stream voltage increases with increasing gas pressure, and the total voltage consumed by the

tube thus gives a minimum at some definite gas pressure. This
pressure of minimum voltage depends on the length of the tube,

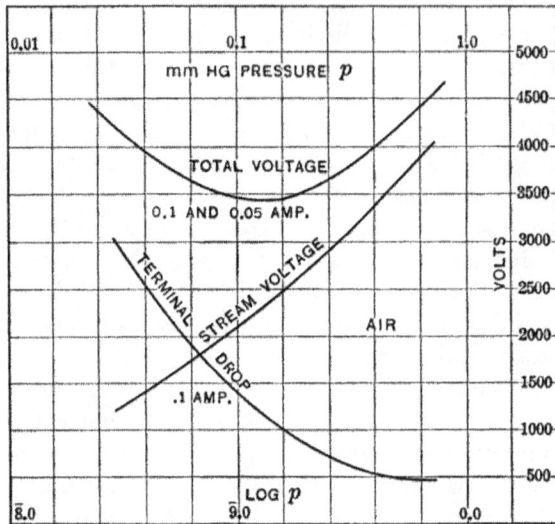

FIG. 16.

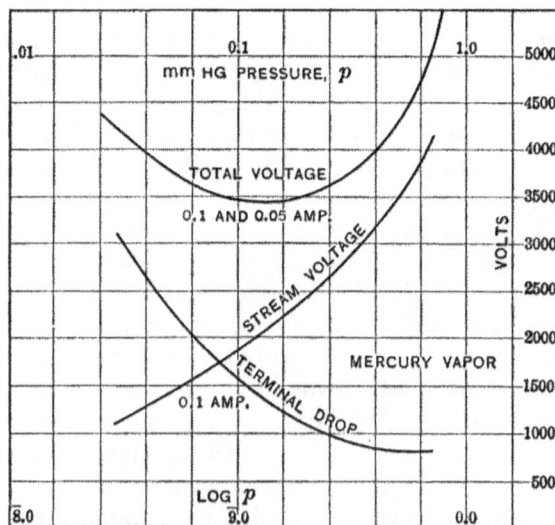

FIG. 17.

and the longer the tube, the lower is the gas pressure which gives
minimum total voltage.

Fig. 16 shows the voltage-pressure characteristic, at constant current of 0.1 amp. and 0.05 amp., of a Geissler tube of 1.3 cm. internal diameter and 200 cm. length, using air as conductor, and Fig. 17 the characteristic of the same tube with mercury vapor as conductor. Figs. 16 and 17 also show the two component voltages, the terminal drop and the stream voltage, separately. As abscissæ are used the log of the gas pressure, in millimeter mercury column. As seen, the terminal drop decreases with increasing gas pressure, and becomes negligible compared with the stream voltage, at atmospheric pressure.

The voltage gradient, per centimeter length of stream, varies from 5 to 20 volts, at gas or vapor pressure from 0.06 to 0.9 mm. At atmospheric pressure (760 mm.) the disruptive voltage gradient, which produces corona, is 21,000 volts effective per centimeter. The specific resistance of the luminous stream is from 65 to 500 ohms per cm.³ in the Geissler tube conduction of Figs. 16 and 17—though this term has little meaning in gas conduction. The specific resistance of the corona in air, as it appears on transmission lines at very high voltages, is still very much higher.

Arc Conduction

20. In the electric arc, the current is carried across the space between the electrodes or arc terminals by a stream of electrode vapor, which issues from a spot on the negative terminal, the so-called cathode spot, as a high-velocity blast (probably of a velocity of several thousand feet per second). If the negative terminal is fluid, the cathode spot causes a depression, by the reaction of the vapor blast, and is in a more or less rapid motion, depending on the fluidity.

As the arc conductor is a vapor stream of electrode material, this vapor stream must first be produced, that is, energy must be expended before arc conduction can take place. The arc, therefore, does not start spontaneously between the arc terminals, if sufficient voltage is supplied to maintain the arc (as is the case with spark conduction) but the arc has first to be started, that is, the conducting vapor bridge be produced. This can be done by bringing the electrodes into contact and separating them, or by a high-voltage spark or Geissler discharge, or by the vapor stream of another arc, or by producing electronic conduction, as by an incandescent filament. Inversely, if the current in the arc stopped even for a moment, conduction ceases, that is, the arc extinguishes and has to be restarted. Thus, arc conduction may also be called *continuous conduction*.

21. The arc stream is conducting only in the direction of its motion, but not in the reverse direction. Any body, which is reached by the arc stream, is conductively connected with it, if positive toward it, but is not in conductive connection, if negative or isolated, since, if this body is negative to the arc stream, an arc stream would have to issue from this body, to connect it conductively, and this would require energy to be expended on the body, before current flows to it. Thus, only if the arc stream is very hot, and the negative voltage of the body impinged by it very high, and the body small enough to be heated to high temperature, an arc spot may form on it by heat energy. If, therefore, a body touched by the arc stream is connected to an alternating voltage, so that it is alternately positive and negative toward the arc stream, then conduction occurs during the half-wave, when this body is positive, but no conduction during the negative half-wave (except when the negative voltage is so high as to give disruptive conduction), and the arc thus rectifies the alternating voltage, that is, permits current to pass in one direction only. The arc thus is a *unidirectional conductor*, and as such extensively used for *rectification* of alternating voltages. Usually vacuum arcs are employed for this purpose, mainly the mercury arc, due to its very great rectifying range of voltage.

Since the arc is a unidirectional conductor, it usually can not exist with alternating currents of moderate voltage, as at the end of every half-wave the arc extinguishes. To maintain an alternating arc between two terminals, a voltage is required sufficiently high to restart the arc at every half-wave by jumping an electrostatic spark between the terminals through the hot residual vapor of the preceding half-wave. The temperature of this vapor is that of the boiling point of the electrode material. The voltage required by the electrostatic spark, that is, by disruptive conduction, decreases with increase of temperature, for a 13-mm. gap about as shown by curve I in Fig. 18. The voltage required to maintain an arc, that is, the direct-current voltage, increases with increasing arc temperature, and therefore increasing radiation, etc., about as shown by curve II in Fig. 18. As seen, the curves I and II intersect at some very high temperature, and materials as carbon, which have a boiling point above this temperature, require a lower voltage for restarting than for maintaining the arc, that is, the voltage required to maintain the arc restarts it at every half-wave of alternating current, and such materials thus give a steady alternating arc. Even materials of a somewhat lower boiling point, in which the starting voltage is not much above the running voltage of the arc, maintain a steady alternating arc, as in starting the voltage consumed by the steadying resistance or reactance is available. Electrode materials of low

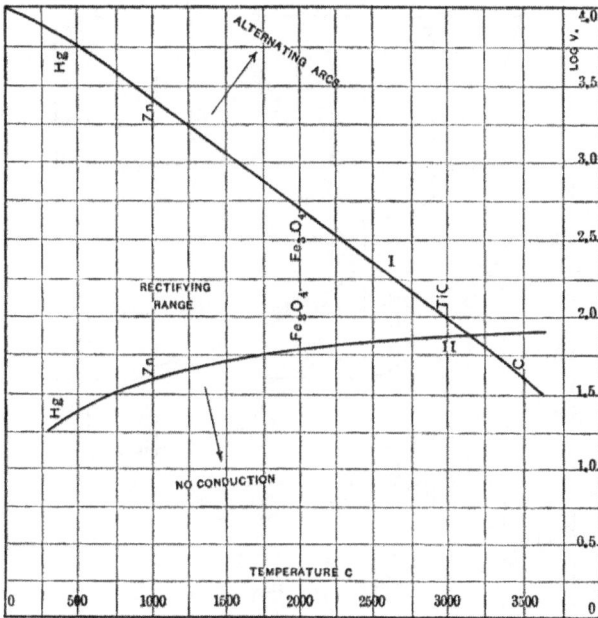

Fig. 18.

boiling point, however, can not maintain steady alternating arcs at moderate voltage.

The range in Fig. 18, above the curve I, thus is that in which alternating arcs can exist; in the range between I and II, an alternating voltage can not maintain the arc, but unidirectional current is produced from an alternating voltage, if the arc conductor is maintained by excitation of its negative terminals, as by an auxiliary arc. This, therefore, is the rectifying range of arc conduction. Below curve II any conduction ceases, as the voltage is insufficient to maintain the conducting vapor stream.

Fig. 18 is only approximate. As ordinates are used the loga-

3

rithm of the voltage, to give better proportions. The boiling points of some' materials are approximately indicated on the curves.

It is essential for the electrical engineer to thoroughly understand the nature of the arc, not only because of its use as illuminant, in arc lighting, but more still because accidental arcs are the foremost cause of instability and troubles from dangerous transients in electric circuits.

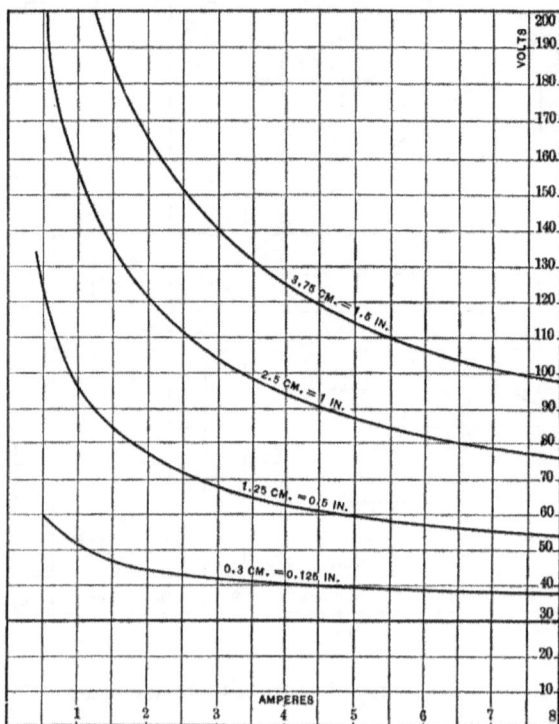

FIG. 19.

22. The voltage consumed by an arc stream, e_1, at constant current, i, is approximately proportional to the arc length, l, or rather to the arc length plus a small quantity, δ, which probably represents the cooling effect of the electrodes.

Plotting the arc voltage, e, as function of the current, i, at constant arc length, gives dropping volt-ampere characteristics, and the voltage increases with decreasing current the more, the longer the arc. Such characteristics are shown in Fig. 19 for the magnetite arcs of 0.3; 1.25; 2.5 and 3.75 cm. length.

These curves can be represented with good approximation by the equation

$$e = a + \frac{c(l + \delta)}{\sqrt{i}} \tag{4}$$

This equation, which originally was derived empirically, can also be derived by theoretical reasoning:

Assuming the amount of arc vapor, that is, the section of the conducting vapor stream, as proportional to the current, and the heat produced at the positive terminal as proportional to the vapor stream and thus the current, the power consumed at the terminals is proportional to the current. As the power equals the current times the terminal drop of voltage, it follows that this terminal drop, a, is constant and independent of current or arc length—similar as the terminal drop at the electrodes in electrolytic conduction is independent of the current.

The power consumed in the arc stream, $p_1 = e_1 i$, is given off from the surface of the stream, by radiation, conduction and convection of heat. The temperature of the arc stream is constant, as that of the boiling point of the electrode material. The power, therefore, is proportional to the surface of the arc stream, that is, proportional to the square root of its section, and therefore the square root of the current, and proportional to the arc length, l, plus a small quantity, δ, which corrects for the cooling effect of the electrodes. This gives

$$p_1 = e_1 i = c \sqrt{i} (l + \delta)$$

or,

$$e_1 = \frac{c(l + \delta)}{\sqrt{i}} \tag{5}$$

as the voltage consumed in the arc stream.

Since a represents the coefficient of power consumed in producing the vapor stream and heating the positive terminal, and c the coefficient of power dissipated from the vapor stream, a and c are different for different materials, and in general higher for materials of higher boiling point and thus higher arc temperature. c, however, depends greatly on the gas pressure in the space in which the arc occurs, and decreases with decreasing gas pressure. It is, approximately, when l is given in centimeter at atmospheric pressure,

$a = 13$ volts for mercury,
  $= 16$ volts for zinc and cadmium (approximately),
  $= 30$ volts for magnetite,
  $= 36$ volts for carbon;
$c = 31$ for magnetite,
  $= 35$ for carbon;
$\delta = 0.125$ cm. for magnetite,
  $= 0.8$ cm. for carbon.

The least agreement with the equation (4) is shown by the carbon arc. It agrees fairly well for arc lengths above 0.75 cm., but for shorter arc lengths, the observed voltage is lower than given by equation (4), and approaches for $l = 0$ the value $e = 28$ volts.

It seems as if the terminal drop, $a = 36$ volts with carbon, consists of an actual terminal drop, $a_0 = 28$ volts, and a terminal drop of $a_1 = 8$ volts, which resides in the space within a short distance from the terminals.

Stability Curves of the Arc

23. As the volt-ampere characteristics of the arc show a decrease of voltage with increase of current, over the entire range of current, the arc is unstable on constant voltage supplied to its terminals, at every current.

Inserting in series to a magnetite arc of 1.8 cm. length, shown as curve I in Fig. 20, a constant resistance of $r = 10$ ohms, the voltage consumed by this resistance is proportional to the current, and thus given by the straight line II in Fig. 20. Adding this voltage II to the arc-voltage curve I, gives the total voltage consumed by the arc and its series resistance, shown as curve III. In curve III, the voltage decreases with increase of current, up to $i_0 = 2.9$ amp. and the arc thus is unstable for currents below 2.9 amp. For currents larger than 2.9 amp. the voltage increases with increase of current, and the arc thus is stable. The point $i_0 = 2.9$ amp. thus separates the unstable lower part of curve III, from the stable upper part.

With a larger series resistance, $r' = 20$ ohms, the stability range is increased down to 1.7 amp., as seen from curve III, but higher voltages are required for the operation of the arc.

With a smaller series resistance, $r'' = 5$ ohms, the stability range is reduced to currents above 4.8 amp., but lower voltages are sufficient for the operation of the arc.

At the stability limit, i_0, in curve III of Fig. 20, the resultant characteristic is horizontal, that is, the slope of the resistance curve II: $r = \dfrac{e'}{i}$, is equal but opposite to that of the arc charac-

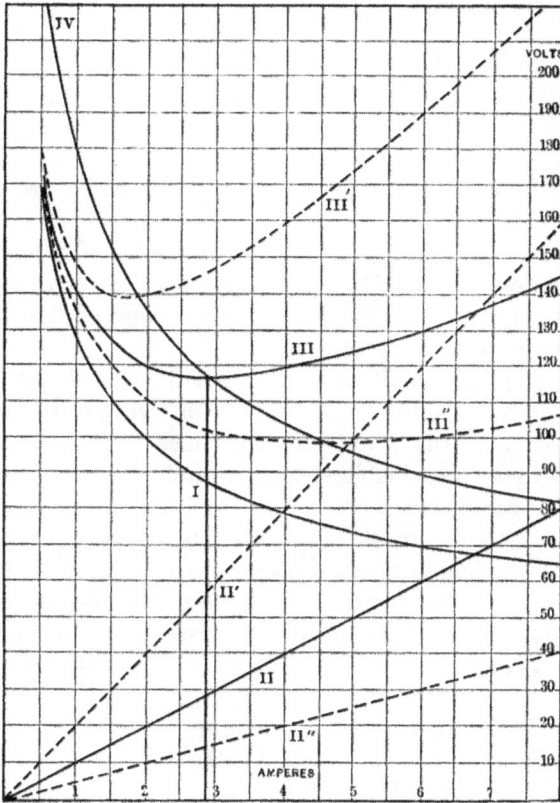

Fig. 20.

teristic I: $\dfrac{de}{di}$. The resistance, r, required to give the stability limit at current, i, thus is found by the condition

$$r = -\frac{de}{di} \tag{6}$$

Substituting equation (4) into (6) gives

$$r = \frac{c(l + \delta)}{2\,i\sqrt{i}} \tag{7}$$

as the minimum resistance to produce stability, hence,

$$ri = \frac{c(l + \delta)}{2\sqrt{i}} = 0.5 \, e_1 \tag{8}$$

where e_1 = arc stream voltage, and

$$E = e + ri$$
$$= a + 1.5 \frac{c(l + \delta)}{\sqrt{i}} \tag{9}$$

is the minimum voltage required by arc and series resistance, to just reach stability.

(9) is plotted as curve IV in Fig. 20, and is called the *stability curve* of the arc. It is of the same form as the arc characteristic I, and derived therefrom by adding 50 per cent. of the voltage, e_1, consumed by the arc stream.

The stability limit of an arc, on constant potential, thus lies at an excess of the supply voltage over the arc voltage $e = a + e_1$, by 50 per cent. of the voltage, e_1, consumed in the arc stream. In general, to get reasonable steadiness and absence of drifting of current, a somewhat higher supply voltage and larger series resistance, than given by the stability curve IV, is desirable.

24. The preceding applies only to those arcs in which the gas pressure an the space surrounding the arc, and thereby the arc vapor pressure and temperature, are constant and independent of the current, as is the case with arcs in air, at "atmospheric pressure."

With arcs in which the vapor pressure and temperature vary with the current, as in vacuum arcs like the mercury arc, different considerations apply. Thus, in a mercury arc in a glass tube, if the current is sufficiently large to fill the entire tube, but not so large that condensation of the mercury vapor can not freely occur in a condensing chamber, the power dissipated by radiation, etc., may be assumed as proportional to the length of the tube, and to the current

$$p = e_1 i = cli$$

thus,

$$e_1 = cl \tag{10}$$

that is, the stream voltage of the tube, or voltage consumed by the arc stream (exclusive terminal drop) is independent of the current. Adding hereto the terminal drop, a, gives as the total voltage consumed by the mercury tube

$$e = a + cl \tag{11}$$

for a mercury arc in a vacuum, it is approximately

$$c = \frac{1.4}{d} \tag{12}$$

where $d =$ diameter of the tube, since the diameter of the tube is proportional to the surface and therefore to the radiation coefficient.

Thus,

$$e = 13 + \frac{1.4 l}{d} \tag{13}$$

At high currents, the vapor pressure rises abnormally, due to incomplete condensation, and the voltage therefore rises, and

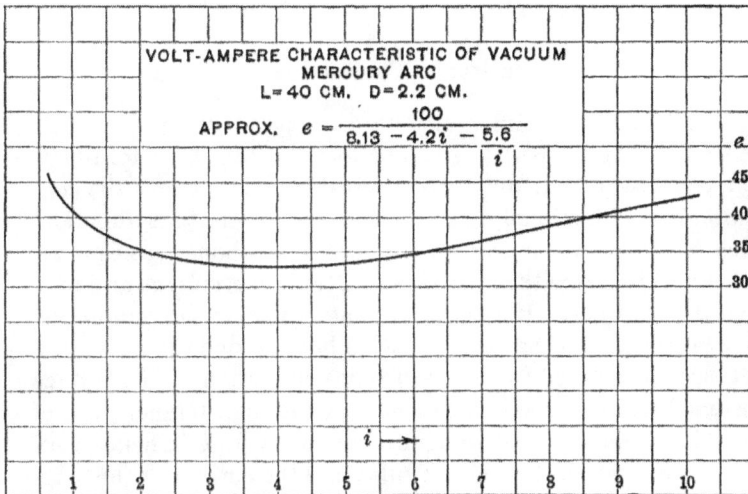

VOLT-AMPERE CHARACTERISTIC OF VACUUM
MERCURY ARC
L= 40 CM. D= 2.2 CM.

APPROX. $e = \dfrac{100}{8.13 - 4.2i - 5.6}$

Fig. 21.

at low currents the voltage rises again, due to the arc not filling the entire tube. Such a volt-ampere characteristic is given in Fig. 21.

25. Herefrom then follows, that the voltage gradient in the mercury arc, for a tube diameter of 2 cm., is about ¾ volts per centimeter or about one-twentieth of what it is in the Geissler tube, and the specific resistance of the stream, at 4 amp., is about 0.2 ohms per cm.[3], or of the magnitude of one one-thousandth of what it is in the Geissler tube.

At higher currents, the mercury arc in a vacuum gives a rising volt-ampere characteristic. Nevertheless it is not stable on constant-potential supply, as the rising characteristic applies only to stationary conditions; the instantaneous characteristic is dropping. That is, if the current is suddenly increased, the voltage drops, regardless of the current value, and then gradually, with the increasing temperature and vapor pressure, increases again, to the permanent value, a lower value or a higher value, whichever may be given by the permanent volt-ampere characteristic.

In an arc at atmospheric pressure, as the magnetite arc, the voltage gradient depends on the current, by equation (1), and at 4 amp. is about 15 to 18 volts per centimeter. The specific resistance of the arc stream is of the magnitude of 1 ohm per cm.[3], and less with larger current arcs, thus of the same magnitude as in vacuum arcs.

Electronic Conduction

26. Conduction occurs at moderate voltages between terminals in a partial vacuum as well as in a perfect vacuum, if the terminals are incandescent. If only one terminal is incandescent, the conduction is unidirectional, that is, can occur only in that direction, which makes the incandescent terminal the cathode, or negative. Such a vacuum tube then rectifies an alternating voltage and may be used as rectifier. If a perfect vacuum exists in the conducting space between the electrodes of such a hot cathode tube, the conduction is considered as true electronic conduction. The voltage consumed by the tube is depending on the high temperature of the cathode, and is of the magnitude of arc voltages, hence very much lower than in the Geissler tube, and the current of the magnitude of arc currents, hence much higher than in the Geissler tube.

27. The complete volt-ampere characteristic of gas and vapor conduction thus would give a curve of the shape in Fig. 22. It consists of three branches separated by ranges of instability or discontinuity. The branch *a*, at very low current, electronic conduction; the branch *b*, discontinuous or Geissler tube conduction; and the branch *c*, arc conduction. The change from *a* to *b* occurs suddenly and abruptly, accompanied by a big rise of current, as soon as the disruptive voltage is reached. The change *b* to *c* occurs suddenly and abruptly, by the formation of a cathode spot, anywhere in a wide range of current, and is accompanied by a sudden drop of voltage. To show the entire range, as abscissæ are used $\sqrt[4]{i}$ and as ordinates $\sqrt{e}$.

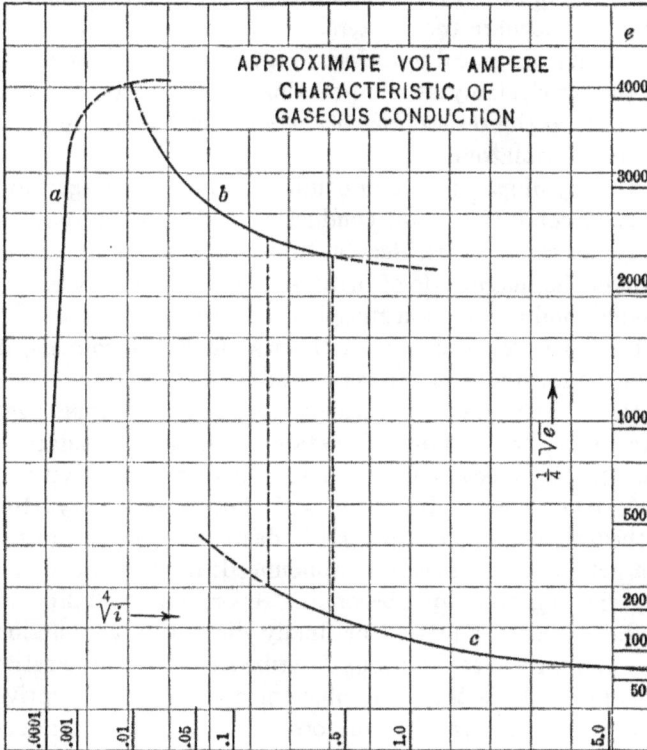

APPROXIMATE VOLT AMPERE CHARACTERISTIC OF GASEOUS CONDUCTION

Fig. 22.

Review

28. The various classes of conduction: metallic conduction, electrolytic conduction, pyroelectric conduction, insulation, gas vapor and electronic conduction, are only characteristic types, but numerous intermediaries exist, and transitions from one type to another by change of electrical conditions, of temperature, etc.

As regards to the magnitude of the specific resistance or resistivity, the different types of conductors are characterized about as follows:

The resistivity of metallic conductors is measured in microhm-centimeters.

The resistivity of electrolytic conductors is measured in ohm-centimeters.

The resistivity of insulators is measured in megohm-centimeters and millions of megohm-centimeters.

The resistivity of typical pyroelectric conductors is of the magnitude of that of electrolytes, ohm-centimeters, but extends from this down toward the resistivities of metallic conductors, and up toward that of insulators.

The resistivity of gas and vapor conduction is of the magnitude of electrolytic conduction: arc conduction of the magnitude of lower resistance electrolytes, Geissler tube conduction and corona conduction of the magnitude of higher-resistance electrolytes.

Electronic conduction at atmospheric temperature is of the magnitude of that of insulators; with incandescent terminals, it reaches the magnitude of electrolytic conduction.

While the resistivities of pyroelectric conductors extend over the entire range, from those of metals to those of insulators, typical are those pyroelectric conductors having a resistivity of electrolytic conductors. In those with lower resistivity, the drop of the volt-ampere characteristic decreases and the instability characteristic becomes less pronounced; in those of higher resistivity, the negative slope becomes steeper, the instability increases, and streak conduction or finally disruptive conduction appears. The streak conduction, described on the pyroelectric conductor, probably is the same phenomenon as the disruptive conduction or breakdown of insulators. Just as streak conduction appears most under sudden application of voltage, but less under gradual voltage rise and thus gradual heating, so insulators of high disruptive strength, when of low resistivity by absorbed moisture, etc., may stand indefinitely voltages applied intermittently—so as to allow time for temperature equalization—while quickly breaking down under very much lower sustained voltage.

CHAPTER III

MAGNETISM

Reluctivity

29. Considering magnetism as the phenomena of a "magnetic circuit," the foremost differences between the characteristics of the magnetic circuit and the electric circuit are:

(a) The maintenance of an electric circuit requires the expenditure of energy, while the maintenance of a magnetic circuit does not require the expenditure of energy, though the starting of a magnetic circuit requires energy. A magnetic circuit, therefore, can remain "remanent" or "permanent."

(b) All materials are fairly good carriers of magnetic flux, and the range of magnetic permeabilities is, therefore, narrow, from 1 to a few thousands, while the range of electric conductivities covers a range of 1 to 10^{18}. The magnetic circuit thus is analogous to an uninsulated electric circuit immersed in a fairly good conductor, as salt water: the current or flux can not be carried to any distance, or constrained in a "conductor," but divides, "leaks" or "strays."

(c) In the electric circuit, current and e.m.f. are proportional, in most cases; that is, the resistance is constant, and the circuit therefore can be calculated theoretically. In the magnetic circuit, in the materials of high permeability, which are the most important carriers of the magnetic flux, the relation between flux, m.m.f. and energy is merely empirical, the "reluctance" or magnetic resistance is not constant, but varies with the flux density, the previous history, etc. In the absence of rational laws, most of the magnetic calculations thus have to be made by taking numerical values from curves or tables.

The only rational law of magnetic relation, which has not been disproven, is Fröhlich's (1882):

"The premeability is proportional to the magnetizability"

$$\mu = a(S - B) \tag{1}$$

where B is the magnetic flux density, S the saturation density,

43

and $S - B$ therefore the magnetizability, that is, the still available increase of flux density, over that existing.

From (1) follows, by substituting,

$$\mu = \frac{B}{H} \tag{2}$$

and rearranging,

$$B = \frac{H}{\alpha + \sigma H} \tag{3}$$

where

$\sigma = \dfrac{1}{S}$ = *saturation coefficient*, that is, the reciprocal of the saturation value, S, of flux density, B, and

$$\alpha = \frac{1}{aS} = \frac{\sigma}{a},$$

for $B = 0$, equation (1) gives

$$\mu_0 = aS = \frac{1}{\alpha}; \quad \alpha = \frac{1}{\mu_0} \tag{4}$$

that is, α is the reciprocal of the magnetic permeability at zero flux density.

A very convenient form of this law has been found by Kennelly (1893) by introducing the reciprocal of the permeability, as reluctivity ρ,

$$\rho = \frac{1}{\mu} = \frac{H}{B},$$

in the form, which can be derived from (3) by transposition.

$$\rho = \alpha + \sigma H \tag{5}$$

As α dominates the reluctivity at lower magnetizing forces, and thereby the initial rate of rise of the magnetization curve, which is characteristic of the "magnetic hardness" of the material, it is called the *coefficient of magnetic hardness*.

30. When investigating flux densities, B, at very high field intensities, H, it was found that B does not reach a finite saturation value, but increases indefinitely; that, however,

$$B_0 = B - H \tag{6}$$

reaches a finite saturation value S, which with iron usually is not far from 20 kilolines per cm.^{2}, and that therefore Fröhlich's and Kennelly's laws apply not to B, but to B_0. The latter, then, is usually called the *metallic magnetic density* or *ferromagnetic density*.

B_0 may be considered as the magnetic flux carried by the molecules of the iron or other magnetic material, in addition to the

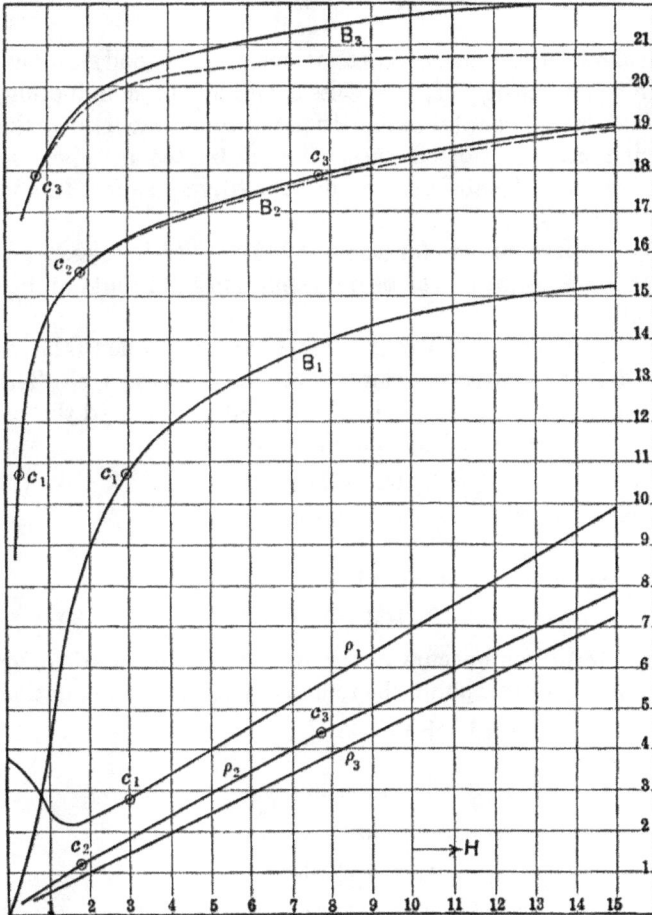

Fig. 23.

space flux, H, or flux carried by space independent of the material in space.

The best evidence seems to corroborate, that with the exception of very low field intensities (where the customary magnetization curve usually has an inward bend, which will be discussed later) in perfectly pure magnetic materials, iron, nickel, cobalt, etc., the linear law of reluctivity (5) and (3) is rigidly obeyed by the metallic induction B_0.

In the more or less impure commercial materials, however, the $\rho - H$ relation, while a straight line, often has one, and occasionally two points, where its slope, and thus the values of α and σ change.

Fig. 23 shows an average magnetization curve, of good standard iron, with field intensity, H, as abscissæ, and magnetic induction, B, as ordinates. The total induction is shown in drawn lines, the metallic induction in dotted lines. The ordinates are given in kilolines per cm.^{2}, the abscissæ in units for B_1, in tens for B_2, and in hundreds for B_3.

The reluctivity curves, for the three scales of abscissæ, are plotted as ρ_1, ρ_2, ρ_3, in tenths of milli-units, in milli-units and in tens of milli-units.

Below $H = 3$, ρ is not a straight line, but curved, due to the inward bend of the magnetization curve, B, in this range. The straight-line law is reached at the point c_1, at $H = 3$, and the reluctivity is then expressed by the linear law

$$\rho_1 = 0.102 + 0.059 H \qquad (7)$$

for

$$3 < H < 18,$$

giving an apparent saturation value,

$$S_1 = 16,950.$$

At $H = 18$, a bend occurs in the reluctivity line, marked by point c_2, and above this point the reluctivity follows the equation

$$\rho_2 = 0.18 + 0.0548 H \qquad (8)$$

for

$$18 < H < 80,$$

giving an apparent saturation value

$$S_2 = 18,250.$$

At $H = 80$, another bend occurs in the reluctivity line, marked by point c_3, and above this point, up to saturation, the reluctivity follows the equation

$$\rho_3 = 0.70 + 0.0477 H \qquad (9)$$

for

$$H > 80$$

giving the true saturation value,

$$S = 20,960.$$

Point c_2 is frequently absent.

Fig. 24 gives once more the magnetization curve (metallic induction) as B, and gives as dotted curves B_1, B_2 and B_3 the magnetization curves calculated from the three linear reluctivity equations (7), (8), (9). As seen, neither of the equations represents

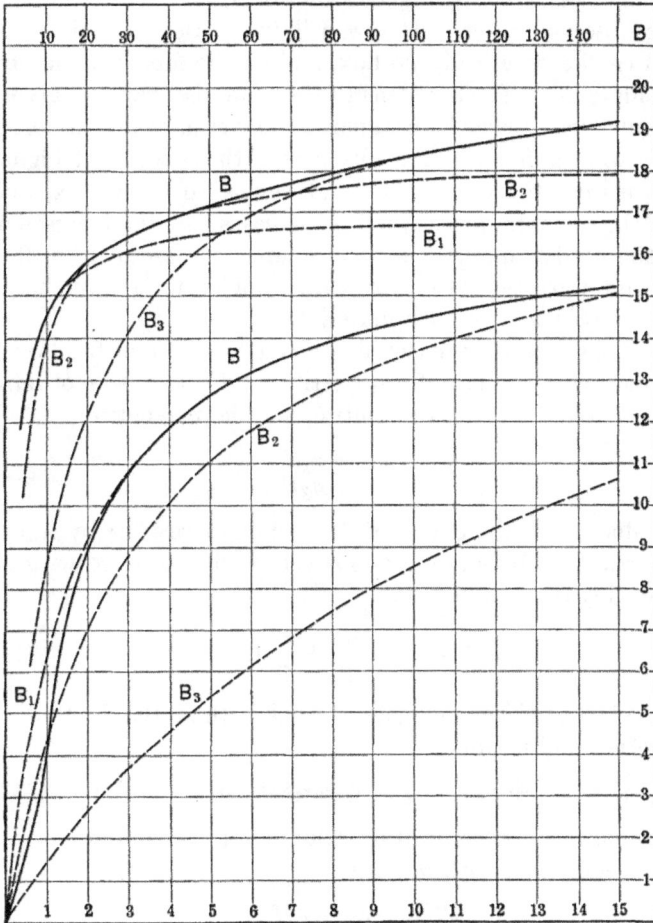

FIG. 24.

B even approximately over the entire range, but each represents it very accurately within its range. The first, equation (7), probably covers practically the entire industrially important range.

37. As these critical points c_2 and c_3 do not seem to exist in perfectly pure materials, and as the change of direction of the reluctivity line is in general the greater, the more impure the material, the cause seems to be lack of homogeneity of the material; that is, the presence, either on the surface as scale, or in the body, as inglomerate, of materials of different magnetic characteristics: magnetite, cementite, silicide. Such materials have a much greater hardness, that is, higher value of α, and thereby would give the observed effect. At low field intensities, H, the harder material carries practically no flux, and all the flux is carried by the soft material. The flux density therefore rises rapidly, giving low α, but tends toward an apparent low saturation value, as the flux-carrying material fills only part of the space. At higher field intensities, the harder material begins to carry flux, and while in the softer material the flux increases less, the increase of flux in the harder material gives a greater increase of total flux density and a greater saturation value, but also a greater hardness, as the resultant of both materials.

Thus, if the magnetic material is a conglomerate of fraction p of soft material of reluctivity ρ_1 (ferrite) and $q = 1 - p$ of hard material of reluctivity, ρ_2 (cementite, silicide, magnetite),

$$\left. \begin{array}{l} \rho_1 = \alpha_1 + \sigma_1 H \\ \rho_2 = \alpha_2 + \sigma_2 H \end{array} \right\} \qquad (10)$$

at low values of H, the part p of the section carries flux by ρ_1, the part q carries flux by ρ_2, but as ρ_2 is very high compared with ρ_1, the latter flux is negligible, and it is

$$\rho^1 = \frac{\rho_1}{p} = \frac{\alpha_1}{p} + \frac{\sigma_1}{p} H \qquad (11)$$

At high values of H, the flux goes through both materials, more or less in series, and it thus is

$$\rho'' = p\rho_1 + q\rho_2 = (p\alpha_1 + q\alpha_2) + (p\sigma_1 + q\sigma_2)H \qquad (12)$$

if we assume the same saturation value, σ, for both materials, and neglect α_1 compared with α_2, it is

$$\rho'' = q\,\alpha_2 + \sigma H \qquad (13)$$

Substituting, as instance, (7) and (9) into (11) and (13) respectively, gives

$$\frac{\alpha_1}{p} = 0.102,$$

$$\frac{\sigma}{p} = 0.059,$$

$$q\alpha_2 = 0.70,$$
$$\sigma = 0.0477,$$

hence,

$$p = 0.80: \quad \rho_1 = 0.082 + 0.0477 \, H,$$
$$q = 0.20: \quad \rho_2 = \quad 3.5 + 0.0477 \, H.$$

However, the saturation coefficients, σ, of the two materials probably are usually not equal.

The deviation of the reluctivity equation from a straight line, by the change of slope at the critical points, c_2 and c_3, thus probably is only apparent, and is the outward appearance of a change of the flux carrier in an unhomogeneous material, that is, the result of a second and magnetically harder material beginning to carry flux.

Such bends in the reluctivity line have been artificially produced by Mr. John D. Ball in combining by superposition two different materials, which separately gave straight-line, ρ, curves, while combined they gave a curve showing the characteristic bend.

Very impure materials, like cast iron, may give throughout a curved reluctivity line.

32. For very low values of field intensity, $H < 3$, however, the straight-line law of reluctivity apparently fails, and the magnetization curve in Fig. 23 has an inward bend, which gives rise of ρ with decreasing H.

This curve is taken by ballistic galvanometer, by the step-by-step method, that is, H is increased in successive steps, and the increase of B observed by the throw of the galvanometer needle. It thus is a "rising magnetization curve."

The first part of this curve is in Fig. 25 reproduced, as B_1, in twice the abscissæ and half the ordinates, so as to give it an average slope of 45°, as with this slope curve shapes such as the inward bend of B_1 below $H = 2$, are best shown ("Engineering Mathematics," p. 286).

Suppose now, at some point, $B_0 = 13.15$, we stop the increase of H, and decrease again, down to 0. We do not return on the same magnetization curve, B_1, but on another curve, B'_1, the "decreasing magnetic characteristic," and at $H = 0$, we are not back to $B = 0$, but a residual or remanent flux is left, in Fig. 25: $R = 7.4$.

Where the magnetic circuit contains an air-gap, as the field circuits of electrical machinery, the decreasing magnetic characteristic, B'_1, is very much nearer to the increasing one, B_1, than in

4

the closed magnetic circuit, Fig. 25, and practically coincides for higher values of H.

There appears no theoretical reason why the rising characteristic, B_1, should be selected as the representative magnetization curve, and not the decreasing characteristic, B'_1, except the incident, that B_1 passes through zero. In many engineering applications, for instance, the calculation of the regulation of a generator, that is, the decrease of voltage under increase of load, it is obviously the decreasing characteristic, B'_1, which is determining.

Suppose we continue B'_1 into negative values of H, to the point A_1, at $H = -1.5$, $B = -4$, and then again reverse, we get a rising magnetization curve, B'', which passes $H = 0$ at a negative remanent magnetism. Suppose we stop at point A_2, at $H = -1.12$, $B = -1.0$: the rising magnetization curve B''' then passes $H = 0$ at a positive remanent magnetism. There must thus be a point, A_0, between A_1 and A_2, such that the rising magnetization curve, B', starting from A_0, passes through the zero point $H = 0$, $B = 0$, and thereby runs into the curve, B_1.

The rising magnetization curve, or standard magnetic characteristic determined by the step-by-step method, B_1, thus is nothing but the rising branch of an unsymmetrical hysteresis cycle, traversed between such limits $+B_0$ and $-A_0$, that the rising branch of the hysteresis cycle passes through the zero point.

33. The characteristic shape of a hysteresis cycle is that it is a loop, pointed at either end and thereby having an inflexion point about the middle of either branch. In the unsymmetrical loop $+B_1$, $-A_0$ of Fig. 25, the zero point is fairly close to one extreme, A_0, and the inflexion point, characteristic of the hysteresis loop, thus lies between 0 and B_0, that is, on that part of the rising branch, which is used as the "magnetic characteristic," B_1, and thereby produces the inward bend in the magnetization curve at low fields, which has always been so puzzling.

If, however, we would stop the increase of H at B''_0, we would get the decreasing magnetization curve, B''_1, and still other curves for other starting points of the decreasing characteristic.

Thus, the relation between magnetic flux density, B, and magnetic field intensity, H, is not definite, but any point between the various rising and decreasing characteristics B'', B_1, B''', B''_1, B'_1, and for some distance outside thereof, is a possible B-H relation. B_1 has the characteristic that it passes through the zero point. But it is not the only characteristic which does this:

if we traverse the hysteresis cycle between the unsymmetrical limits $+A_0$ and $-B_0$, as shown in Fig. 26, its decreasing branch B_3 passes through the zero point, that is, has the same feature as B_1. It is interesting to note, that B_3 does not show an inward bend, and the reluctivity curve of B_3, given as ρ_3 in Fig. 28, apparently is a straight line.

Magnetic characteristics are frequently determined by the method of reversals, by reversing the field intensity, H, and observing the voltage induced thereby by ballistic galvanometer,

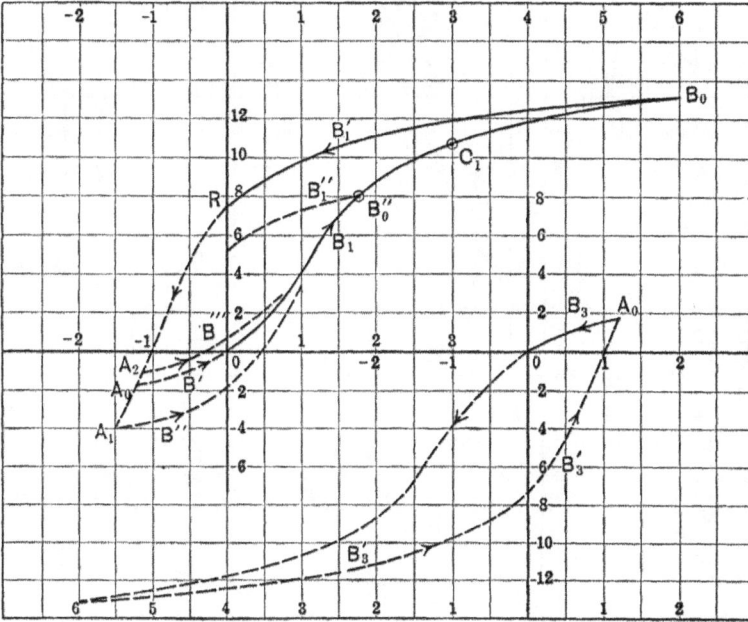

FIGS. 25 AND 26.

or using an alternating current for field excitation, and observing the induced alternating voltage, preferably by oscillograph to eliminate wave-shape error.

This "alternating magnetic characteristic" is the one which is of consequence in the design of alternating-current apparatus. It differs from the "rising magnetic characteristic," B_1 by giving lower values of B, for the same H, materially so at low values of H. It shows the inward bend at low fields still more pronounced than B_1 does. It is shown as curve B_2 in Fig. 27, and its reluctivity line given as ρ_2 in Fig. 28. At higher values of H: from $H = 3$ upward, B_1 and B_2 both coincide with the curve, B_0, representing the straight-line reluctivity law.

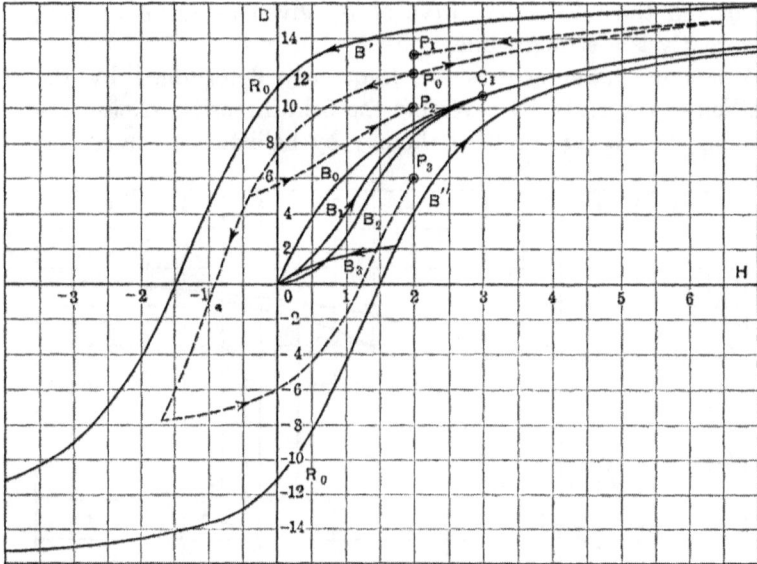

Fig. 27.

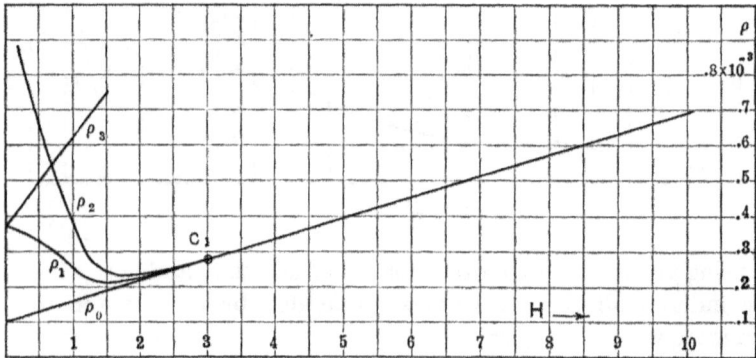

Fig. 28.

The alternating characteristic, B_2, is not a branch of any hysteresis cycle. It is reproducible and independent of the previous history of the magnetic circuit, except perhaps at extremely low values of H, and in view of its engineering importance as representing the conditions in the alternating magnetic field, it would appear the most representative magnetic characteristic, and is commonly used as such.

It has, however, the disadvantage that it represents an unstable condition.

Thus in Fig. 27, an alternating field $H = 1$ gives an alternating flux density, $B_2 = 2.6$. If, however, this field strength $H = 1$ is left on the magnetic circuit, the flux does not remain at $B_2 = 2.6$, but gradually creeps up to higher values, especially in the presence of mechanical vibrations or slight pulsations of the magnetizing current. To a lesser extent, the same occurs with the values of curve, B_1, to a greater extent with B_3. At very low densities, this creepage due to instability of the B-H relation may amount to hundreds of per cent. and continue to an appreciable extent for minutes, and with magnetically hard materials for many years. Thus steel structures in the terrestrial magnetic field show immediately after erection only a small part of the magnetization, which they finally assume, after many years.

Thus the alternating characteristic, B_2, however important in electrical engineering, can, due to its instability, not be considered as representing the true physical relation between B and H any more than the branches of hysteresis cycles B_1 and B_3.

34. Correctly, the relation between B and H thus can not be expressed by a curve, but by an area.

Suppose a hysteresis cycle is performed between infinite values of field intensity: $H = \pm \infty$, that is, practically, between very high values such as are given for instance by the isthmus method of magnetic testing (where values of H of over 40,000 have been reached. Very much lower values probably give practically the same curve). This gives a magnetic cycle shown in Fig. 5 as B', B''. Any point, H, B, within the area of this loop between B' and B'' of Fig. 27 then represents a possible condition of the magnetic circuit, and can be reached by starting from any other point, H_0, B_0, such as the zero point, by gradual change of H.

Thus, for instance, from point P_0, the points P_1, P_2, P_3, etc., are reached on the curves shown in the dotted lines in Fig. 27.

As seen from Fig. 27, a given value of field intensity, such as $H = 1$, may give any value of flux density between $B = -4.6$ and $B = +13.6$, and a given value of flux density, such as $B = 10$, may result from any value of field intensity, between $H = -0.25$ to $H = +3.4$

The different values of B, corresponding to the same value of H in the magnetic area, Fig. 27, are not equally stable, but the values near the limits B' and B'' are very unstable, and become more stable toward the interior of the area. Thus, the relation of point P_1, Fig. 27: $H = 2$, $B = 13$, would rapidly change, by the flux density decreasing, to P_0, slower to P_2 and then still slower, while from point P_3 the flux density would gradually creep up.

If thus follows, that somewhere between the extremes B' and B'', which are most unstable, there must be a value of B, which is stable, that is, represents the stationary and permanent relation between B and H, and toward this stable value, B_0, all other values would gradually approach. This, then, would give the true magnetic characteristic: the stable physical relation between B and H.

At higher field intensities, beyond the first critical point, c_1, this stable condition is rapidly reached, and therefore is given by all the methods of determining magnetic characteristics. Hence, the curves B_1, B_2, B_0 coincide there, and the linear law of reluctivity applies. Below c_1, however, the range of possible, B, values is so large, and the final approach to the stable value so slow, as to make it difficult of determination.

35. For $H = 0$, the magnetic range is from $-R_0 = -11.2$ to $+R_0 = 11.2$; the permanent value is zero. The method of reaching the permanent value, whatever may be the remanent magnetism, is well known; it is by "demagnetizing" that is, placing the material into a powerful alternating field, a demagnetizing coil, and gradually reducing this field to zero. That is, describing a large number of cycles with gradually decreasing amplitude.

The same can be applied to any other point of the magnetization curve. Thus for $H = 1$, to reach permanent condition, an alternating m.m.f. is superimposed upon $H = 1$, and gradually decreased to zero, and during these successive cycles of decreasing amplitude, with $H = 1$, as mean value, the flux density gradually approaches its permanent or stable value. (The only requirement is, that the initial alternating field must be higher than any unidirectional field to which the magnetic circuit had been exposed.)

This seems to be the value given by curve B_0, that is, by the straight-line law of reluctivity. In other words, it is probable that:

Fröhlich's equation, or Kennelly's linear law of reluctivity represent the permanent or stable relation between B and H, that is, the true magnetic characteristic of the material, over the entire range down to $H = 0$, and the inward bend of the magnetic characteristic for low field intensities, and corresponding increase of reluctivity ρ, is the persistence of a condition of magnetic instability, just as remanent and permanent magnetism are.

In approaching stable conditions by the superposition of an alternating field, this field can be applied at right angles to the unidirectional field, as by passing an alternating current lengthwise, that is, in the direction of the lines of magnetic force, through the material of the magnetic circuit. This superimposes a circular alternating flux upon the continuous-length flux, and permits observations while the circular alternating flux exists, since the latter does not induce in the exploring circuit of the former. Some 20 years ago Ewing has already shown, that under these conditions the hysteresis loop collapses, the inward bend of the magnetic characteristic practically vanishes, and the magnetic characteristic assumes a shape like curve B_0.

To conclude, then, it is probable that:

In pure homogeneous magnetic materials, the stable relation between field intensity, H, and flux density, B, is expressed, over the entire range from zero to infinity, by the linear equation of reluctivity

$$\rho = a + \sigma H,$$

where ρ applies to the metallic magnetic induction, $B - H$.

In unhomogeneous materials, the slope of the reluctivity line changes at one or more critical points, at which the flux path changes, by a material of greater magnetic hardness beginning to carry flux.

At low field intensities, the range of unstable values of B is very great, and the approach to stability so slow, that considerable deviation of B from its stable value can persist, sometimes for years, in the form of remanent or permanent magnetism, the inward bend of the magnetic characteristic, etc.

CHAPTER IV

MAGNETISM

Hysteresis

36. Unlike the electric current, which requires power for its maintenance, the maintenance of a magnetic flux does not require energy expenditure (the energy consumed by the magnetizing current in the ohmic resistance of the magnetizing winding being an electrical and not a magnetic effect), but energy is required to produce a magnetic flux, is then stored as potential energy in the magnetic flux, and is returned at the decrease or disappearance of the magnetic flux. However, the amount of energy returned at the decrease of magnetic flux is less than the energy consumed at the same increase of magnetic flux, and energy is therefore dissipated by the magnetic change, by conversion into heat, by what may be called *molecular magnetic friction*, at least in those materials, which have permeabilities materially higher than unity.

Thus, if a magnetic flux is periodically changed, between $+B$ and $-B$, or between B_1 and B_2, as by an alternating or pulsating current, a dissipation of energy by molecular friction occurs during each magnetic cycle. Experiment shows that the energy consumed per cycle and cm.^{3} of magnetic material depends only on the limits of the cycle, B_1 and B_2, but not on the speed or wave shape of the change.

If the energy which is consumed by molecular friction is supplied by an electric current as magnetizing force, it has the effect that the relations between the magnetizing current, i, or magnetic field intensity, H, and the magnetic flux density, B, is not reversible, but for rising, H, the density, B, is lower than for decreasing H; that is, the magnetism lags behind the magnetizing force, and the phenomenon thus is called *hysteresis*, and gives rise to the *hysteresis loop*.

However, hysteresis and molecular magnetic friction are not

the same thing, but the hysteresis loop is the measure of the molecular magnetic friction only in that case, when energy is supplied to or abstracted from the magnetic circuit only by the magnetizing current, but not otherwise. Thus, if mechanical work is done by the magnetic cycle—as when attracting and dropping an armature—the hysteresis loops enlarge, representing not only the energy dissipated by molecular magnetic friction, but also that converted into mechanical work. Inversely, if mechanical energy is supplied to the magnetic circuit as by vibrating it mechanically, the hysteresis loop collapses or overturns, and its area becomes equal to the molecular magnetic friction minus the mechanical energy absorbed. The reaction machine, as synchronous motor and as generator, is based on this feature. See "Reaction Machine," "Theory and Calculation of Electrical Apparatus."

In general, when speaking of hysteresis, molecular magnetic friction is meant, and the hysteresis cycle assumed under the condition of no other energy conversion, and this assumption will be made in the following, except where expressly stated otherwise.

The hysteresis cycle is independent of the frequency within commercial frequencies and far beyond this range. Even at frequencies of hundred thousand cycles, experimental evidence seems to show that the hysteresis cycle is not materially changed, except in so far as eddy currents exert a demagnetizing action and thereby require a change of the impressed m.m.f., to get the same resultant m.m.f., and cause a change of the magnetic flux distribution by their screening effect.

A change of the hysteresis cycle occurs only at very slow cycles —cycles of a duration from several minutes to years—and even then to an appreciable extent only at very low magnetic densities. Thus at low values of B—below 1000—hysteresis cycles taken by ballistic galvanometer are liable to become irregular and erratic, by "magnetic creepage." For most practical purposes, however, this may be neglected.

37. As the industrially most important varying magnetic fields are the alternating magnetic fields, the hysteresis loss in alternating magnetic fields, that is, in symmetrical cycles, is of most interest.

In general, if a magnetic flux changes from the condition H_1, B_1: point P_1 of Fig. 29, to the condition H_2, B_2: point P_2, and we assume this magnetic circuit surrounded by an electric circuit of

n turns, the change of magnetic flux induces in the electric circuit the voltage, in absolute units,

$$e = n \frac{d\Phi}{dt} \tag{1}$$

it is, however,

$$\Phi = sB \tag{2}$$

where s = section of magnetic circuit. Hence

$$e = ns \frac{dB}{dt} \tag{3}$$

If i = current in the electric circuit, the m.m.f. is

$$F = ni \tag{4}$$

and the magnetizing force

$$f = \frac{ni}{l} \tag{5}$$

where l = length of the magnetic circuit.

And the field intensity

$$H = 4\pi f \tag{6}$$

hence, substituting (5) into (6) and transposing,

$$i = \frac{lH}{4\pi n} \tag{7}$$

is the magnetizing current in the electric circuit, which produces the flux density, B.

The power consumed by the voltage induced in the electric circuit thus is

$$p = ei = \frac{slH}{4\pi} \frac{dB}{dt} \tag{8}$$

or, per cm.³ of the magnetic circuit, that is, for $s = 1$ and $l = 1l$,

$$p = \frac{H}{4\pi} \frac{dB}{dt} \tag{9}$$

and the energy consumed by the change from H_1, B_1 to H_2, B_2, which is transferred from the electric into the magnetic circuit, or inversely,

$$w_{1,2} = \frac{1}{4\pi} \int_1^2 H dB \text{ ergs} \tag{10}$$

$$= \frac{A_{1,2}}{4\pi},$$

where $A_{1,2}$ is the area shown shaded in Fig. 29.

The energy consumed during a cycle, from H_0, B_0 to $- H_0$, $- B_0$ and back to H_0, B_0, thus is

$$w = \frac{1}{4\,\pi}\int_0^0 H dB \text{ ergs} \tag{11}$$

$$= \frac{A}{4\,\pi} \text{ ergs} \tag{12}$$

where

$$\int_0^0 H dB = A \text{ is the area of the hysteresis loop, shown shaded}$$

in Fig. 30.

As the magnetic condition at the end of the cycle is the same as

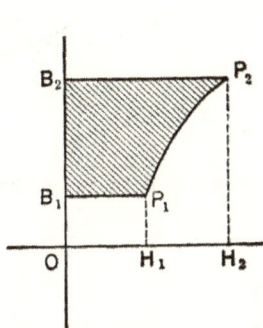

Fig. 29.

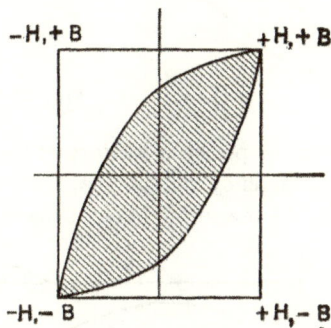

Fig. 30.

at the beginning, all this energy, w, is dissipated as heat, that is, is the hysteresis energy which measures the molecular magnetic friction.

38. If in Fig. 30 the shaded area represents the hysteresis loop between $+ H$, $+ B$, and $- H$, $- B$, giving with a sinusoidal alternating flux the voltage and current waves, Fig. 31, the maximum area, which the hysteresis loop could theoretically assume, is given by the rectangle between $+ H$, $+ B$; $- H$, $+ B$; $- H$, $- B$; $+ H$, $- B$. This would mean, that the magnetic flux does not appreciably decrease with decreasing field intensity, until the field has reversed to full value. It would give the theoretical wave shape shown as Fig. 32. As seen, this is the extreme exaggeration of wave shape, Fig. 31.

The total energy of this rectangle, or maximum available magnetic energy, is

$$w_0 = \frac{4\,HB}{4\,\pi} = \frac{HB}{\pi} \tag{12}$$

or, if μ = permeability, thus $H = \dfrac{B}{\mu}$, it is

$$w_0 = \frac{B^2}{\pi\mu} \tag{13}$$

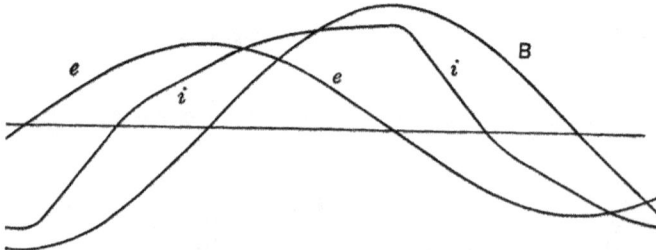

Fig. 31.

the maximum possible hysteresis loss.

The inefficiency of the magnetic cycle, or percentage loss of energy in the magnetic cycle, thus is

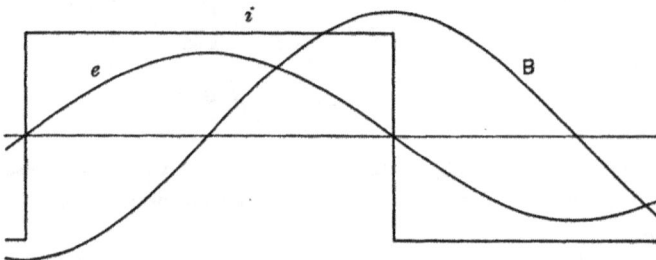

Fig. 32.

$$
\begin{aligned}
\zeta &= \frac{w}{w_0} = \frac{\pi\mu w}{B^2} \\
&= \frac{\mu}{4\,B^2}\int_0^0 H\,dB \\
&= \frac{A\mu}{4\,B^2}
\end{aligned}
\tag{14}
$$

39. Experiment shows that for medium flux density, that is, thoses values of B which are of the most importance industrially,

from $B = 1000$ to $B = 12,000$, the hysteresis loss can with sufficient accuracy for most practical purposes be approximated by
the empirical equation,

$$w = \eta B^{1.6} \qquad (15)$$

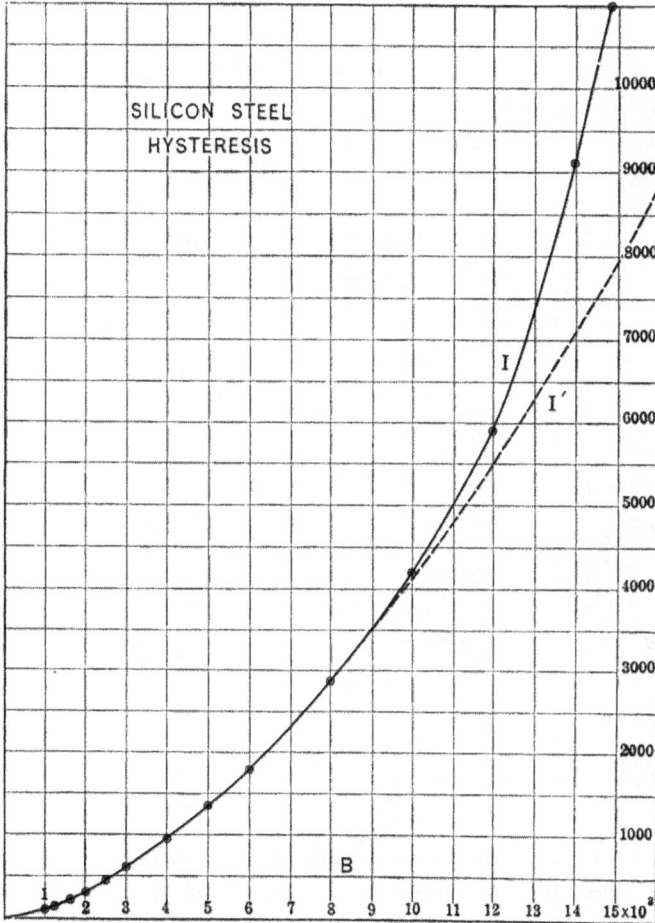

SILICON STEEL
HYSTERESIS

Fig. 33.

where η, the "coefficient of hysteresis," is of the magnitude
of 1×10^{-3} to 2×10^{-3} for annealed soft sheet steel, if B is
given in lines of force per cm.², and w is ergs per cm.³ and cycle.

Very often w is given in joules, or watt-seconds per cycle and
per kilogram or pound of iron, and B in lines per square inch,
or w is given in watts per kilogram or per pound at 60 cycles.

In Fig. 33 is shown, with B as abscissæ, the hysteresis loss, w, of a sample of silicon steel. The observed values are marked by circles. In dotted lines is given the curve calculated by the equation

$$w = 0.824 \times 10^{-3} B^{1.6} \qquad (16)$$

As seen, the agreement the curve of 1.6^{th} power with the test values is good up to $B = 10,000$, but above this density, the observed values rise above the curve.

40. In Fig. 34 is plotted, with field intensity, H, as abscissæ, the magnetization curve of ordinary annealed sheet steel, in

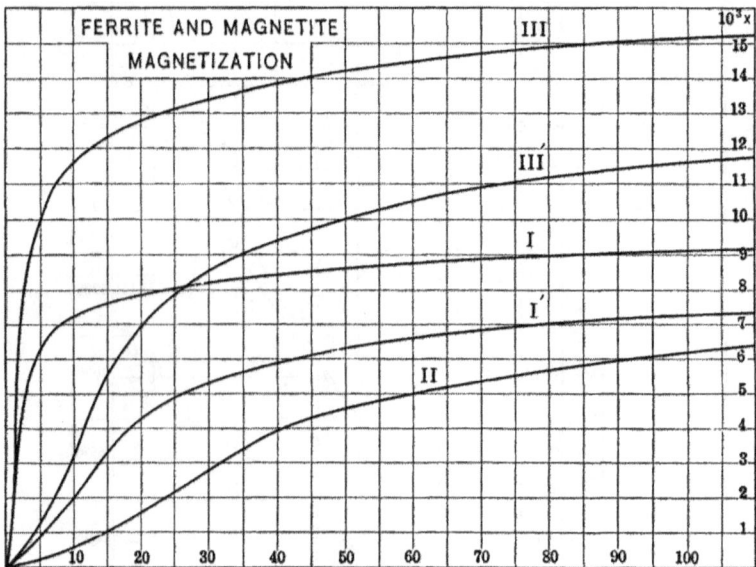

FERRITE AND MAGNETITE
MAGNETIZATION

Fig. 34.

half-scale, as curve I, and the magnetization curve of magnetite, Fe_3O_4—which is about the same as the black scale of iron—in double-scale, as curve II. As III then is plotted, in full-scale, a curve taking 0.8 of I and 0.2 of II. This would correspond to the average magnetic density in a material containing 80 per cent. of iron and 20 per cent. (by volume) of scale. Curves I' and III' show the initial part of I and III, with ten times the scale of abscissæ and the same scale of ordinates.

Fig. 35 then shows, with the average magnetic flux density, B, taken from curve III of Fig. 34, as abscissæ, the part of the magnetic flux density which is carried by the magnetite, as curve I. As seen, the magnetite carries practically no flux up to $B = 10$, but beyond $B = 12$, the flux carried by the magnetite rapidly increases.

As curve II of Fig. 35 is shown the hysteresis loss in this inhomogeneous material consisting of 80 per cent. ferrite (iron) and 20 per cent. magnetite (scale) calculated from curves I and II of Fig.

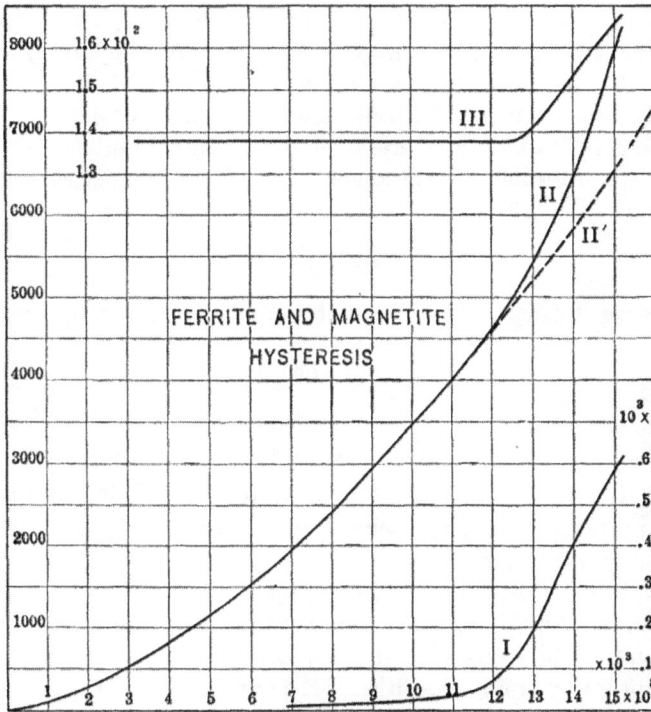

Fig. 35.

34 under the assumption that either material rigidly follows the 1.6^{th} power law up to the highest densities, by the equation,

Iron:

$$w_1 = 1.2 \; B_I{}^{1.6} \times 10^{-3}.$$

Scale:

$$w_2 = 23.5 \; B_{II}{}^{1.6} \times 10^{-3},$$

As curve II' is shown in dotted lines the 1.6^{th} power equation,

$$w = 1.38 \;\; B^{1.6} \times 10^{-3}.$$

As seen, while either constituent follows the 1.6^{th} power law, the combination deviates therefrom at high densities, and gives an increase of hysteresis loss, of the same general characteristic as shown with the silicon steel in Fig. 33, and with most similar materials.

As curve III in Fig. 35 is then shown the increase of the hysteresis coefficient η, at high densities, over the value 1.38×10^{-3}, which it has at medium densities.

Thus, the deviation of the hysteresis loss at high densities, from the 1.6^{th} power law, may possibly be only apparent, and the result of lack of homogeneity of the material.

41. At low magnetic densities, the law of the 1.6^{th} power must cease to represent the hysteresis loss even approximately.

The hysteresis loss, as fraction of the available magnetic energy, is, by equation (14),

$$\zeta = \frac{\pi \mu w}{B^2} \tag{14}$$

Substituting herein the parabolic equation of the hysteresis loss,

$$w = \eta B^n \tag{17}$$

where $n = 1.6$, it is

$$\zeta = \mu \pi \eta \quad B^{n-2} \tag{18}$$
$$= \pi \mu \eta \quad B^{.4}$$

With decreasing density B, B^{n-2} steadily increases, if $n < 2$, and as the permeability μ approaches a constant value, ζ, steadily increases in this case, thus would become unity at some low density, B, and below this, greater than unity. This, however, is not possible, as it would imply more energy dissipated, than available, and thus would contradict the law of conservation of energy. Thus, for low magnetic densities, if the parabolic law of hysteresis (17) applies, the exponent must be: $n \gtrless 2$.

In the case of Fig. 33, for $\eta = 0.824 \times 10^{-3}$, assuming the permeability for extremely low density as

$$\mu = 1500,$$

ζ becomes unity, by equation (18), at

$$B = 30.$$

If $n > 2$, B^{n-2} steadily decreases with decreasing B, and the percentage hysteresis loss becomes less, that is, the cycle approaches reversibility for decreasing density; in other words, the hysteresis loss vanishes. This is possible, but not probable, and the probability is that for very low magnetic densities, the hysteresis losses approach proportionality with the square of the magnetic density, that is, the percentage loss approaches constancy.

From equation (17) follows

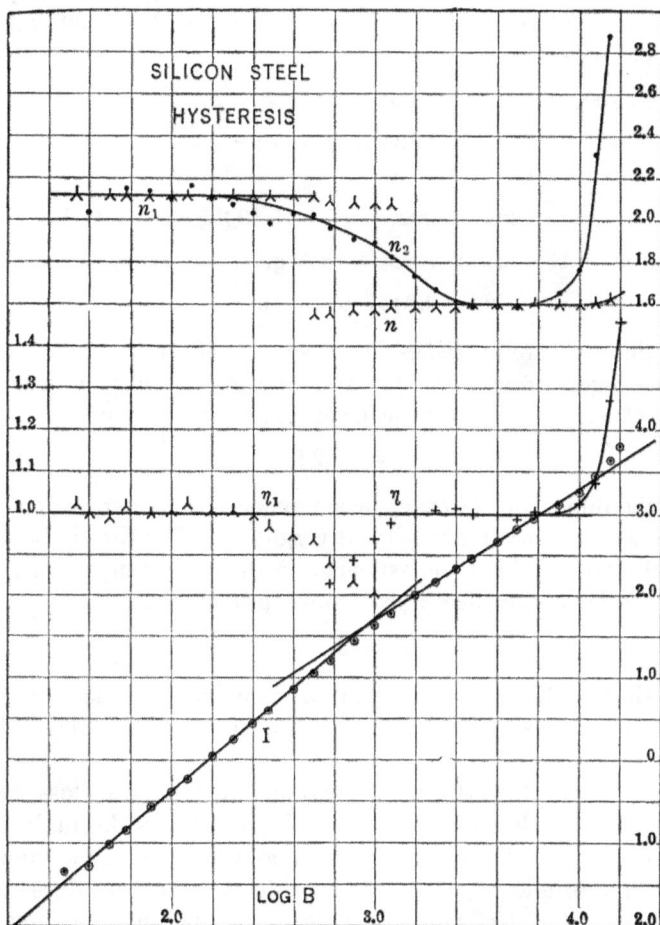

SILICON STEEL

HYSTERESIS

Fɪɢ. 36.

$$\log w = \log \eta + n \log B \qquad (19)$$

That is:

"If the hysteresis loss follows a parabolic law, the curve plotted with log w against log B is a straight line, and the slope of this straight line is the exponent, n."

5

Thus, to investigate the hysteresis law, $\log w$ is plotted against $\log B$. This is done for the silicon steel, Fig. 33, over the range from $B = 30$ to $B = 16{,}000$, in Fig. 36, as curve I.

Curve I contains two straight parts, for medium densities, from $\log B = 3$; $B = 1000$, to $\log B = 4$; $B = 10{,}000$, with slope 1.6006, and for low densities, up to $\log B = 2.6$; $B = 400$, with slope 2.11. Thus it is

For $1000 \leqq \mathrm{B} \leqq 10{,}000$:

$$w = 0.824 \ B^{1 \cdot 6} \times 10^{-3}$$

For $B \leqq 400$:

$$w = 0.00257 \ B^{2 \cdot 11} \times 10^{-3}$$

However, in this lower range, $n = 2$ gives a curve:

$$w = 0.0457 \ B^2 \times 10^{-3}$$

which still fairly well satisfies the observed values.

As the logarithmic curve for a sample of ordinary annealed sheet steel, Fig. 37, gives for the lower range the exponent,

$$n = 1.923,$$

and as the difficulties of exact measurements of hysteresis losses increase with decreasing density, it is quite possible that in both, Figs. 36 and 37 the true exponent in the lower range of magnetic densities is the theoretically most probable one,

$$n = 2,$$

that is, that at about $B = 500$, in iron the point is reached, below which the hysteresis loss varies with the square of the magnetic density.

42. As over most of the magnetic range the hysteresis loss can be expressed by the parabolic law (17), it appears desirable to adapt this empirical law also to the range where the logarithmic curve, Figs. 36 and 37, is curved, and the parabolic law does not apply, above $B = 10{,}000$, and between $B = 500$ and $B = 1000$, or thereabouts. This can be done either by assuming the coefficient η as variable, or by assuming the exponent n as variable.

(a) Assuming η as constant,

$$\eta = 0.824 \times 10^{-3} \text{ for the medium range, where } n \cong 1.6$$
$$\eta_1 = 0.0457 \times 10^{-3} \text{ for the low range, where } n_1 \cong 2$$

The coefficients n and n_1 calculated from the observed values of w, then, are shown in Fig. 36 by the three-cornered stars in the upper part of the figure.

(b) Assuming n as constant,

$n = 1.6$ for the medium range, where $\eta \cong 0.0824 \times 10^{-3}$
$n_1 = 2$ for the low range, where $\eta_1 \cong 0.0457 \times 10^{-3}$

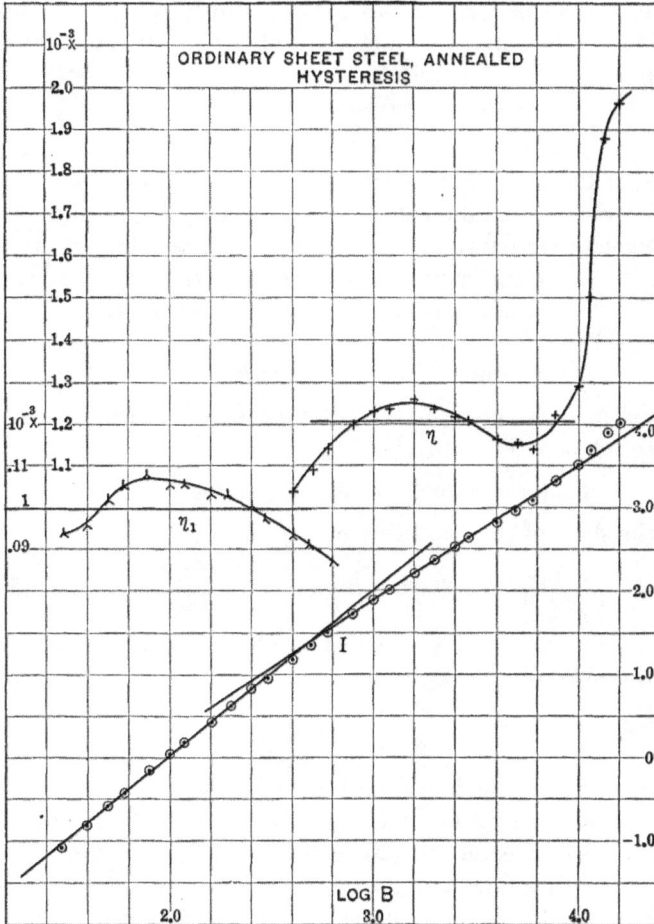

FIG. 37.

The variation of η and η_1, from the values in the constant range, then, are best shown in per cent., that is, the loss w calculated from the parabolic equation and a correction factor applied for values of B outside of the range.

Fig. 37 shows the values of η and η_1, as calculated from the parabolic equations with $n = 1.6$ and $n_1 = 2$, and Fig. 36 shows the percentual variation of η and η_1.

The latter method, (b), is preferable, as it uses only one exponent, 1.6, in the industrial range, and uses merely a correction factor. Furthermore, in the method (a), the variation of the exponent is very small, rising only to 1.64, or by 2.5 per cent., while in method (b) the correction factor is 1.46, or 46 per cent., thus a much greater accuracy possible.

43. If the parabolic law applies,

$$w = \eta B^n \tag{17}$$

the slope of the logarithmic curve is the exponent n.

If, however, the parabolic law does not rigidly apply, the slope of the logarithmic curve is not the exponent, and in the range, where the logarithmic curve is not straight, the exponent thus can not even be approximately derived from the slope.

From (17) follows

$$\log w = \log \eta + n \log B, \tag{19}$$

differentiating (19), gives, in the general case, where the parabolic law does not strictly apply,

$$d \log w = d \log \eta + nd \log B + \log B dn,$$

hence, the slope of the logarithmic curve is

$$\frac{d \log w}{d \log B} = n + \left(\log B \frac{dn}{d \log B} + \frac{d \log \eta}{d \log B} \right) \tag{20}$$

If $n = $ constant, and $\eta = $ constant, the second term on the right-hand side disappears, and it is

$$\frac{d \log w}{d \log B} = n \tag{21}$$

that is, the slope of the logarithmic curve is the exponent.

If, however, η and n are not constant, the second term on the right-hand side of equation (20) does not in general disappear, and the slope thus does not give the exponent.

Assuming in this latter case the slope as the exponent, it must be

$$\log B \frac{dn}{d \log B} + \frac{d \log \eta}{d \log B} = 0.$$

Or,

$$\frac{d \log \eta}{dn} = - \log B \tag{22}$$

In this case, n and much more still η show a very great variation, and the variation of η is so enormous as to make this representation valueless.

As illustration is shown, in Fig. 36, the slope of the curve as n_2. As seen, n_2 varies very much more than n or n_1.

To show the three different representations, in the following table the values of n and η are shown, for a different sample of iron.

TABLE

$B \ 10^3$	(a) η = const. = 1.254	(b) n = const. = 1.6	(c) $n_2 = \dfrac{d \log w}{d \log B}$	η_2
below 10.00	$n = 1.6$	$\eta = 1.254 \times 100^{-3}$	$n_2 = 1.6$	$\eta_2 = 1.254 \times 10^{-6}$
10.00	= 1.601	= 1.268	= 1.79	230.00
11.23	= 1.604	= 1.302	= 2.23	3.68
12.63	= 1.617	= 1.468	= 2.66	0.0488
13.30	= 1.624	= 1.570	= 2.83	0.0133
14.00	= 1.630	= 1.668	= 2.98	0.0032
14.65	= 1.634	= 1.738	= 3.15	0.00069

As seen, to represent an increase of hysteresis loss by $\dfrac{1.738}{1.254} = 1.39$, or 39 per cent., under (c), n_2 is nearly doubled, and η_2 reduced to $\dfrac{1}{1,800,000}$ of its initial value.

44. The equation of the hysteresis loss at medium densities,

$$W = \eta B^n; \quad n = 1.6$$

is entirely empirical, and no rational reason has yet been found why this approximation should apply. Calculating the coefficient n from test values of B and W, shows usually values close to 1.6, but not infrequently values of n are found, as low as 1.55, and even values below 1.5, and values up to 1.7 and even above 1.9 In general, however, the more accurate tests give values of n which do not differ very much from 1.6, so that the losses can still be represented by the curve with the exponent $n = 1.6$, without serious error. This is desirable, as it permits comparing different materials by comparing the coefficients η. This would not be the case, if different values of n were used, as even a small change of n makes a very large change of η: a change of n by 1 per cent., at $B = 10,000$, changes η by about 16 per cent.

Thus in Fig. 37 is represented as I the logarithmic curve of a sample of ordinary annealed sheet steel, which at medium density gives the exponent $n = 1.556$, at low densities the exponent $n_1 = 1.923$. Assuming, however, $n = 1.6$ and $n_1 = 2.0$, gives the average values $\eta = 1.21 \times 10^{-3}$ and $\eta_1 = 0.10 \times 10^{-3}$, and the

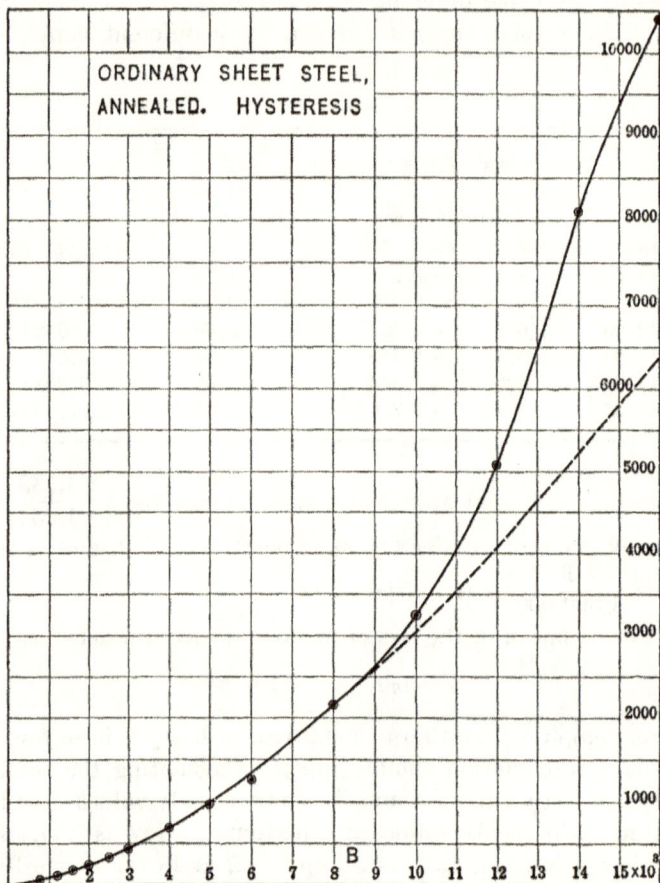

ORDINARY SHEET STEEL, ANNEALED. HYSTERESIS

Fig. 38.

individual calculated values of η and η_1 are then shown on Fig. 37 by crosses and three-pointed stars, respectively.

Fig. 38 then shows the curve of observed loss, in drawn line, and the 1.6^{th} power curve calculated in dotted line, and Fig. 39 the lower range of the calculated curve, with the observations marked by circles. Fig. 40 shows, for the low range, the curve

of $\eta_1 B^2$, in two different scales, with the observed values marked
by cycles. As seen, although in this case the deviation of n from
1.6 respectively 2 is considerable, the curves drawn with $n =$
1.6 and $n_1 = 2$ still represent the observed values fairly well in

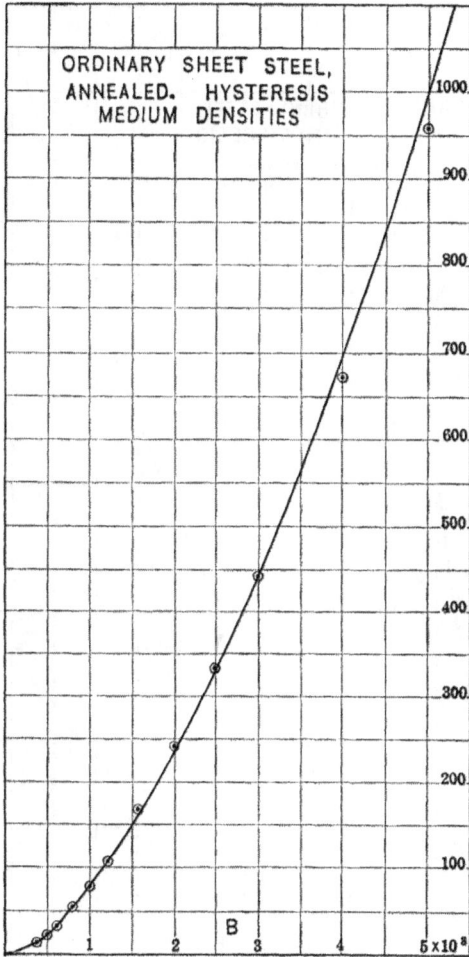

ORDINARY SHEET STEEL,
ANNEALED. HYSTERESIS
MEDIUM DENSITIES

FIG. 39.

the range of B from 500 to 10,000, and below 500, respectively, so
that the 1.6^{th} power equation for the medium, and the quadratic
equation for the low values of B can be assumed as sufficiently
accurate for most purposes, except in the range of high densities

in those materials, where the increase of hysteresis loss occurs there.

While the measurement of the hysteresis loss appears a very simple matter, and can be carried out fairly accurately over a

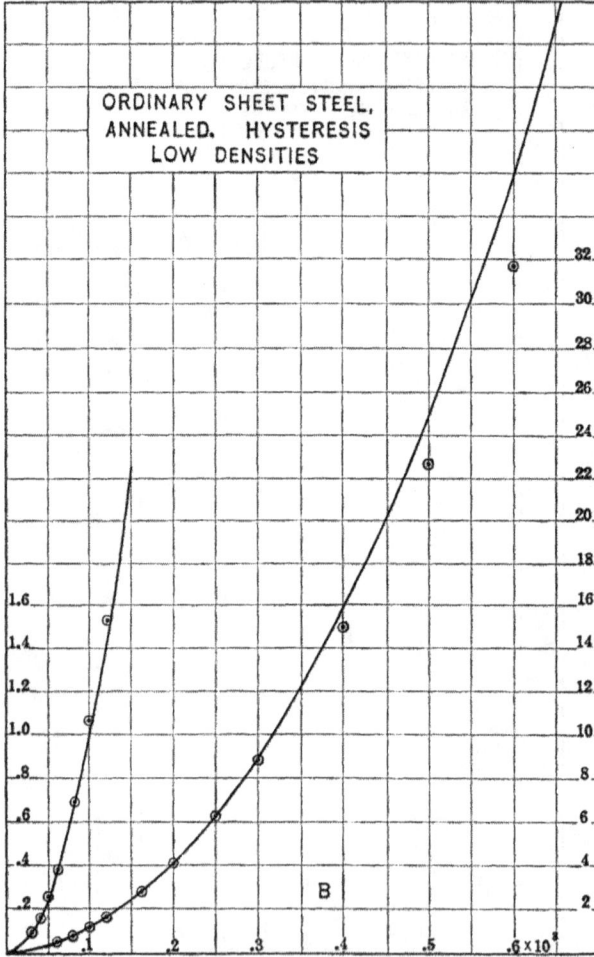

FIG. 40.

narrow range of densities, it is one of the most difficult matters to measure the hysteresis loss over a wide range of densities with such accuracy as to definitely determine the exact value of the exponent *n*, due to varying constant errors, which are beyond control. While true errors of observations can be eliminated by multiplying data, with a constant error this is not the case, and if the constant error changes with the magnetic density, it results in an apparent change of n. Such constant errors, which increase or decrease, or even reverse with changing B, are in the Ballistic galvanometer method the magnetic creepage at lower B, and at higher B the sharp-pointed shape of the hysteresis loop, which makes the area between rising and decreasing characteristic difficult to determine. In the wattmeter method by alternating current, varying constant errors are the losses in the instruments, the eddy-current losses which change with the changing flux distribution by magnetic screening in the iron, with the temperature, etc., by wave-shape distortion, the unequality of the inner and outer length of the magnetic circuit, etc.

45. Symmetrical magnetic cycles, that is, cycles performed between equal but opposite magnetic flux densities, $+B$ and $-B$, are industrially the most important, as they occur in practically all alternating-current apparatus. Unsymmetrical cycles, that is, cycles between two different values of magnetic flux density, B_1 and B_2, which may be of different, or may be of the same sign, are of lesser industrial importance, and therefore have been little investigated until recently.

However, unsymmetrical cycles are met in many cases in alternating- and direct-current apparatus, and therefore are of importance also.

In most inductor alternators the magnetic flux in the armature does not reverse, but pulsates between a high and a low value in the same direction, and the hysteresis loss thus is that of an unsymmetrical non-reversing cycle.

Unsymmetrical cycles occur in transformers and reactors by the superposition of a direct current upon the alternating current, as discussed in the chapter "Shaping of Waves," or by the equivalent thereof, such as the suppression of one-half wave of the alternating current. Thus, in the transformers and reactors of many types of rectifiers, as the mercury-arc rectifier, the magnetic cycle is unsymmetrical.

Unsymmetrical cycles occur in certain connections of transformers (three-phase star-connection) feeding three-wire synchronous converters, if the direct-current neutral of the converter is connected to the transformer neutral.

They may occur and cause serious heating, if several transformers with grounded neutrals feed the same three-wire distribution circuit, by stray railway return current entering the three-wire a ternating distribution circuit over one neutral and leaving it over another one.

Two smaller unsymmetrical cycles often are superimposed on an alternating cycle, and then increase the hysteresis loss. Such occurs in transformers or reactors by wave shapes of impressed voltage having more than two zero values per cycle, such as that shown in Fig. 51 of the chapter on "Shaping of Waves."

They also occur sometimes in the armatures of direct-current motors at high armature reaction and low field excitation, due to the flux distortion, and under certain conditions in the armatures of regulating pole converters.

A large number of small unsymmetrical cycles are sometimes superimposed upon the alternating cycle by high-frequency pulsation of the alternating flux due to the rotor and stator teeth, and then may produce high losses. Such, for instance, is the case in induction machines, if the stator and rotor teeth are not proportioned so as to maintain uniform reluctance, or in alternators or direct-current machines, in which the pole faces are slotted to receive damping windings, or compensating windings, etc., if the proportion of armature and pole-piece slots is not carefully designed.

46. The hysteresis loss in an unsymmetrical cycle, between limits B_1 and B_2, that is, with the amplitude of magnetic variation $B = \dfrac{B_1 - B_2}{2}$, follows the same approximate law of the 1.6^{th} power,

$$w_0 = \eta_0 B^{1.6}$$

as long as the average value of the magnetic flux variation,

$$B_0 = \frac{B_1 + B_2}{2},$$

is constant.

With changing B_0, however, the coefficient η_0 changes, and increases with increasing average flux density, B_0.

John D. Ball has shown, that the hysteresis coefficient of the unsymmetrical cycle increases with increasing average density, B_0, and approximately proportional to a power of B_0. That is,

$$\eta_0 = \eta + \beta\eta \, B_0^{1.9}.$$

Thus, in an unsymmetrical cycle between limits B_1 and B_2 of magnetic flux density, it is

$$w = \left\{ \eta + \beta \left(\frac{B_1 + B_2}{2} \right)^{1.9} \right\} \left(\frac{B_1 - B_2}{2} \right)^{1.6} \tag{23}$$

where η is the coefficient of hysteresis of the alternating-current cycle, and for $B_2 = -B_1$, equation (23) changes to that of the symmetrical cycle.

Or, if we substitute,

$$B_0 = \frac{B_1 + B_2}{2} \tag{24}$$

$\qquad\qquad$ = average value of flux density, that is, average of maximum and minimum.

$$B = \frac{B_1 - B_2}{2} \tag{25}$$

$\qquad\qquad$ = amplitude of unsymmetrical cycle,

it is

$$w = (\eta + \beta B_0^{1.9}) B^{1.6} \tag{26}$$

or,

$$w = \eta_0 B^{1.6} \tag{27}$$

where

$$\eta_0 = \eta + \beta B_0^{1.9} \tag{28}$$

or, more general,

$$w = \eta_0 B^n \tag{29}$$

$$\eta_0 = \eta + \beta B_0^m \tag{30}$$

For a good sample of ordinary annealed sheet steel, it was found,

$$\eta = 1.06 \times 10^{-3} \tag{31}$$

$$\beta = 0.344 \times 10^{-10}$$

For a sample of annealed medium silicon steel,

$$\eta = 1.05 \times 10^{-3}$$
$$\tag{32}$$
$$\beta = 0.32 \times 10^{-10}$$

Fig. 41 shows, with B_0 as abscissæ, the values of η_0, by equations (30) and (32).

As seen, in a moderately unsymmetrical cycle, such as between $B_1 = +12,000$ and $B_2 = -4000$, the increase of the hysteresis loss over that in a symmetrical cycle of the same amplitude, is moderate, but the increase of hysteresis loss becomes very large

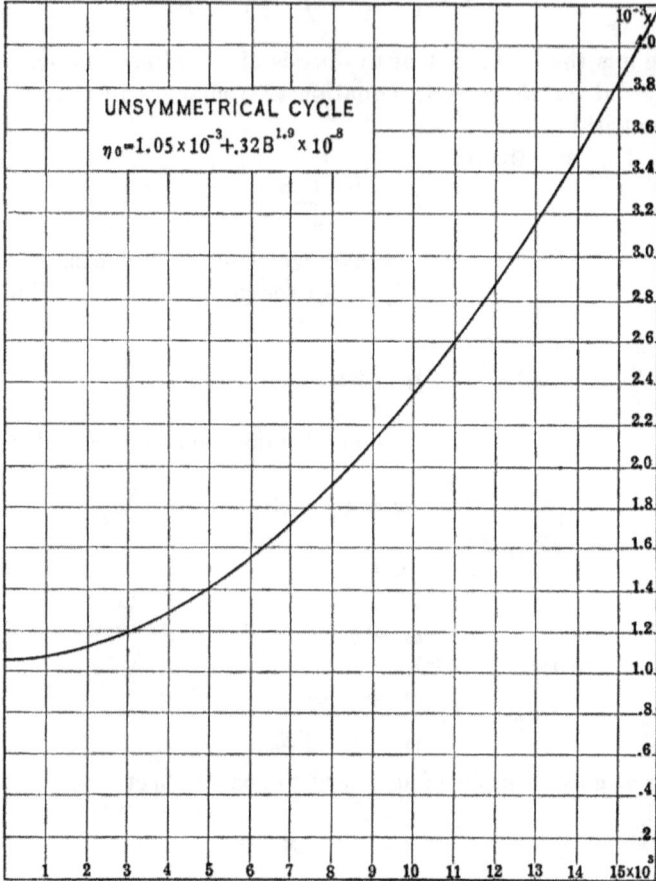

UNSYMMETRICAL CYCLE

$$\eta_0 = 1.05 \times 10^{-3} + .32 B^{1.9} \times 10^{-8}$$

FIG. 41.

in highly unsymmetrical cycles, such as between $B_1 = 16,000$ and $B_2 = 12,000$.

CHAPTER V

MAGNETISM

Magnetic Constants

47. With the exception of a few ferromagnetic substances, the magnetic permeability of all materials, conductors and dielectrics, gases, liquids and solids, is practically unity for all industrial purposes. Even liquid oxygen, which has the highest permeability, differs only by a fraction of a per cent. from non-magnetic materials.

Thus the permeability of neodymium, which is one of the most paramagnetic metals, is $\mu = 1.003$; the permeability of bismuth, which is very strongly diamagnetic, is $\mu = 1 - 0.00017 = 0.99983$.

The magnetic elements are iron, cobalt, nickel, manganese and chromium. It is interesting to note that they are in atomic weight adjoining each other, in the latter part of the first half of the first large series of the periodic system:

	Ti	V	Cr	Mn	Fe	Co	Ni	Cu	Zn
Atomic weight	48	51	52	55	56	58	59	61	65

The most characteristic, because relatively most constant, is the metallic magnetic saturation, S, or its reciprocal, the saturation coefficient, σ, in the reluctivity equation. The saturation density seems to be little if any affected by the physical condition of the material. By the chemical composition, such as by the presence of impurities, it is affected only in so far as it is reduced approximately in proportion to the volume occupied by the non-magnetic materials, except in those cases where new compounds result.

It seems, that the saturation value is an absolute limit of the element, and in any mixture, alloy or compound, the saturation value reduced to the volume of the magnetic metal contained therein, can not exceed that of the magnetic metal, but may be lower, if the magnetic metal partly or wholly enters a compound of lower intrinsic saturation value. Thus, if $S = 21 \times 10^3$ is the saturation value of iron, an alloy or compound containing

72 per cent. by volume of iron can have a maximum saturation value of $S = 0.72 \times 21 \times 10^3 = 15.1 \times 10^3$ only, or a still lower saturation value.

The only known exception herefrom seems to be an iron-cobalt alloy, which is alleged to have a saturation value about 10 per cent. higher than that of iron, though cobalt is lower than iron.

The coefficient of magnetic hardness, α, however, and the co-efficient of hysteresis, η, vary with the chemical, and more still with the physical characteristic of the magnetic material, over an enormous range.

Thus, a special high-silicon steel, and the chilled glass hard tool steel in the following tables, have about the same percentage of non-magnetic constituents, 4 per cent., and about the same saturation value, $S = 19.2 \times 10^3$, but the coefficient of hardness of chilled tool steel, $\alpha = 8 \times 10^{-3}$, is 200 times that of the special silicon steel, $\alpha = 0.04 \times 10^{-3}$, and the coefficient of hysteresis of the chilled tool steel, $\eta = 75 \times 10^{-3}$, is 125 times that of the silicon steel, $\eta = 0.6 \times 10^{-3}$. Hardness and hysteresis loss seem to depend in general on the physical characteristics of the material, and on the chemical constitution only as far as it affects the physical characteristics.

Chemical compounds of magnetic metals are in general not ferromagnetic, except a few compounds as magnetite, which are ferromagnetic.

With increasing temperature, the magnetic hardness α, decreases, that is, the material becomes magnetically softer, and the saturation density, S, also slowly decreases, until a certain critical temperature is reached (about 760°C. with iron), at which the material suddenly ceases to be magnetizable or ferromagnetic, but usually remains slightly paramagnetic.

As the result of the increasing magnetic softness and decreasing saturation density, with increasing temperature the density, B, at low field intensities, H, increases, at high field intensities decreases. Such B-temperature curves at constant H, however, have little significance, as they combine the effect of two changes, the increase of softness, which predominates at low H, and the decrease of saturation, which predominates at high H.

Heat treatment, such as annealing, cooling, etc., very greatly changes the magnetic constants, especially α and η—more or less in correspondence with the change of the physical constants brought about by the heat treatment.

Very extended exposure to moderate temperature—100 to 200°C.—increases hardness and hysteresis loss with some materials, by what is called ageing, while other materials are almost free of ageing.

48. The most important, and therefore most completely investigated magnetic metal is iron.

Its saturation value is probably between $S = 21.0 \times 10^3$ and $S = 21.5 \times 10^3$, the saturation coefficient thus $\sigma = 0.047$.

As all industrially used iron contains some impurities, carbon, silicon, manganese, phosphorus, sulphur, etc., usually saturation values between 20×10^3 and 21×10^3 are found on sheet steel or cast steel, etc., lower values, 19 to 19.5×10^3, in silicon steels containing several per cent. of Si, and still much lower values, 12 to 15×10^3, in very impure materials, such as cast iron.

Two types of iron alloys seem to exist:

1. Those in which the alloying material does not directly affect the magnetic qualities, but only indirectly, by reducing the volume of the iron and thereby the saturation value, and by changing the physical characteristics and thereby the hardness and hysteresis loss.

Such apparently are the alloys with carbon, silicon, titanium, chromium, molybdenum and tungsten, etc., as cast iron, silicon steel, magnet steel, etc.

2. Those in which the alloying material changes the magnetic, characteristics.

Such apparently are the alloys with nickel, manganese, mercury, copper, cobalt, etc.

In this class also belong the chemical compounds of the magnetic materials.

Thus, a manganese content of 10 to 15 per cent. makes the iron practically non-magnetic, lowers the permeability to $\mu = 1.4$. However, even here it is not certain whether this is not an extreme case of magnetic hardness, and at extremely high magnetic fields the normal saturation value of the iron would be approached.

Some nickel steels (25 per cent. Ni) may be either magnetic, or non-magnetic. However, pure iron, when heated to high incandescence, becomes non-magnetic at a certain definite temperature, and when cooling down, becomes magnetizable again at another definite, though lower temperature, and between these two temperatures, iron may be magnetic or unmagnetic, depending whether it has reached this temperature from lower, or from higher temperatures. Apparently, for these nickel steels, the critical temperature range, within which they can be magnetic or unmagnetic, is within the range of atmospheric temperature, and thus, after heating, they become non-magnetic, after cooling to sufficiently low temperature, they become magnetizable again. Thus, a steel containing 17 per cent. nickel, 4.5 per cent. chromium, 3 per cent. manganese, has permeability 1.004, that is, is almost completely unmagnetic.

Heterogeneous mixtures, such as powdered iron incorporated in resin, or iron filings in air, seem to give saturation densities not far different from those corresponding to their volume percentage of iron, but give an enormous increase of hardness, α, and hysteresis, η, as is to be expected.

Most chemical compounds of iron are non-magnetic. Ferromagnetic is only magnetite, which is the intermediate oxide and may be considered as ferrous ferrite. There also is an alleged magnetic sulphide of iron, though I have never seen it, magnetkies, Fe_7S_8 or Fe_8S_9.

As magnetite, Fe_3O_4, contains 72 per cent. of Fe, by weight, and has the specific weight 5.1, its volume per cent. of iron would be 48 per cent., and the saturation density $S = 10 \times 10^3$.

Observations on the magnetic constants of magnetite give a saturation density of 4.7×10^3 to 5.91×10^3, so that magnetite would fall in the second class of iron compounds, those in which the saturation density is affected, and lowered, by the composition.

Not only *magnetite*, which may be considered as ferrous ferrite, but numerous other ferrites, that is, salts of the acid $Fe_2O_4H_2$, are to some extent ferromagnetic, such as copper and cobalt ferrite, calcium ferrite, etc.

49. *Cobalt*, next adjoining to iron in the periodic system of elements, is the magnetic metal which has been least investigated. Its saturation value probably is between $S = 12 \times 10^3$ and $S = 14 \times 10^3$, and its magnetic characteristic looks very similar to that of cast iron. Partly this is due to the similar saturation value, partly probably due to the feature that most of the available data were taken on cast cobalt.

It is interesting to note that Cobalt retains its magnetizability up to much higher temperatures than iron or any other material, so that above 800 degrees C., Cobalt is the only magnetic material.

More information is available on *nickel*, the metal next adjoining to cobalt in the periodic system of elements. Its saturation density is the lowest of the magnetic metals, probably between $S = 6 \times 10^3$ and $S = 7 \times 10^3$.

Some data on nickel and nickel alloys are given in the following table. In general, nickel seems to show characteristics very similar to those of iron, except that all the magnetic densities are reduced in proportion to the lower saturation density; but the effect of the physical characteristics on the magnetic constants appears to be the same. Interesting is, that nickel seems to be least sensitive to impurities in their effect on the reluctivity curve.

Nickel ceases to be magnetizable already below red heat.

The next metal beyond nickel, in the periodic system of elements, is copper, and this is non-magnetic, as far as known.

On the other side of iron, in the periodic system, is *manganese*.

This is very interesting in so far as it has never been observed in a strongly magnetic state, but many of the alloys of manganese are more or less strongly magnetic, and estimating from the saturation values of manganese alloys, the saturation value of manganese as pure metal should be about $S = 30 \times 10^3$. This would make it the most magnetic metal.

In favor of manganese as magnetic metal also is the unusual behavior of its alloys with iron: the alloys of nickel, and of cobalt with iron also show unusual characteristics, and this seems to be a characteristic of alloys between magnetic metals.

The best known magnetic manganese alloys are the Heusler alloys, of manganese with copper and aluminum, and the characteristics of three such alloys are given in the following table. The most magnetic shows about the same saturation value as magnetite, but higher saturation values, equal to those of nickel, have been observed.

A curious feature of some Heusler alloys is, that when slowly cooled from high temperatures, they are very little magnetic, and have low saturation values. The quicker they are cooled, the higher their permeability and their saturation value, and the best values have been reached by dropping the molten alloy into water, so suddenly chilling it.

In general, the Heusler alloys are especially sensitive to heat treatment, and some of them show the ageing in a most pro-

6

nounced degree, so that maintaining the alloy for a considerable time at moderate temperature, increases hardness and hysteresis loss more than tenfold.

Magnetic alloys of manganese also are known with antimony, arsenic, phosphorus, bismuth, boron, with zinc and with tin, etc. Usually, the best results are given by alloys containing 20 to 30 per cent. of manganese. Little is known of these magnetic alloys, except that they may be in a magnetic state, or in an unmagnetic stage. They are most conveniently produced by dissolving manganese metal in the superheated alloying metal, or in this metal with the addition of some powerful reducing metal, as sodium or aluminum, but the alloy is only sometimes magnetic, sometimes practically unmagnetic, and the conditions of the formation of the magnetic state are unknown.

Apparently, there also exists an intermediary oxide of manganese, or a compound oxide of manganese with that of the other metal, which is strongly magnetic. The black slag, appearing in the fusion of manganese with other metals such as antimony, zinc, tin, without flux, often is strongly magnetic, more so than the alloy itself.

A mixture of about 25 per cent. powdered manganese metal, and 75 per cent. powdered antimony metal, heated together to a moderate temperature—in a test-tube—gives a strongly magnetic black powder, which can be used like iron filings, to show the lines of forces of the magnetic field, but has not further been investigated.

A considerable number of such magnetic manganese alloys have been investigated by Heusler and others, and their constants are given in the following table.

It is supposed that these magnetic manganese alloys are chemical compounds, similar as magnetite or magnetkies. Thus the copper-aluminum-manganese alloy of Heusler is a compound of 1 atom of aluminum with 3 atoms of copper or manganese: $Al(Mn$ or $Cu)_3$, usually $AlMnCu_2$. Other magnetic manganese compounds then are:

With antimony	$MnSb$ and Mn_2Sb
With bismuth	$MnBi$
With arsenic	$MnAs$
With boron	MnB
With phosphorus	MnP
With tin	Mn_4Sn and Mn_2Sn

Next adjacent to manganese in the periodic system of elements is *chromium*. Neither the metal, nor any of its alloys (except those with magnetic metals) have ever been observed in the magnetic state. There is, however, an intermediary oxide of chromium, alleged to be Cr_5O_9 (a basic chromic chromate?) which is strongly magnetic. It forms, in black scales, in a narrow range of temperature, by passing CrO_2Cl_2 with hydrogen through a heated tube.

A second strongly magnetic chromium oxide is Cr_4O_9 (a basic chromic bichromate?). It is easily produced by rapidly heating CrO_3, but the product is not always the same. Their magnetic characteristics have never been investigated, and they are the only indication which would point to chromium having potentially magnetic qualities.

The metal next to chromium in the periodic system of elements, vanadium, is non-magnetic, as far as known.

50. On attached tables are given the magnetic constants of the better known magnetic materials, metals, alloys, mixtures and compounds:

The first tables give the saturation density, S, and the demagnetization temperature, that is, temperature at which the material ceases to be ferromagnetic, and its specific gravity.

It is interesting to note that with some magnetic materials the demagnetization temperature is very close to, or within the range of, atmospheric temperature.

The second table gives more complete data of those materials, of which such data are available. It gives:

S = saturation density, or value of $B - H$ for infinitely high H;

α = coefficient of magnetic hardness;

σ = coefficient of magnetic saturation.

Where the reluctivity line shows a bend at some critical point, α and σ are given for the lower range—which is the one industrially most useful—together with the range of field intensity, for which this value applies, and are given also for the highest range observed, together with the value of field intensity H, above which the latter values of α and σ apply.

η = coefficient of hysteresis, in the 1.6^{th} power law.

β = coefficient of unsymmetrical cycle, for the two cases where this is known.

Demagnetization temperature, that is, temperature at which ferromagnetism ceases.

ρ = electrical resistivity of the material—which refers to the eddy-current losses in magnetic cycles.

Sp. gr. = specific gravity of the material.

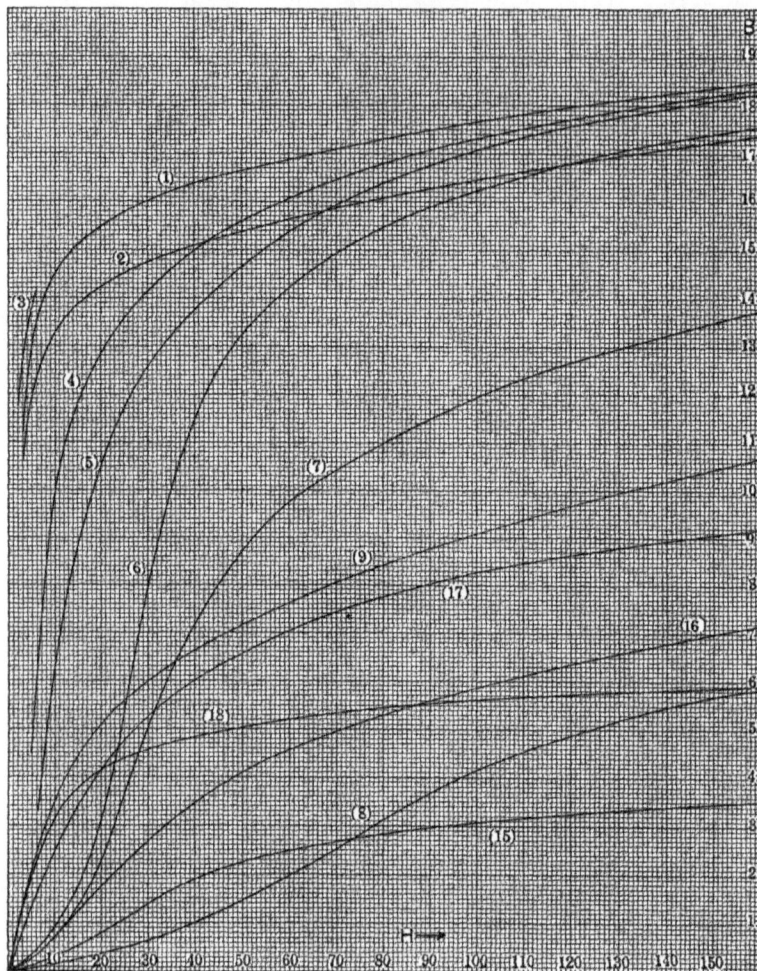

Fig. 42.

Fig. 42 gives the magnetic characteristics, up to $H = 160$ (beyond this, the linear law of reluctivity usually applies), for a number of magnetic materials of higher values of saturation densities.

Fig. 43 gives, with twice the scale of ordinates, but the same abscissæ, the magnetic characteristic of some materials of low saturation density.

Fig. 44 gives, with ten times the scale of abscissæ, and the same scale of ordinates, the initial part of the magnetic characteristic,

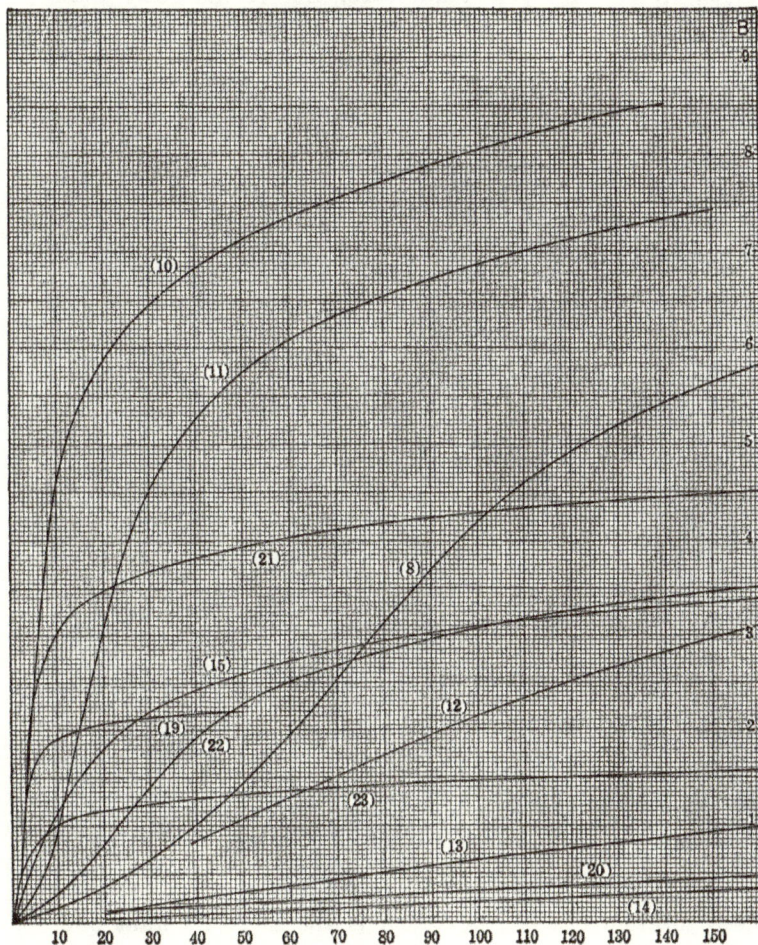

Fig. 43.

up to $H = 16$, for the magnetically soft materials of Fig. 42, that is, materials with low value of α, which rise so rapidly to high values of density that the initial part of their characteristic is not well shown in the scale of Fig. 42.

The magnetic characteristics in Figs. 42, 43 and 44 are denoted by numbers, and these numbers refer to the materials given in the table of "Magnetic Constants" under the same numbers.

With regards to the magnetic data, it must be realized, however, that the numerical values, especially of the less-investigated materials, are to some extent uncertain, due to the great difficulty of exact magnetic measurements.

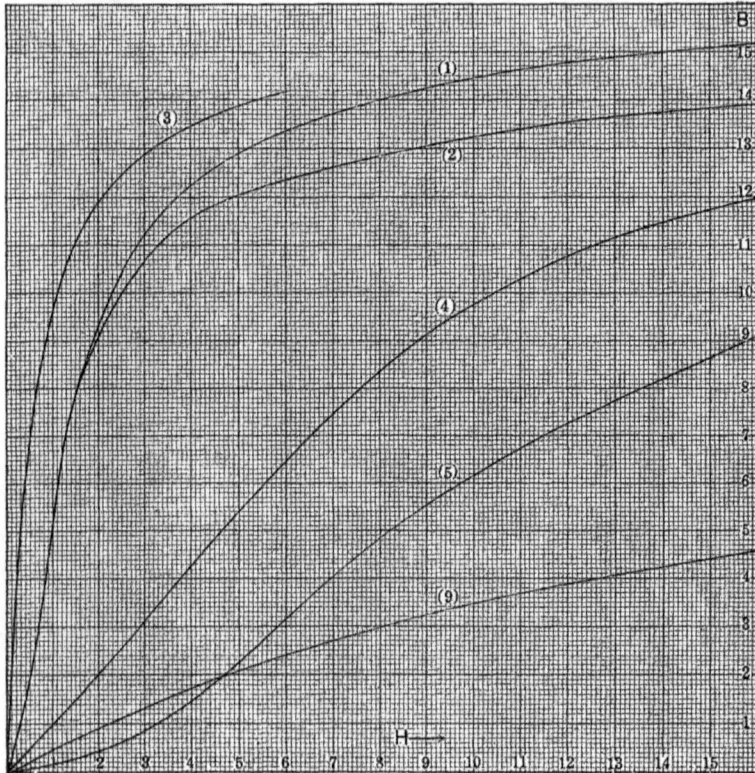

FIG. 44.

The saturation density, S, which is the most constant and most definite and permanent magnetic quantity, can be measured either directly, by measuring B in such very high fields, $-H = 10,000$ and over—that $B-H$ does not further increase, or indirectly, by observing the B, H curve up to moderately high fields, therefrom derive the reluctivity curve: ρ, H, and from the straight-line law of the latter curve determine σ and therewith B.

Table I.—Saturation Density

$$S = (B - H)_{H = \infty}$$

Material	Authority	Saturation density, S	Demagnetization temperature, °C.	Specific gravity
		$\times 10^{+3}$		
Iron:				
Most probable value......................	1916	21.0–21.5	760	7.70
Best standard sheet steel, annealed.........	1915–16	20.70		
Average standard sheet steel, annealed......	1915–16	20.20		
Pure iron...............................	Wedekind	21.10	765	
Swedish wrought iron....................	Ewing	21.25		
Iron, 99.88 per cent.....................	Hatfield	21.15		
Iron....................................	Gumlich	21.60?		
Electrolytic iron........................	Gumlich	21.70?		
Commercial steel........................	Williams	22.00?		
Vacuum-melted electrolytic iron...........	Williams	22.60?		
Pure iron...............................	DuBois	23.20?	756	7.86
Average sheet iron.......................	1892	20.10		
Average medium silicon steel, annealed, 2.5 per cent...............................	1915–16	19.25		
Average soft steel castings................	1915–16	20.20		
Special tungsten steel....................	....	20.30		
Magnet steel............................	....	18.50		
Average cast iron.......................	1915–16	15.00		
Fe_2Co, cobalt iron.....................	Williams	22.50?	520	
Fe_2Co, cobalt iron, vacuum-melted and forged.	Williams	25.80?		
Fe_2Co, cobalt-iron, probable value, about.....	....	23.50		
Scale of silicon steel.......................	....	9–10		
Iron amalgam, 11 per cent.................	1892	0.90		
Magnetite, Fe_3O_4......................	1892	4.70	....	5.10
Magnetite, Fe_3O_4......................	DuBois	5.46–5.91	536–589	
Magnetkies, Fe_7S_8 or FeS_2..............	DuBois	0.88	....	4.60
$CuFe_2O_4$, copper ferrite.................	....		280	
$CoFe_2O_4$, cobalt ferrite.................	....	....	280–290	
Cobalt:				
Probable value...........................	1916	12–14		
Cast cobalt.............................	....	11.10		
Cobalt.................................	....	12.10		
Cobalt.................................	Wedekind	13.30		
Cobalt, 1.66 per cent. Fe................	Ewing	16.45?		
Cobalt.................................	DuBois	17.20?	1075	8.70
Cobalt, pure............................	Stifler	17.85?		
Cobalt, vacuum-melted..................	Williams	18.85?		
Nickel:				
Probable value...........................	1916	6–7		
Nickel, 99 per cent. pure.................	....	6.15		
Nickel wire, soft........................	1892	5.88		
Nickel, cast............................	....	6.52		
Nickel.................................	DuBois	7.27?	340–376	8.93
Nickel.................................	....	8.17?		
Monel metal..............................	....	2.22		
Binel metal..............................	....	0.26		
Manganese, Heusler alloys:				
$AlMnCu_2$, soft, high permeability...........		4.67		
$AlMnCu_2$, hard, high permeability..........		3.92		
$AlMnCu_2$, soft, low permeability...........		1.56		
$AlMnCu_2$, highest values.................		7.00	310	
Manganese-antimony, MnSb................		7.00	310–330	
Manganese-antimony, Mn_2Sb.............		3.85		
Manganese-boron, MnB...................		3.10		
Manganese-phosphorus, MnP..............		0.70	18–26	
Manganese-bismuth, MnB		....	360–380	
Manganese-arsenic, MnAs.................			40–50	
Manganese-tin, Mn_4Sn				
Chromium:				
Cr_4O_9, chromic bichromate				
Cr_8O_9, chromic chromate				

TABLE II.—MAGNETIC CONSTANTS OF MATERIALS

Reference number	Material	Saturation density, S ×10³	High range Hardness, α ×10⁻³	High range Saturation, σ ×10⁻³	High range Range, H	Low range Hardness, α ×10⁻³	Low range Saturation, σ ×10⁻³	Low range Range, H	Hysteresis Coefficient, η ×10⁻³	Hysteresis Asymmetry coefficient, β ×10⁻¹⁰	Demagnetization temperature, °C	Electrical resistivity, ρ ×10⁻⁶	Specific gravity, Sp. gr.
(1)	*Iron:* S = 21.0 to 21.5 Standard sheet steel, annealed, average 1915–16	20.2	0.76	0.0494	>100	0.10	0.0588	2.5–50	1.2	0.344		12	7.7
	Standard sheet steel, annealed, best	20.7	0.74	0.0483	>100	0.06	0.0583	2.5–50	2.0				
	Soft iron wire (Ewing)					0.2	0.0635						
	Pure iron (Wedekind)	21.1	1.5	0.0474	150–845	0.3	0.05		3.5	0.32	765		7.86
(2)	Average sheet iron, 1892	20.1											
	Medium silicon sheet steel, annealed, 2.5 per cent. Si, average 1915–16	19.25	0.93	0.0519	>100	0.10	0.0639	2.5–50	0.8			40	7.5
(3)	Special alloy steel, best values	20.2	0.85	0.0495	>100	0.04	0.064	7–6	0.3			60	
(4)	Soft steel castings, average 1915–16	20.6	1.0	0.0486	>100	0.465	0.0547	10–50	9–12				
	Soft steel castings, 1892	20.3	1.2	0.0493	>60	0.44–0.9	0.051–0.059						
(5)	Steel shafting												
(6)	Tungsten steel, special	18.5	2.92	0.0541	>60				48.0				
(7)	Magnet steel								75.0				
(8)	Chilled tool steel, 1892					0.876	0.0509		8–20				
(9)	Cast iron, average 1915–16	15.0	4.2	0.0667	>80	1.93	0.102	20–50	12–16			80–85	6.7
	Cast iron, average 1892					2.0–2.9	0.094–0.098	20–50					
(10)	Scale of silicon steel, vacuum-annealed, 5.5–5.7 per cent. Si; 78.5 per cent. Fe	9.7	2.15	0.103	>90	1.01	0.121	40–90	7.0			30–80	6.0
(11)	Scale of silicon steel, furnace-annealed, 5.3–5.5 per cent. Si; 76 per cent. Fe	8.9	3.30	0.112	>90	2.89	0.117	40–90	21.6			100–160	5.64
(12)	Powdered iron in resin 10 ÷ 1 by weight	10.1	39.0	0.0993					24.0			80×10⁴	3.2–3.6
(13)	Powdered iron in resin, 3 ÷ 1 by weight, 1 ÷ 1 by volume	8.4	173.0	0.119					56.0				
(14)	Iron filings in air, 30 per cent. volume, 1892	3.5	70.0	0.290					231.0				
	Iron amalgam, 11 per cent. (Hg₃Fe₃), 1892	0.9	500.0	1.12					23.5				
(15)	Magnetite, Fe₃O₄, 1892	4.7	8.9	0.213							536–589		5.1–5.2
	Magnetite, (DuBois)	5.46											
	Magnetite, (DuBois)	5.91											

TABLE II.—MAGNETIC CONSTANTS OF MATERIALS—*Continued*

Reference number	Material	High range				Low range			Hysteresis		Demagnetization temperature, °C	Electrical resistivity, ρ	Specific gravity, Sp. gr.
		Saturation density, S	Hardness, α	Saturation, σ	Range, H	Hardness, α	Saturation, σ	Range, H	Coefficient, η	Asymmetry coefficient, β			
		$\times 10^3$	$\times 10^{-3}$	$\times 10^{-3}$		$\times 10^{-8}$	$\times 10^{-3}$		$\times 10^{-3}$	$\times 10^{-10}$		10^{-6}	
	Cobalt: Probably S = 12 to 14												
(16)	Cast cobalt	11.1	8.7	0.09	>40				10–20		1075		8.7
	Cast cobalt	13.6	3.0	0.073	>30								
(17)	Soft cobalt	11.1	5.0	0.09									
	Cobalt (Wedekind)	13.3	35.0	0.075	370–785								
	Nickel: Probably S = 6 to 7												
(18)	Nickel, 99 per cent	6.15	0.61	0.163	>5				20–25				8.93
	Soft nickel wire, 1892	5.85	1.0	0.17					12.5		340–376		
	Hard nickel wire, 1892		3.5						38.5				
	Cast nickel	6.52		0.153									
(19)	Monel metal	2.22	1.51	0.45	>70	0.855	0.455	5–70	44.0			44–53	
(20)	Binel metal, Fe, 70; Ni, 23.8; Cr, 3.68; Mn, 2.47; P, 0.06; Si, 0.04; S, 0.019	0.26	0.564	3.84	>140	0.35	0.522	60–140				88	
	Manganese: (In alloys, S = 28 to 30)												
(21)	Heusler alloy, Al(Mn or Cu)₂, soft, high permeability	4.67	2.39	0.214	>40	0.9	0.242	5–40	3– 30.0		615 (310)		
(22)	Heusler alloy, Al(Mn or Cu)₂, hard, high permeability	3.92	6.4	0.256									
(23)	Heusler alloy, Al(Mn or Cu)₂, soft low permeability	1.56	8.0	0.643	>40	2.4	0.765	5–40					
	Heusler alloy, Al(Mn or Cu)₂, best values	7.0											
	Manganese-antimony, MnSb	7.0	73.0	0.143	180–530						310–330		
	Manganese-antimony, Mn₂Sb	3.85	208.0	0.26	190–790								
	Manganese-boron, MnB	3.1	182.0	0.32	85–775								
	Manganese-phosphorus, MnP	0.7	420.0	1.44	335–775						18–26		

Such extremely high fields, as to reach complete magnetic saturation, are produced only between the conical pole faces of a very powerful large electromagnet. The area of the field then is very small, and it is difficult to get perfect uniformity of the field. The tendency is to underestimate the field, and this gives too high values of S. Thus, in the following table those values of S, which appear questionable for this case, have been marked by the interrogation sign.

The indirect method, from the straight-line reluctivity curve, gives more accurate values of S, as S is derived from a complete curve branch, and this method thus is preferable. However, the value derived in this manner is based on the assumption that there is no further critical point in the reluctivity curve beyond the observed range. This is correct with iron, as the best tests by the direct method check. With cobalt, there may be a critical point in the reluctivity curve beyond the observed range, as there are several observations by the direct method, which give very much higher, though erratic, values of saturation, S.

The value of the magnetic hardness, a, also is difficult to determine for very soft materials, especially where the method of observation requires correction for joints, etc., and the extremely high values of permeability—over 15,000—therefore appear questionable.

CHAPTER VI

MAGNETISM

MECHANICAL FORCES

1. General

51. Mechanical forces appear wherever magnetic fields act on electric currents. The work done by all electric motors is the result of these forces. In electric generators, they oppose the driving power and thereby consume the power which finds its equivalent in the electric power output. The motions produced by the electromagnet are due to these forces. Between the primary and the secondary coils of the transformer, between conductor and return conductor of an electric circuit, etc., such mechanical forces appear.

The electromagnet, and all electrodynamic machinery, are based on the use of these mechanical forces between electric conductors and magnetic fields. So also is that type of transformer which transforms constant alternating voltage into constant alternating current. In most other cases, however, these mechanical forces are not used, and therefore are often neglected in the design of the apparatus, under the assumption that the construction used to withstand the ordinary mechanical strains to which the apparatus may be exposed, is sufficiently strong to withstand the magnetic mechanical forces. In the large apparatus, operating in the modern, huge, electric generating systems, these mechanical forces due to magnetic fields may, however, especially under abnormal, though not infrequently occurring, conditions of operation (as short-circuits), assume such formidable values, so far beyond the normal mechanical strains, as to require consideration. Thus generators and large transformers on big generating systems have been torn to pieces by the magnetic mechanical forces of short-circuits, cables have been torn from their supports, disconnecting switches blown open, etc.

In the following, a general study of these forces will be given. This also gives a more rational and thereby more accurate design of the electromagnet, and permits the determination of what may be called the efficiency of an electromagnet.

Investigations and calculations dealing with one form of energy only, as electromagnetic energy, or mechanical energy, usually are relatively simple and can be carried out with very high accuracy. Difficulties, however, arise when the calculation involves the relation between several different forms of energy, as electric energy and mechanical energy. While the elementary relations between different forms of energy are relatively simple, the calculation involving a transformation from one form of energy to another, usually becomes so complex, that it either can not be carried out at all, or even only approximate calculation becomes rather laborious and at the same time gives only a low degree of accuracy. In most calculations involving the transformation between different forms of energy, it is therefore preferable not to consider the relations between the different forms of energy at all, but to use the *law of conservation of energy* to relate the different forms of energy, which are involved.

Thus, when mechanical motions are produced by the action of a magnetic field on an electric circuit, energy is consumed in the electric circuit, by an induced e.m.f. At the same time, the stored magnetic energy of the system may change. By the law of conservation of energy, we have:

Electric energy consumed by the induced e.m.f. = mechanical energy produced, + increase of the stored magnetic energy. (1) The consumed electric energy, and the stored magnetic energy, are easily calculated, as their calculation involves one form of energy only, and this calculation then gives the mechanical work done, $= Fl$, where $F =$ mechanical force, and $l =$ distance over which this force moves.

Where mechanical work is not required, but merely the mechanical forces, which exist, as where the system is supported against motions by the mechanical forces—as primary and secondary coils of a transformer, or cable and return cable of a circuit—the same method of calculation can be employed, by assuming some distance l of the motion (or dl); calculating the mechanical energy $w_0 = Fl$ by (1), and therefrom the mechanical force as $F = \dfrac{w_0}{l}$, or $F = \dfrac{dw_0}{dl}$.

Since the induced e.m.f., which consumes (or produces) the electric energy, and also the stored magnetic energy, depend on the current and the inductance of the electric circuit, and in alternating-current circuits the impressed voltage also depends on the inductance of the circuit, the inductance can frequently be expressed by supply voltage and current; and by substituting this in equation (1), the mechanical work of the magnetic forces can thus be expressed, in alternating-current apparatus, by supply voltage and current.

In this manner, it becomes possible, for instance, to express the mechanical work and thereby the pull of an alternating electromagnet, by simple expressions of voltage and current, or to give the mechanical strains occurring in a transformer under short-circuits, by an expression containing only the terminal voltage, the short-circuit current, and the distance between primary and secondary coils, without entering into the details of the construction of the apparatus.

This general method, based on the law of conservation of energy, will be illustrated by some examples, and the general equations then given.

2. The Constant-current Electromagnet

52. Such magnets are most direct-current electromagnets, and also the series operating magnets of constant-current arc lamps on alternating-current circuits.

Let i_0 = current, which is constant during the motion of the armature of the electromagnet, from its initial position 1, to its final position 2, l = the length of this motion, or the stroke of the electromagnet, in centimeters, and n = number of turns of the magnet winding.

The magnetic flux Φ, and the inductance

$$L = \frac{n\Phi}{i_0} 10^{-8} \qquad (2)$$

of the magnet, vary during the motion of its armature, from a minimum value,

$$\Phi_1 = \frac{i_0 L_1}{n} 10^8 \qquad (3)$$

in the initial position, to a maximum value,

$$\Phi_2 = \frac{i_0 L_2}{n} 10^8 \qquad (4)$$

in the end position of the armature.

Hereby an e.m.f. is induced in the magnet winding,

$$e' = n \frac{d\Phi}{dt} 10^{-8} = i_0 \frac{dL}{dt} \qquad (5)$$

This consumes the power

$$p = i_0 e' = i_0^2 \frac{dL}{dt} \qquad (6)$$

and thereby the energy

$$w = \int_2^1 p \, dt = n i_0^2 (L_2 - L_1) \qquad (7)$$

Assuming that the inductance, in any fixed position of the armature, does not vary with the current, that is, that magnetic saturation is absent,[1] the stored magnetic energy is:

In the initial position, 1,

$$w_1 = \frac{i_0^2 L_1}{2} \qquad (8)$$

in the end position, 2,

$$w_2 = \frac{i_0^2 L_2}{2} \qquad (9)$$

The increase of the stored magnetic energy, during the motion of the armature, thus is

$$w' = w_2 - w_1 = \frac{i_0^2}{2} (L_2 - L_1) \qquad (10)$$

The mechanical work done by the electromagnet thus is, by the law of conservation of energy,

$$w_0 = w - w'$$
$$= \frac{i_0^2}{2} (L_2 - L_1) \text{ joules.} \qquad (11)$$

If l = length of stroke, in centimeters, F = average force, or pull of the magnet, in gram weight, the mechanical work is

$$Fl \text{ gram-cm.}$$

Since

$$g = 981 \text{ cm.-sec.} \qquad (12)$$

= acceleration of gravity, the mechanical work is, in absolute units,

$$Flg$$

---

[1] If magnetic saturation is reached, the stored magnetic energy is taken from the magnetization curve, as the area between this curve and the vertical axis, as discussed before.

and since 1 joule $= 10^7$ absolute units, the mechanical work is

$$w_0 = Flg \ 10^{+7} \ \text{joules.} \tag{13}$$

From (11) and (12) then follows,

$$Fl = \frac{i_0{}^2}{2g} (L_2 - L_1) 10^{+7} \ \text{gram-cm.} \tag{14}$$

as the *mechanical work of the electromagnet*, and

$$F = \frac{i_0{}^2}{2g} \frac{L_2 - L_1}{l} 10^7 \ \text{grams} \tag{15}$$

as the average force, or pull of the electromagnet, during its stroke l.

Or, if we consider only a motion element dl,

$$F = \frac{i_0{}^2}{2g} \frac{dL}{dl} 10^{+7} \ \text{grams} \tag{16}$$

as the force, or pull of the electromagnet in any position l.

Reducing from gram-centimeters to foot-pounds, that is, giving the stroke l in feet, the pull F in pounds, we divide by

$$454 \times 30.5 = 13,850$$

which gives, after substituting for g from (12)

$$(14): Fl = 3.68 \ i_0{}^2 (L_2 - L_1) \ \text{ft.-lb} \tag{17}$$

$$(15): \ F = 3.68 \ i_0{}^2 \frac{L_2 - L_1}{l} \ \text{lb.} \tag{18}$$

$$(16): \ F = 3.68 \ i_0{}^2 \frac{dL}{dl} \ \text{lb.} \tag{19}$$

These equations apply to the direct-current electromagnet as well as to the alternating-current electromagnet.

In the alternating-current electromagnet, if i_0 is the effective value of the current, F is the effective or average value of the pull, and the pull or force of the electromagnet pulsates with double frequency between 0 and $2F$.

53. In the alternating-current electromagnet usually the voltage consumed by the resistance of the winding, $i_0 r$, can be neglected compared with the voltage consumed by the reactance of the winding, $i_0 x$, and the latter, therefore, is practically equal to the terminal voltage, e, of the electromagnet. We have then, by the general equation of self-induction,

$$e = 2\pi \ fL i_0 \tag{20}$$

where f = frequency, in cycles per second.

From which follows,

$$i_0 L = \frac{e}{2\,\pi f} \tag{21}$$

and substituting (21) in equations (14) to (19), gives as the equation of the *mechanical work, and the pull of the alternating-current electromagnet*.

In the metric system:

$$Fl = \frac{i_0(e_2 - e_1)10^7}{4\,\pi f g} \text{ gram-cm.} \tag{22}$$

$$F = \frac{i_0(e_2 - e_1)\,10^7}{4\,\pi f g l} = \frac{i_0}{4\,\pi f g}\,\frac{de}{dl}\,10^7 \text{ grams} \tag{23}$$

In foot-pounds:

$$Fl = \frac{0.586\,i_0(e_2 - e_1)}{f} \text{ ft.-lb.} \tag{24}$$

$$F = \frac{0.586\,i_0(e_2 - e_1)}{fl} = \frac{0.586 i_0}{f}\,\frac{de}{dl} \text{ lb.} \tag{25}$$

Example.—In a 60-cycle alternating-current lamp magnet, the stroke is 3 cm., the voltage, consumed at the constant alternating current of 3 amp. is 8 volts in the initial position, 17 volts in the end position. What is the average pull of the magnet?

$$
\begin{aligned}
l &= 3 \text{ cm.} \\
e_1 &= 8 \\
e_2 &= 17 \\
f &= 60 \\
i_0 &= 3
\end{aligned}
$$

hence, by (23),

$$F = 122 \text{ grams } (= 0.27 \text{ lb.})$$

The work done by an electromagnet, and thus its pull, depend, by equation (22), on the current i_0 and the difference in voltage between the initial and the end position of the armature, $e_2 - e_1$; that is, depend upon the difference in the volt-amperes consumed by the electromagnet at the beginning and at the end of the stroke. With a given maximum volt-amperes, $i_0 e_2$, available for the electromagnet, the maximum work would thus be done, that is, the greatest pull produced, if the volt-amperes at the beginning of the stroke were zero, that is, $e_1 = 0$, and the theoretical maximum output of the magnet thus would be

$$F_m l = \frac{i_0 e_2 10^7}{4\,\pi f g} \tag{26}$$

and the ratio of the actual output, to the theoretically maximum output, or the efficiency of the electromagnet, thus is, by (22) and (26),

$$\eta = \frac{F}{F_m} = \frac{e_2 - e_1}{e_2} \tag{27}$$

or, using the more general equation (14), which also applies to the direct-current electromagnet,

$$\eta = \frac{L_2 - L_1}{L_2} \tag{28}$$

The efficiency of the electromagnet, therefore, is the difference between maximum and minimum voltage, divided by the maximum voltage; or the difference between maximum and minimum volt-ampere consumption, divided by the maximum volt-ampere consumption; or the difference between maximum and minimum inductance, divided by the maximum inductance.

As seen, this expression of efficiency is of the same form as that of the thermodynamic engine,

$$\frac{T_2 - T_1}{T_2}$$

From (26) it also follows, that the maximum work which can be derived from a given expenditure of volt-amperes, $i_0 e_2$, is limited. For $i_0 e_2 = 1$, that is, for 1 volt-amp. the maximum work, which could be derived from an alternating electromagnet, is, from (26),

$$F_m l = \frac{10^7}{4 \pi f g} = \frac{810}{f} \text{ gram-cm.} \tag{29}$$

That is, a 60-cycle electromagnet can never give more than 13.5 gram-cm., and a 25-cycle electromagnet never more than 32.4 gram-cm. pull per volt-ampere supplied to its terminals.

Or inversely, for an average pull of 1 gram over a distance of 1 cm., a minimum of $\frac{1}{13.5}$ volt-amp. is required at 60 cycles, and a minimum of $\frac{1}{32.4}$ volt-amp. at 25 cycles.

Or, reduced to pounds and inches:

For an average pull of 1 lb. over a distance of 1 in., at least 86 volt-amp. are required at 60 cycles, and at least 36 volt-amp. at 25 cycles.

This gives a criterion by which to judge the success of the design of electromagnets.

7

3. The Constant-potential Alternating Electromagnet

54. If a constant alternating potential, e_0, is impressed upon an electromagnet, and the voltage consumed by the resistance, ir, can be neglected, the voltage consumed by the reactance, x, is constant and is the terminal voltage, e_0, thus the magnetic flux, Φ, also is constant during the motion of the armature of the electromagnet. The current, i, however, varies, and decreases from a maximum, i_1, in the initial position, to a minimum, i_2, in the end position of the armature, while the inductance increases from L_1 to L_2.

The voltage induced in the electric circuit by the motion of the armature,

$$e' = n\frac{d\Phi}{dt} 10^8 \tag{30}$$

then is zero, and therefore also the electrical energy expended,

$$w = 0.$$

That is, the electric circuit does no work, but the mechanical work of moving the armature is done by the stored magnetic energy.

The increase of the stored magnetic energy is

$$w' = \frac{i_2{}^2L_2 - i_1{}^2L_1}{2} \tag{31}$$

and since the mechanical energy, in joules, is by (13),

$$w_0 = Flg\,10^7$$

the equation of the law of conservation of energy,

$$w = w' + w_0 \tag{32}$$

then becomes

$$0 = \frac{i_2{}^2L_2 - i_1{}^2L_1}{2} + Flg\,10^{-7},$$

or

$$Fl = \frac{i_1{}^2L_1 - i_2{}^2L_2}{2g} 10^7 \text{ gram-cm.} \tag{33}$$

Since, from the equation of self-induction, in the initial position,

$$e_0 = 2\,\pi f L_1 i_1 \tag{34}$$

in the end position

$$e_0 = 2\,\pi f L_2 i_2 \tag{35}$$

substituting (34) and (35) in (33), gives the equation of the constant-potential alternating electromagnet.

$$Fl = \frac{e_0(i_1 - i_2)}{4 \pi f g} 10^7 \text{ gram-cm.} \tag{36}$$

and

$$F = \frac{e_0(i_1 - i_2)}{4 \pi f g l} 10^7 = \frac{e_0}{4 \pi f g} \frac{di}{dl} 10^7 \text{ grams} \tag{37}$$

or, in foot-pounds,

$$Fl = \frac{0.586 \, e_0(i_1 - i_2)}{f} \text{ ft.-lb.} \tag{38}$$

$$F = \frac{0.586 \, e_0(i_1 - i_2)}{fl} = \frac{0.586 \, e_0}{f} \frac{di}{dl} \text{ lb.} \tag{39}$$

Substituting $Q = ei =$ volt-amperes, in equations (36) to (39) of the constant-potential alternating electromagnet, and equations (22) to (25) of the constant-current alternating magnet, gives the same expression of mechanical work and pull:

In metric system:

$$Fl = \frac{\Delta Q}{4 \pi f g} 10^7 \text{ gram-cm.} \tag{40}$$

$$F = \frac{\Delta Q}{4 \pi f g l} 10^7 = \frac{1}{4 \pi f g} \frac{dQ}{dl} 10^7 \text{ grams} \tag{41}$$

In foot-pounds:

$$Fl = \frac{0.586 \, \Delta Q}{f} \text{ ft.-lb.} \tag{42}$$

$$F = \frac{0.586 \, \Delta Q}{fl} = \frac{0.586}{f} \frac{dQ}{dl} \text{ lb.} \tag{43}$$

where $\Delta Q =$ difference in volt-amperes consumed by the magnet in the initial position, and in the end position of the armature.

Both types of alternating-current magnet, then, give the same expression of efficiency,

$$\eta = \frac{\Delta Q}{Q_m} \tag{44}$$

where Q_m is the maximum volt-amperes consumed, corresponding to the end position in the constant-current magnet, to the initial position in the constant-potential magnet.

4. Short-circuit Stresses in Alternating-current Transformers

55. At short-circuit, no magnetic flux passes through the secondary coils of the transformer, if we neglect the small voltage consumed by the ohmic resistance of the secondary coils. If the supply system is sufficiently large to maintain constant voltage at the primary terminals of the transformer even at short-circuit, full magnetic flux passes through the primary coils.[1] In this case the total magnetic flux passes between primary coils and secondary coils, as self-inductive or leakage flux. If then x = self-inductive or leakage reactance, e_0 = impressed e.m.f., $i_0 = \dfrac{e_0}{x}$ is the short-circuit current of the transformer. Or, if as usual the reactance is given in per cent., that is, the ix (where i = full-load current of the transformer) given in per cent. of e, the short-circuit current is equal to the full-load current divided by the percentage reactance. Thus a transformer with 4 per cent. reactance would give a short-circuit current, at maintained supply voltage, of 25 times full-load current.

To calculate the force, F, exerted by this magnetic leakage flux on the transformer coils (which is repulsion, since primary and secondary currents flow in opposite direction) we may assume, at constant short-circuit current, i_0, the secondary coils moved against this force, F, and until their magnetic centers coincide with those of the primary coils; that is, by the distance, l, as shown diagrammatically in Fig. 45, the section of a shell-type transformer. When brought to coincidence, no magnetic flux passes between primary and secondary coils, and during this motion, of length, l, the primary coils thus have cut the total magnetic flux, Φ, of the transformer.

Hereby in the primary coils a voltage has been induced,

$$e' = n \frac{d\Phi}{dt} 10^{-8}$$

where n = effective number of primary turns.

The work done or rather absorbed by this voltage, e', at current, i_0, is

$$w = \int e' i_0 dt = n i_0 \Phi \, 10^{-8} \text{ joules.} \tag{45}$$

[1] If the terminal voltage drops at short-circuit on the transformer secondaries, the magnetic flux through the transformer primaries drops in the same proportion, and the mechanical forces in the transformer drop with the square of the primary terminal voltage, and with a great drop of the terminal voltage, as occurs for instance with large transformers at the end of a transmission line or long feeders, the mechanical forces may drop to a small fraction of the value, which they have on a system of practically unlimited power.

If L = leakage inductance of the transformer, at short-circuit, where the entire flux, Φ, is leakage flux, we have

$$\Phi = \frac{Li_0}{n} 10^8 \qquad (46)$$

hence, substituted in (45)

$$w = i_0{}^2 L \qquad (47)$$

The stored magnetic energy at short-circuit is

$$w_1 = \frac{i_0{}^2 L}{2} \qquad (48)$$

and since at the end of the assumed motion through distance, l, the leakage flux has vanished by coincidence between primary and secondary coils, its stored magnetic energy also has vanished, and the change of stored magnetic energy therefore is

$$w' = w_1 = \frac{i_0{}^2 L}{2} \qquad (49)$$

Hence, the mechanical work of the magnetic forces of the short-circuit current is

$$w_0 = w - w' = \frac{i_0{}^2 L}{2} \qquad (50)$$

It is, however, if F is the force, in grams, l, the distance between the magnetic centers of primary and secondary coils,

$$w_1 = Flg \, 10^{-7} \text{ joules.}$$

Hence,

$$Fl = \frac{i_0{}^2 L}{2 \, g} 10^7 \text{ gram-cm.} \qquad (51)$$

and

$$F = \frac{i_0{}^2 L}{2 \, gl} 10^7 \text{ grams} \qquad (52)$$

the mechanical force existing between primary and secondary coils of a transformer at the short-circuit current, i_0.

Since at short-circuit, the total supply voltage, e_0, is consumed by the leakage inductance of the transformer, we have

$$e_0 = 2 \, \pi f L i_0 \qquad (53)$$

hence, substituting (53) in (52), gives

$$F = \frac{e_0 i_0 \, 10^7}{4 \, \pi f g l} \text{ grams}$$

$$= \frac{810 \, e_0 i_0}{f l} \text{ grams} \qquad (54)$$

Example —Let, in a 25-cycle 1667-kw. transformer, the supply voltage, $e_0 = 5200$, the reactance $= 4$ per cent. The transformer contains two primary coils between three secondary coils, and the distance between the magnetic centers of the adjacent coils or half coils is 12 cm., as shown diagrammatically in Fig. 45. What force is exerted on each coil face during short-circuit, in a system which is so large as to maintain constant terminal voltage?

At 5200 volts and 1667 kw., the full-load current is 320 amp. At 4 per cent. reactance the short-circuit current therefore,

$$i_0 = \frac{320}{0.04} = 8000 \text{ amp.} \quad \text{Equation (54) then gives, for } f = 25,$$

$l = 12$,

$$F = 112 \times 10^6 \text{ grams}$$
$$= 112 \text{ tons.}$$

This force is exerted between the four faces of the two primary coils, and the corresponding faces of the secondary coils, and on every coil face thus is exerted the force

$$\frac{F}{4} = 28 \text{ tons}$$

This is the average force, and the force varies with double frequency, between 0 and 56 tons, and is thus a large force.

56. Substituting $i_0 = \frac{e_0}{x}$ in (54), gives as the short-circuit force

of an alternating-current transformer, at maintained terminal voltage, e_0, the value

$$F = \frac{e_0^2 \, 10^7}{4 \, \pi f g l x} = \frac{810 \, e_0^2}{f l x} \text{ grams} \tag{55}$$

That is, the short-circuit stresses are inversely proportional to the leakage reactance of the transformer, and to the distance, l, between the coils.

In large transformers on systems of very large power, safety therefore requires the use of as high reactance as possible.

High reactance is produced by massing the coils of each circuit.

Let in a transformer

$$n = \text{number of coil groups}$$

(where one coil is divided into two half coils, one at each end of the coil stack, as one secondary coil in Fig. 45, where $n = 2$) the mechanical force per coil face then is, by (55),

$$F_0 = \frac{F}{2n} = \frac{e_0{}^2\,10^7}{8\,\pi f g n l x} = \frac{810\,e_0{}^2}{2\,f n l x}\ \text{grams} \qquad (56)$$

Let x = leakage reactance of transformer;

l_0 = distance between coil surfaces;

l_1 = thickness of primary coil;

l_2 = thickness of secondary coil.

Between two adjacent coils, P and S in Fig. 45, the leakage flux density is uniform for the width l_0 between the coil surfaces,

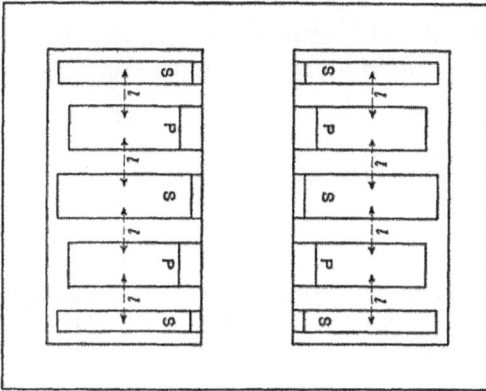

and then decreases toward the interior of the coils, over the distance $\frac{l_1}{2}$ respectively $\frac{l_2}{2}$, to zero at the coil centers. All the coil turns are interlinked with the leakage flux in the width, l_0, but toward the interior of the coils, the number of turns interlinked with the leakage flux decreases, to zero at the coil center, and as the leakage flux density also decreases, proportional to the distance from the coil center, to zero in the coil center, the interlinkages between leakage flux and coil turns decrease over the space $\frac{l_1}{2}$ respectively $\frac{l_2}{2}$, proportional to the square of the distance from the coil center, thus giving a total interlinkage distance,

$$\int_0^{\frac{l_1}{2}} u^2\,du = \frac{l_1}{6},$$

where u is the distance from the coil center.

Thus the total interlinkages of the leakage flux with the coil turns are the same as that of a uniform leakage flux density over the width $l_0 + \dfrac{l_1}{6} + \dfrac{l_2}{6}$. This gives the effective distance between coil centers, for the reactance calculation,

$$l = l_0 + \frac{l_1 + l_2}{6} \tag{57}$$

Assuming now we regroup the transformer coils, so as to get m primary and m secondary coils, leaving, however, the same iron structure.

The leakage flux density between the coils is hereby changed in proportion to the changed number of ampere-turns per coil, that is, by the factor $\dfrac{n}{m}$.

The effective distance between the coils, l, is changed by the same factor $\dfrac{n}{m}$.

The number of interlinkages between leakage flux and electric circuits, and thus the leakage reactance, x, of the transformer, thus is changed by the factor

$$\left(\frac{n}{m}\right)^2.$$

That is, by regrouping the transformer winding within the same magnetic circuit and without changing the number of turns of the electric circuit, the leakage reactance, x, changes inverse proportional to the square of the number of coil groups.

As by equation (56) the mechanical force is inverse proportional to x, l and n, and x changes proportional to $\left(\dfrac{n}{m}\right)^2$, l proportional to $\dfrac{n}{m}$, the mechanical force per coil thus changes proportional to

$$\left(\frac{n}{m}\right)^2 \times \frac{n}{m} \times \frac{m}{n} = \left(\frac{n}{m}\right)^2$$

That is, regrouping the transformer winding in the same winding space changes the mechanical force inverse proportional to the square of the coil groups, thus inverse proportional to the change of leakage reactance.

However, the distance l_0 between the coils is determined by insulation and ventilation. Thus its decrease, when increasing the number of coil groups, would usually not be permissible, but more winding space would have to be provided by changing the magnetic circuit, and inversely, with a reduction of the number of coil groups, the winding space, and with it the magnetic circuit, would be reduced.

Assuming, then, that at the change from n to m coil groups, the distance between the coils, l_0, is left the same.

The effective leakage space then changes from

$$l = l_0 + \frac{l_1 + l_2}{6},$$

to

$$l' = l_0 + \frac{n}{m}\frac{l_1 + l_2}{6} = l\frac{l_0 + \dfrac{n}{m}\dfrac{l_1 + l_2}{6}}{l_0 + \dfrac{l_1 + l_2}{6}},$$

and the leakage reactance thus changes from

$$x$$

to

$$x' = \frac{n}{m}\frac{l'}{l}x;$$

hence the mechanical force per coil, from

$$F_0 = \frac{F}{2\,n} = \frac{e_0{}^2\,10^7}{8\,\pi fnglx},$$

to

$$F'_0 = \frac{F'}{2m} = \frac{e_0{}^2\,10^7}{8\,\pi fngl'x'}$$

$$= F_0\,\frac{nlx}{ml'x'}$$

$$= F_0\left(\frac{l}{l'}\right)^2$$

$$= F_0\left(\frac{l_0 + \dfrac{n}{m}\dfrac{l_1 + l_2}{6}}{l_0 + \dfrac{l_1 + l_2}{6}}\right)^2 \qquad (58)$$

Thus, if $\dfrac{l_1 + l_2}{6}$ is large compared with l_0,

$$F'_0 = \left(\frac{n}{m}\right)^2 F_0,$$

that is, the mechanical forces vary with the square of the number of coil groups.

If $\dfrac{l_1 + l_2}{6}$ is small compared with l_0,

$$F_0{}^1 = F_0$$

that is, the mechanical forces are not changed by the change of the number of coil groups.

In actual design, decreasing the number of coil groups usually materially decreases the mechanical forces, but materially less than proportional to the square of the number of coil groups.

5. Repulsion between Conductor and Return Conductor

57. If i_0 is the current flowing in a circuit consisting of a conductor and the return conductor parallel thereto, and l the distance between the conductors, the two conductors repel each other by the mechanical force exerted by the magnetic field of the circuit, on the current in the conductor.

As this case corresponds to that considered in section 2, equation (16) applies, that is,

$$F = \frac{i_0{}^2}{2\,g}\frac{dL}{dl}\,10^7 \text{ grams},$$

The inductance of two parallel conductors, at distance l from each other, and conductor diameter l_d is, per centimeter length of conductor,

$$L = \left(4 \log \frac{2\,l}{l_d} + \mu\right) 10^{-9} \text{ henrys} \tag{59}$$

Hence, differentiated,

$$\frac{dL}{dl} = \frac{4 \times 10^{-9}}{l}$$

and, substituted in (16),

$$F = \frac{i_0{}^2}{50\,gl} \text{ grams} \tag{60}$$

or substituting (12),

$$F = \frac{20.4\ i_0^2\ 10^{-6}}{l}\ \text{grams} \qquad (61)$$

If $l = 150$ cm. (5 ft.)

$$i_0 = 200\ \text{amp.}$$

this gives

$F = 0.0054$ grams per centimeter length of circuit, hence it is inappreciable.

If, however, the conductors are close together, and the current very large, as the momentary short-circuit current of a large alternator, the forces may become appreciable.

For example, a 2200-volt 4000-kw. quarter-phase alternator feeds through single conductor cables having a distance of 15 cm. (6 in.) from each other. A short-circuit occurs in the cables, and the momentary short-circuit current is 12 times full-load current. What is the repulsion between the cables?

Full-load current is, per phase, 910 amp. Hence, short-circuit current, $i_0 = 12 \times 910 = 10,900$ amp. $l = 15$. Hence,

$$F = 160\ \text{grams per centimeter.}$$

Or multiplied by $\dfrac{30.5}{454}$

$$F = 10.8\ \text{lb. per feet of cable.}$$

That is, pulsating between 0 and 21.6 lb. per foot of cable. Hence sufficient to lift the cable from its supports and throw it aside.

In the same manner, similar problems, as the opening of disconnecting switches under short-circuit, etc., can be investigated.

6. General Equations of Mechanical Forces in Magnetic Fields

58. In general, in an electromagnetic system in which mechanical motions occur, the inductance, L, is a function of the position, l, during the motion. If the system contains magnetic material, in general the inductance, L, also is a function of the current, i, especially if saturation is reached in the magnetic material.

Let, then, L = inductance, as function of the current, i, and position, l;

L_1 = inductance, as function of the current, i, in the initial position 1 of the system;

L_2 = inductance, as function of the current, i, in the end position 2 of the system.

If then Φ = magnetic flux, n = number of turns interlinked with the flux, the induced e.m.f. is

$$e' = n\frac{d\Phi}{dt} 10^{-8} \tag{62}$$

We have, however,

$$n\Phi = iL\ 10^8;$$

hence,

$$e' = \frac{d(iL)}{dt} \tag{63}$$

the power of this induced e.m.f. is

$$p = ie' = i\frac{d(iL)}{dt},$$

and the energy

$$w = \int_1^2 pdt = \int_1^2 id(iL)$$

$$= \int_1^2 i^2 dL + \int_1^2 iL di \tag{64}$$

The stored magnetic energy in the initial position 1 is

$$w_1 = \int_0^1 id(iL_1) \tag{65}$$

In the end position 2,

$$w_2 = \int_0^2 id(iL_2) \tag{66}$$

and the mechanical work thus is, by the law of conservation of energy

$$w_0 = w - w_2 + w_1$$

$$= \int_1^2 id(iL) + \int_0^1 id(iL_1) - \int_0^2 id(iL_2) \tag{67}$$

and since the mechanical work is

$$w_0 = Flg\ 10^{-7} \tag{68}$$

We have:

$$Fl = \frac{10^7}{g}\left\{\int_1^2 id(iL) + \int_0^1 id(iL_1) - \int_0^2 id(iL_2)\right\} \text{ gram-cm.} \tag{69}$$

If L is not a function of the current, i, but only of the position, that is, if saturation is absent, L_1 and L_2 are constant, and equation (69) becomes,

$$Fl = \frac{10^7}{g} \left\{ \int_1^2 id(iL) + \frac{i_1{}^2L_1 - i_2{}^2L_2}{2} \right\} \text{ gram-cm.} \quad (70)$$

(a) If $i =$ constant, equation (70) becomes,

$$Fl = \frac{10^7}{g} \frac{i^2(L_2 - L_1)}{2}$$

(Constant-current electromagnet.)

(b) If $L =$ constant, equation (70) becomes,

$$Fl = 0.$$

That is, mechanical forces are exerted only where the inductance of the circuit changes with the mechanical motion which would be produced by these forces.

(c) If $iL =$ constant, equation (70) becomes,

$$Fl = \frac{10^7}{g} \frac{iL(i_1 - i_2)}{2}$$

(Constant-potential electromagnet.)

In the general case, the evaluation of equation (69) can usually be made graphically, from the two curves, which give the variation of L_1 with i in the initial position, of L_2 with i in the final position, and the curve giving the variation of L and i with the motion from the initial to the final position.

In alternating magnetic systems, these three curves can be determined experimentally by measuring the volts as function of the amperes, in the fixed initial and end position, and by measuring volts and amperes, as function of the intermediary positions, that is, by strictly electrical measurement.

As seen, however, the problem is not entirely determined by the two end positions, but the function by which i and L are related to each other in the intermediate positions, must also be given. That is, in the general case, the mechanical work and thus the average mechanical force, are not determined by the end positions of the electromagnetic system. This again shows an analogy to thermodynamic relations.

If then in case of a cyclic change, the variation from position

1 to 2 is different from that from position 2 back to 1, such a cyclic change produces or consumes energy.

$$w = \int_1^2 id(iL) + \int_2^1 id(iL) = \int_1^1 id(iL)$$

Such a case is the hysteresis cycle. The reaction machine (see Theory and Calculation of Electrical Apparatus) is based on such cycle.

SECTION II

CHAPTER VII

SHAPING OF WAVES: GENERAL

59. In alternating-current engineering, the sine wave, as shown in Fig. 46, is usually aimed at as the standard. This is not due to any inherent merit of the sine wave.

For all those purposes, where the energy developed by the current in a resistance is the object, as for incandescent lighting, heating, etc., any wave form is equally satisfactory, as the energy of the wave depends only on its effective value, but not on its shape.

With regards to insulation stress, as in high-voltage systems, a flat-top wave of voltage and current, such as shown in Fig. 47, would be preferable, as it has a higher effective value, with the same maximum value and therefore with the same strain on the insulation, and therefore transmits more energy than the sine wave, Fig. 46.

Inversely, a peaked wave of voltage, such as Fig. 48, and such as the common saw-tooth wave of the unitooth alternator, is superior in transformers and similar devices, as it transforms the energy with less hysteresis loss. The peaked voltage wave, Fig. 48, gives a flat-topped wave of magnetism, Fig. 47, and thereby transforms the voltage with a lesser maximum magnetic flux, than a sine wave of the same effective value, that is, the same power. As the hysteresis loss depends on the maximum value of the magnetic flux, the reduction of the maximum value of the magnetic flux, due to a peaked voltage wave, results in a lower hysteresis loss, and thus higher efficiency of transformation. This reduction of loss may amount to as much as 15 to 25 per cent. of the total hysteresis loss, in extreme cases.

Inversely, a peaked voltage wave like Fig. 48 would be objectionable in high-voltage transmission apparatus, by giving an unnecessary high insulation strain, and a flat-top wave of voltage like Fig. 47, when impressed upon a transformer, would give a peaked wave of magnetism and thereby an increased hysteresis loss.

111

The advantage of the sine wave is, that it remains unchanged in shape under most conditions, while this is not the case with any other wave shape, and any other wave shape thus introduces the danger, that under certain conditions, or in certain parts of the circuit, it may change to a shape which is undesirable or even

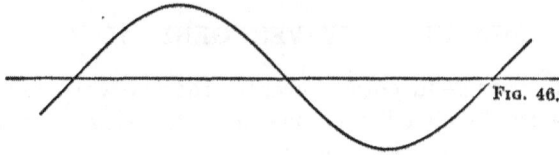

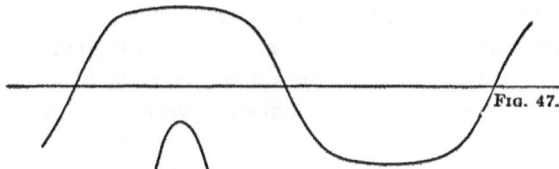

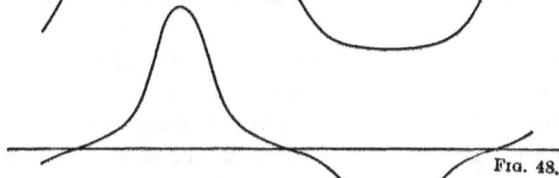

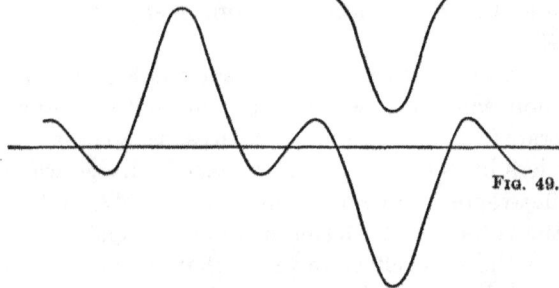

Figs. 46 to 49.

dangerous. Voltage, e, and current, i, are related to each other by proportionality, by differentiation and by integration, with resistance, r, inductance, L, and capacity, C, as factors,

$$e = ri,$$

$$e = L \frac{di}{dt},$$

$$e = C \int idt,$$

and as the differentials and integrals of sines are sines, as long as r, L and C are constant—which is mostly the case—sine waves of voltage produce sine waves of current and inversely, that is, the sine wave shape of the electrical quantities remains constant.

A flat-topped current wave like Fig. 47, however, would by differentiation give a self-inductive voltage wave, which is peaked, like Fig. 48. A voltage wave like Fig. 48, which is more efficient in transformation, may by further distortion, as by intensification of the triple harmonic by line capacity, assume the shape,

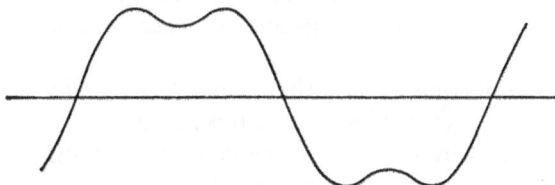

FIG. 50.

Fig. 49, and the latter then would give, when impressed upon a transformer, a double-peaked wave of magnetism, Fig. 50, and such wave of magnetism gives a magnetic cycle with two small

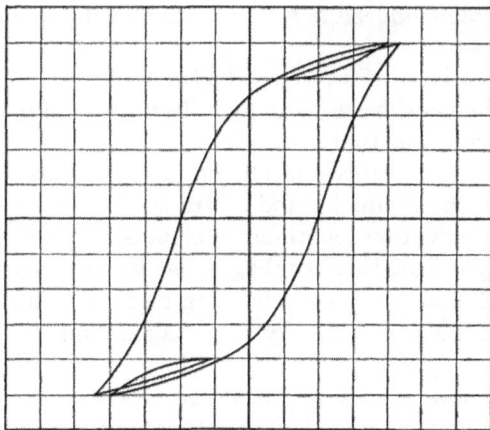

FIG. 51.

secondary loops at high density, as shown in Fig. 51, and an additional energy loss by hysteresis in these two secondary loops, which is considerable due to the high mean magnetic density, at which the secondary loop is traversed, so that in spite of the reduced maximum flux density, the hysteresis loss may be increased.

Therefore, in alternating-current engineering, the aim gener-

8

ally is to produce and use a wave which is a sine wave or nearly so.

60. In an alternating-current generator, synchronous or induction machine, commutating machine, etc., the wave of voltage induced in a single armature conductor or "face conductor" equals the wave of field flux distribution around the periphery of the magnet field, modified, however, by the reluctance pulsations of the magnetic circuit, where such exist. As the latter produce higher harmonics, they are in general objectionable and to be avoided as far as possible.

By properly selecting the length of the pole arc and the length of the air-gap between field and armature, a sinusoidal field flux distribution and thereby a sine wave of voltage induced in the armature face conductor could be produced. In this direction, however, the designer is very greatly limited by economic consideration: length of pole arc, gap length, etc., are determined within narrow limits by the requirement of the economic use of the material, questions of commutation, of pole-face losses, of field excitation, etc., so that as a rule the field flux distribution and with it the voltage induced in a face conductor differs materially from sine shape.

The voltage induced in a face conductor may contain even harmonics as well as odd harmonics, and often, as in most inductor alternators, a constant term.

The constant term cancels in all turn windings, as it is equal and opposite in the conductor and return conductor of each turn. Direct-current induction (continuous, or pulsating current) thus is possible only in half-turn windings, that is, windings in which each face conductor has a collector ring at either end, so-called unipolar machines (see "Theory and Calculation of Electrical Apparatus").

In every winding, which repeats at every pole or 180 electrical degrees, as is almost always the case, the even harmonics cancel, even if they existed in the face conductor. In any machine in which the flux distribution in successive poles is the same, and merely opposite in direction, that is, in which the poles are symmetrical, no even harmonics are induced, as the field flux distribution contains no even harmonics. Even harmonics would, however, exist in the voltage wave of a machine designed as shown diagrammatically in Fig. 52, as follows:

The south poles S have about one-third the width of the north poles N, and the armature winding is a unitooth 50 per cent. pitch winding, shown as A in Fig. 52.

Assuming sinusoidal field flux distribution in the air-gaps under the poles N and S of Fig. 52, curve I in Fig. 53 shows the field flux distribution and thus the voltage induced in a single-face conductor. Curve II shows the voltage wave in a 50 per cent. pitch turn and therewith that of the winding A. As seen, this contains a pronounced second harmonic in addition to the fundamental. If, then, a second 50 per cent. pitch winding is located on the arma-

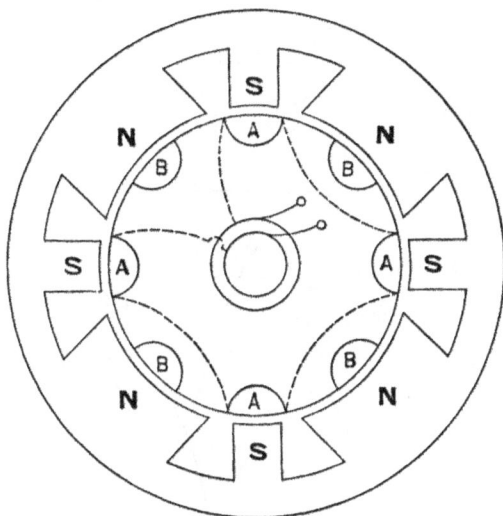

FIG. 52.

ture, shown as B in Fig. 52, by connecting B and A in series with each other in such direction that the fundamentals cancel (that is, in opposition for the fundamental wave), we get voltage wave III of Fig. 53, which contains only the even harmonics, that is, is of double frequency. Connecting A and B in series so that the fundamentals add and the second harmonics cancel, gives the wave IV. If the machine is a three-phase Y-connected alternator, with curve IV as the voltage per phase, or Y voltage, the delta or terminal voltage, derived by combination of two Y voltages under 60°, then is given by the curve V of Fig. 53. Fig. 54 shows the corresponding curves for the flux distribution of uniform density under the pole and tapering off at the pole corners, curve I, such as would approximately correspond to actual conditions. As seen, curve III as well as V are approximately sine waves, but the one of twice the frequency of the other. Thus, such a machine, by reversing connections between the two windings A and B, could be made to give two frequencies, one double the other, or as synchronous motor could run at two speeds, one one-half the other.

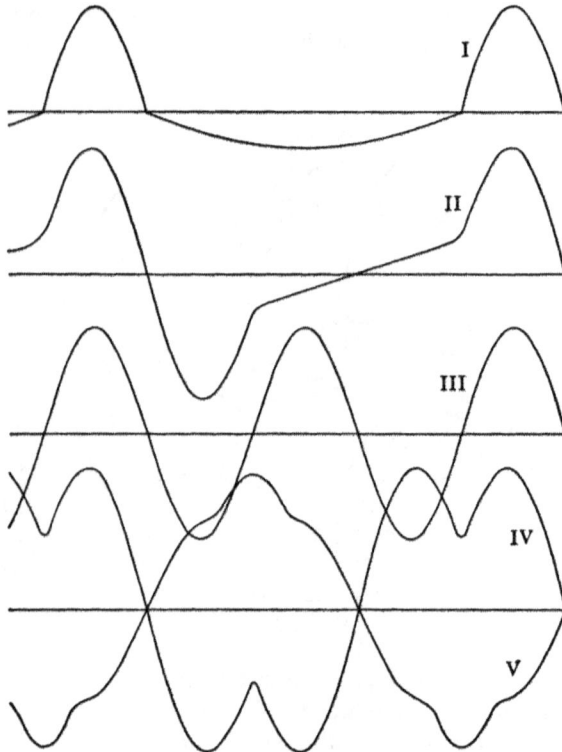

FIG. 53.

61. Distribution of the winding over an arc of the periphery of the armature eliminates or reduces the higher harmonics, so that the terminal voltage wave of an alternator with distributed winding is less distorted, or more nearly sine-shaped, than that of a single turn of the same winding (or that of a unitooth alternator). The voltage waves of successive turns are slightly out of phase with each other, and the more rapid variations due to higher harmonics thus are smoothed out. In two armature turns different in position on the armature circumference by δ electrical degrees ("electrical degrees" means counting the pitch of two poles as 360°), the fundamental waves are δ degrees out of phase, the third harmonics 3δ degrees, the fifth harmonics 5δ degrees, and so on, and their resultants thus get less and less, and becomes zero for that harmonic n, where $n\delta = 180°$.

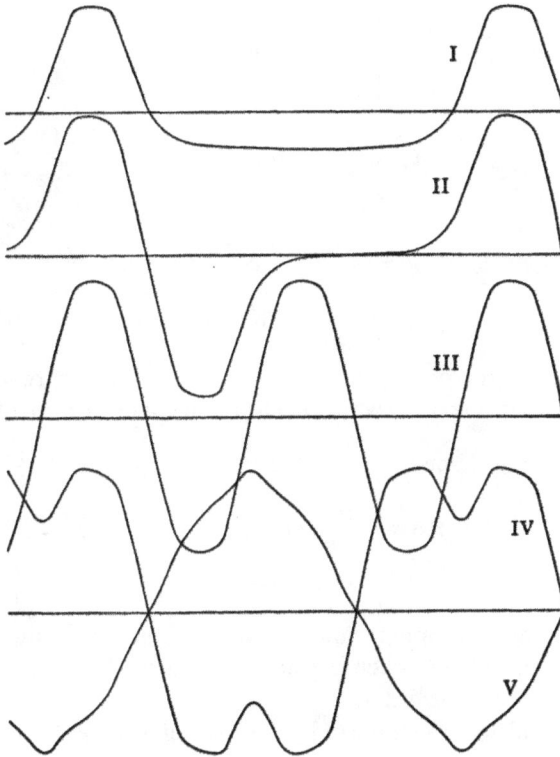

FIG. 54.

If

$$e = e_1 \sin \Phi + e_3 \sin 3\ (\Phi - \alpha_3) + e_5 \sin 5\ (\Phi - \alpha_5)$$
$$+ e_7 \sin 7\ (\Phi - \alpha_7) + \ldots \tag{1}$$

is the voltage wave of a single turn, and the armature winding of m turns covers an arc of ω electrical degrees on the armature periphery (per phase), the coefficients of the harmonics of the resultant voltage wave are

$$E_n = me_n \text{ avg. cos} \begin{cases} +\dfrac{n\omega}{2} \\ \\ -\dfrac{n\omega}{2} \end{cases} \qquad (2)$$

or, since

$$\text{avg. cos} \begin{cases} +\dfrac{n\omega}{2} \\ \\ -\dfrac{n\omega}{2} \end{cases} = \dfrac{2}{n\omega} \sin \dfrac{n\omega}{2}$$

$$E_n = \frac{2m}{n\omega} e_n \sin \frac{n\omega}{2} \qquad (3)$$

and

$$E = \frac{2m}{\omega} \left\{ e_1 \sin \frac{\omega}{2} \sin \phi + \frac{e_3}{3} \sin \frac{3\omega}{2} \sin 3(\phi - a_3) \right.$$
$$\left. + \frac{e_5}{5} \sin \frac{5\omega}{2} \sin 5(\phi - \alpha_5) + \dots \right\} \qquad (4)$$

Thus, in a three-phase winding like that of the three-phase synchronous converter, in which each phase covers an arc of 120° $= \dfrac{2\pi}{3}$, it is $\dfrac{\omega}{2} = \dfrac{\pi}{3}$, hence,

$$E = \frac{3m\sqrt{3}}{2\pi} \left\{ e_1 \sin \phi - \frac{e_5}{5} \sin 5(\phi - \alpha_5) \right.$$
$$\left. + \frac{e_7}{7} \sin 7(\phi - \alpha_7) - + \dots \right\} \qquad (5)$$

that is, the third harmonic and all its multiples, the ninth, fifteenth, etc., cancel, all other harmonics are greatly reduced, the more, the higher their order.

In a three-phase Y-connected winding, in which each phase covers 60° $= \dfrac{\pi}{3}$ of the periphery, as commonly used in induction and synchronous machines, it is $\dfrac{\omega}{2} = \dfrac{\pi}{6}$, hence,

$$E = \frac{3m}{\pi} \left\{ e_1 \sin \phi + \frac{2}{3} e_3 \sin 3(\phi - \alpha_3) + \frac{1}{5} e_5 \sin 5(\phi - \alpha_5) \right.$$
$$- \frac{1}{7} e_7 \sin 7(\phi - \alpha_7) - \frac{2}{9} e_9 \sin 9(\phi - \alpha_9)$$
$$\left. - \frac{1}{11} e_{11} \sin 11(\phi - \alpha_{11}) + \frac{1}{13} e_{13} \sin 13(\phi - \alpha_{13}) + - \dots \right\} \qquad (6)$$

Here the third harmonics do not cancel, but are especially large. Thus in a Y-connected three-phase machine of the usual 60° winding, the Y voltage may contain pronounced third harmonics, which, however, cancel in the delta voltage.

Thus with the distributed armature winding, which is now almost exclusively used, the wave-shape distortion due to the nonsinusoidal distribution of the field flux is greatly reduced, that is, the higher harmonics in the voltage wave decreased, the more so, the higher their order, and very high harmonics, such as the seventeenth, thirty-fifth, etc., therefore do not exist in such machines to any appreciable extent, except where produced by other causes. Such are a pulsation of the magnetic reluctance of the field due to the armature slots, or a pulsation of the armature reactance, as discussed in Chapter XXV of "Theory and Calculation of Alternating-current Phenomena," or a space resonance of the armature conductors with some of the harmonics. The latter may occur if the field flux distribution contains a harmonic of such order, that the voltages induced by it are in phase in the successive armature conductors, and therefore add, that is, when the spacing of the armature conductors coincides with a harmonic of the field flux, and the armature turn pitch and winding pitch are such that this harmonic does not cancel.

Inversely, if two turns are displaced from each other on the armature periphery by $\frac{1}{n}$ of the pole pitch, or $\frac{\pi}{n}$, and are connected in series, then in the resultant voltage of these two turns, the nth harmonics are out of phase by n times $\frac{\pi}{n}$, or by $\pi = 180°$, that is, are in opposition and so cancel.

Thus in a unitooth Y-connected three-phase alternator, while each phase usually contains a strong third harmonic, the terminal voltage can contain no third harmonic or its multiples: the two phases, which are in series between each pair of terminals, are one-third pole pitch, or 60 electrical degrees displaced on the armature periphery, and their third harmonic voltages therefore $3 \times 60 = 180°$ displaced, or opposite, that is, cancel, and no third harmonic can appear in the terminal voltage wave, or delta voltage, but a pronounced third harmonic may exist—and give trouble—in the voltage between each terminal and the neutral, or the Y voltage.

62. By the use of a fractional-pitch armature winding, higher harmonics can be eliminated. Assume the two sides of the armature turn, conductor and return conductor, are not separated from each other by the full pitch of the field pole, or 180 electrical degrees, but by less (or more); that is, each armature turn or coil covers not the full pitch of the pole, but the part p less (or more), that is, covers $(1 \pm p)$ 180°. The coil then is said to be $(1 \pm p)$ fractional pitch, or has the pitch deficiency p. The voltages induced in the two sides of the coil then are not equal and in phase, but are out of phase by 180 p for the fundamental, and by 180 np for the nth harmonic. Thus, if $np = 1$, for this n^{th} harmonic the voltages in the two sides of the coil are equal and opposite, thus cancel, and this harmonic is eliminated.

Therefore, two-thirds pitch winding eliminates the third harmonic, four-fifths pitch winding the fifth harmonic, etc.

Peripherally displacing half the field poles against the other half by the fraction q of the pole pitch, or by 180 q electrical degrees, causes the voltages induced by the two sets of field poles to be out of phase by 180 nq for the nth harmonic, and thereby eliminates that harmonic, for which $nq = 1$.

By these various means, if so desired, a number of harmonics can be eliminated. Thus in a Y-connected three-phase alternator with the winding of each phase covering 60 electrical degrees, with four-fifths pitch winding and half the field poles offset against the other by one-seventh of the pole pitch, the third, fifth, and seventh harmonic and their multiples are eliminated, that is, the lowest harmonic existing in the terminal voltage of such a machine is the eleventh, and the machine contains only the eleventh, thirteenth, seventeeth, ninteenth, twenty-third, twenty-ninth, thirty-first, thirty-seventh, etc. harmonics. As by the distributed winding these harmonics are greatly decreased, it follows that the terminal voltage wave would be closely a sine, irrespective of the field flux distribution, assuming that no slot harmonics exist.

63. In modern machines, the voltage wave usually is very closely a sine, as the pronounced lower harmonics, caused by the field flux distribution, which gave the saw-tooth, flat-top, peak or multiple-peak effects in the former unitooth machines, are greatly reduced by the distributed winding and the use of fractional pitch. Individual high harmonics, or pairs of high harmonics, are occasionally met, such as the seventeenth and ninteenth, or the thirty-fifth and thirty-seventh, etc. They are due to the pulsation of the magnetic field flux caused by the pulsation of the field reluctance by the passage of the armature slots, and occasionally, under load, by magnetic saturation of the armature self-inductive flux, that is, flux produced by the current in an armature slot and surrounding this slot, in cases where very many ampere conductors are massed in one slot, and the slot opening bridged or nearly so.

The low harmonics, third, fifth, seventh, are relatively harmless, except where very excessive and causing appreciable increase of the maximum voltage, or the maximum magnetic flux and thus hysteresis loss. The very high harmonics as a rule are relatively harmless in all circuits containing no capacity, since they are necessarily fairly small and still further suppressed by the inductance of the circuit. They may become serious and even dangerous, however, if capacity is present in the circuit, as the current taken by capacity is proportional to the frequency, and even small voltage harmonics, if of very high order, that is, high frequency, produce very large currents, and these in turn may cause dangerous voltages in inductive devices connected in series into the circuit, such as current transformers, or cause resonance effects in transformers, etc. With the increasing extent of very high-voltage transmission, introducing capacity into the systems, it thus becomes increasingly important to keep the very high harmonics practically out of the voltage wave.

Incidentally it follows herefrom, that the specifications of wave shape, that it should be within 5 per cent. of a sine wave, which is still occasionally met, has become irrational: a third harmonic of 5 per cent. is practically negligible, while a thirty-fifth harmonic of 5 per cent., in the voltage wave, would hardly be permissible. This makes it necessary in wave-shape specifications, to discriminate against high harmonics. One way would be, to specify not the wave shape of the voltage, but that of the current taken by a small condenser connected across the voltage. In the condenser current, the voltage harmonics are multiplied by their order. That is, the third harmonic is increased three times, the fifth harmonic five times, the thirty-fifth harmonic 35 times, etc. However, this probably overemphasizes the high harmonics, gives them too much weight, and a better way appears to be, to specify the current wave taken by a small condenser having a specified amount of non-inductive resistance in series.

Thus for instance, if $x = 1000$ ohms = capacity reactance of the condenser, at fundamental frequency, $r = 100$ ohms = resistance in series to the condenser, the impedance of this circuit, for the nth harmonic, would be

$$Z_n = r - j\frac{x}{n} = 100 - \frac{1000}{n}j \qquad (7)$$

or, absolute, the impedance,

$$z_n = 1000\sqrt{\frac{1}{n^2} + 0.01} \qquad (8)$$

and, the admittance,

$$y_n = \frac{0.001\, n}{\sqrt{1 + 0.01\, n^2}} \qquad (9)$$

and therefore, the multiplying factor,

$$f = \frac{y_n}{y_1} = \frac{1.005\, n}{\sqrt{1 + 0.01\, n^2}} \qquad (10)$$

this gives, for

n	f	n	f
1	1.0	13	8.0
3	2.9	15	8.4
5	4.5	25	9.3
7	5.8	35	9.6
9	6.7	45	9.8
11	7.4	∞	10.0

Thus, with this proportion of resistance and capacity, the maximum intensification is tenfold, for very high harmonics. By using a different value of the resistance, it can be made anything desired.

A convenient way of judging on the joint effect of all harmonics of a voltage wave is by comparing the current taken by such a condenser and resistance, with that taken by the same condenser and resistance, at a sine wave of impressed voltage, of the same effective value.

Thus, if the voltage wave

$$e = 600 + 18_3 + 12_5 + 9_7 + 4_9 + 2_{11} + 3_{13} + 30_{23} + 24_{25}$$
$$= 600\ \{\ 1 + 0.03_3 + 0.02_5 + 0.015_7 + 0.0067_9 + 0.0033_{11}$$
$$+ 0.005_{13} + 0.05_{23} + 0.04_{25}\ \}$$

(where the indices indicate the order of the harmonics) of effect-ive value

$$e = \sqrt{600^2 + 18^2 + 12^2 + 9^2 + 4^2 + 2^2 + 3^2 + 30^2 + 24^2}$$
$$= 601.7$$

is impressed upon the condenser resistance of the admittance, y_n, the current wave is

$$i = 0.603 \{ 1 + 0.087_3 + 0.09_5 + 0.087_7 + 0.0445_9 + 0.0247_{11}$$
$$+ 0.04_{13} + 0.46_{23} + 0.37_{25} \}$$
$$= 0.603 \times 1.173$$
$$= 0.707$$

while with a sine wave of voltage, of $e_0 = 601.7$, the current would be

$$i_0 = 0.599,$$

giving a ratio

$$\frac{i}{i_0} = 1.18,$$

or 18 per cent. increase of current due to wave-shape distortion by higher harmonics.

64. While usually the sine wave is satisfactory for the purpose for which alternating currents are used, there are numerous cases where waves of different shape are desirable, or even necessary for accomplishing the desired purpose. In other cases, by the internal reactions of apparatus, such as magnetic saturation, a wave-shape distortion may occur and requires consideration to avoid harmful results.

Thus in the regulating pole converter (so-called "split-pole converter") variations of the direct-current voltage are produced at constant alternating-current voltage input, by superposing a third harmonic produced by the field flux distribution, as discussed under "Regulating Pole Converter" in "Theory and Calculation of Electrical Apparatus." In this case, the third harmonic must be restricted to the local or converter circuit by proper transformer connections: either three-phase connection of the converter, or Y or double-delta connections of the transformers with a six-phase converter.

The appearance of a wave-shape distortion by the third har-monic and its multiples, in the neutral voltage of Y-connected transformers, and its intensifications by capacity in the secondary circuit, and elimination by delta connection, has been discussed in Chapter XXV of "Theory and Calculation of Alternating-current Phenomena."

In the flickering of incandescent lamps, and the steadiness of arc lamps at low frequencies, a difference exists between the flat-top wave of current with steep zero, and the peaked wave with flat zero, the latter showing appreciable flickering already at a somewhat higher frequency, as is to be expected.

In general, where special wave shapes are desirable, they are usually produced locally, and not by the generator design, as with the increasing consolidation of all electric power supply in large generating stations, it becomes less permissible to produce a desired wave shape within the generator, as this is called upon to supply power for all purposes, and therefore the sine wave as the standard is preferable.

One of the most frequent causes of very pronounced wave-shape distortion, and therefore a very convenient means of producing certain characteristic deviations from sine shape, is magnetic saturation, and as instance of a typical wave-shape distortion, its causes and effects, this will be more fully discussed in the following.

CHAPTER VIII

SHAPING OF WAVES BY MAGNETIC SATURATION

65. The wave shapes of current or voltage produced by a closed magnetic circuit at moderate magnetic densities, such as are commonly used in transformers and other induction apparatus, have

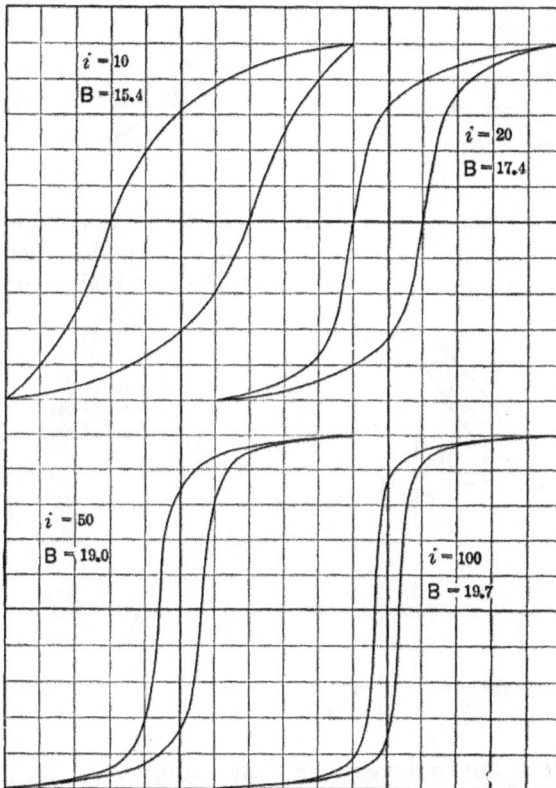

$i = 10$
$B = 15.4$

$i = 20$
$B = 17.4$

$i = 50$
$B = 19.0$

$i = 100$
$B = 19.7$

FIG. 55.

been discussed in "Theory and Calculation of Alternating-current Phenomena."

The characteristic of the wave-shape distortion by magnetic

saturation in a closed magnetic circuit is the production of a high peak and flat zero, of the current with a sine wave of impressed voltage, of the voltage with a sine wave of current traversing the circuit.

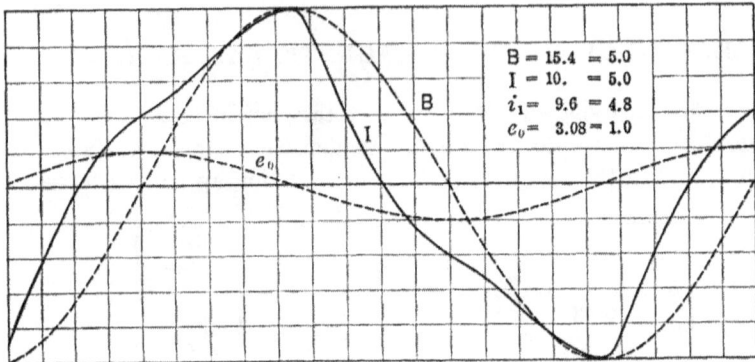

$$B = 15.4 = 5.0$$
$$I = 10. = 5.0$$
$$i_1 = 9.6 = 4.8$$
$$e_0 = 3.08 = 1.0$$

FIG. 56.

In Fig. 55 are shown four magnetic cycles, corresponding respectively to beginning saturation: $B = 15.4$ kilolines per cm.[2], $H = 10$; moderate saturation: $B = 17.4$, $H = 20$; high saturation:

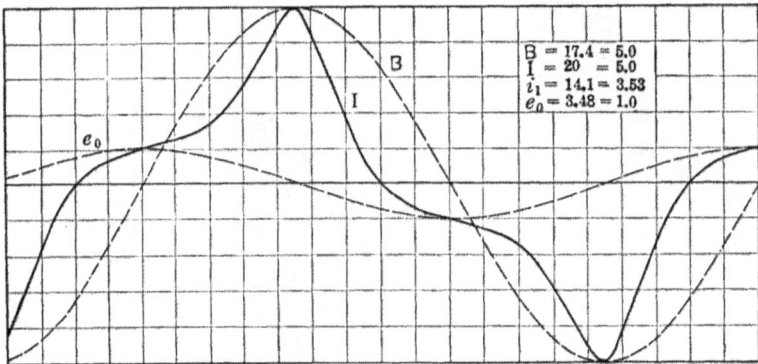

$$B = 17.4 = 5.0$$
$$I = 20 = 5.0$$
$$i_1 = 14.1 = 3.53$$
$$e_0 = 3.48 = 1.0$$

FIG. 57.

$B = 19.0$, $H = 50$; and very high saturation: $B = 19.7$, $H = 100$. Figs. 56, 57, 58 and 59 show the four corresponding current waves I, at a sine wave of impressed voltage e_0, and therefore sine wave of magnetic flux, B (neglecting ir drop in the winding, or rather, e_0 is the voltage induced by the alternating magnetic flux density B). In these four figures, the maximum values of e_0, B and I are chosen of the same scale, for wave-shape comparison, though in reality, in Fig. 59, very high saturation, the maximum of current, I, is ten times as high as in Fig. 56, beginning saturation. As seen, in Fig. 56 the current is the usual saw-tooth wave of transformer-exciting current, but slightly peaked, while in Fig. 59 a high peak exists. The numerical values are given in Table I.

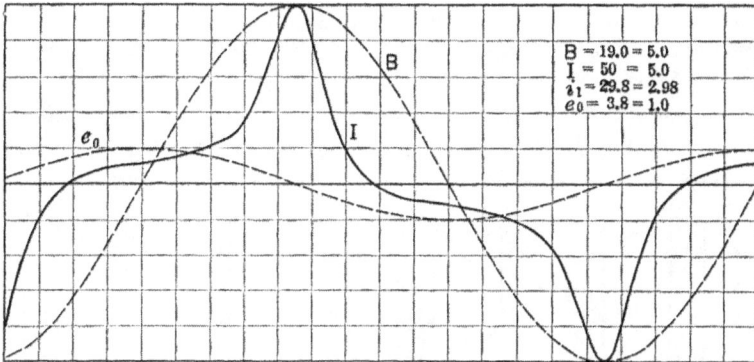

$$B = 19.0 = 5.0$$
$$I = 50 = 5.0$$
$$i_1 = 29.8 = 2.98$$
$$e_0 = 3.8 = 1.0$$

Fɪɢ. 58.

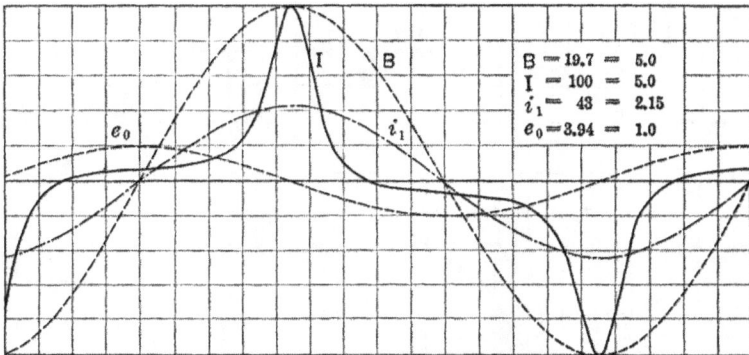

$$B = 19.7 = 5.0$$
$$I = 100 = 5.0$$
$$i_1 = 43 = 2.15$$
$$e_0 = 3.94 = 1.0$$

Fɪɢ. 59.

That is, at beginning saturation, the maximum value of the saw-tooth wave of current differs little from what it would be with a sine wave of the same effective value, being only 4 per cent. higher. At moderate saturation, however, the current peak is already 42 per cent. higher than in a sine wave of the same effective value, and becomes 132 per cent. higher than in a sine wave, at the very high saturation of Fig. 59.

Inversely, while the maximum values of current at the higher

TABLE I

	Begin- ning sat- uration, $B = 15.4$	Moder- ate sat- uration, $B = 17.4$	High satura- tion, $B = 19.0$	Very high satura- tion, $B = 19.7$
Sine wave of voltage, e_0, maximum.......	3.08	3.48	3.80	3.94
Maximum value of current, I...........	10.00	20.00	50.00	100.00
Effective value of current, $\times \sqrt{2} : i_1$.....	9.6	14.1	29.8	43.0
Form factor of current wave $\frac{I}{i_1}$	1.04	1.42	1.68	2.32
Ratio of effective currents.............	1.00	1.47	3.11	4.48

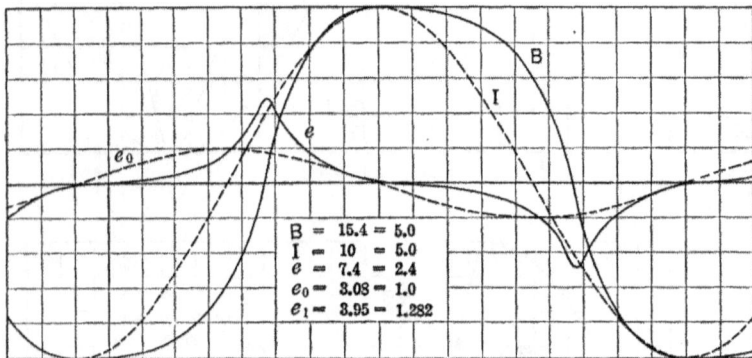

B = 15.4 = 5.0
I = 10 = 5.0
e = 7.4 = 2.4
e_0 = 3.08 = 1.0
e_1 = 3.95 = 1.282

FIG. 60.

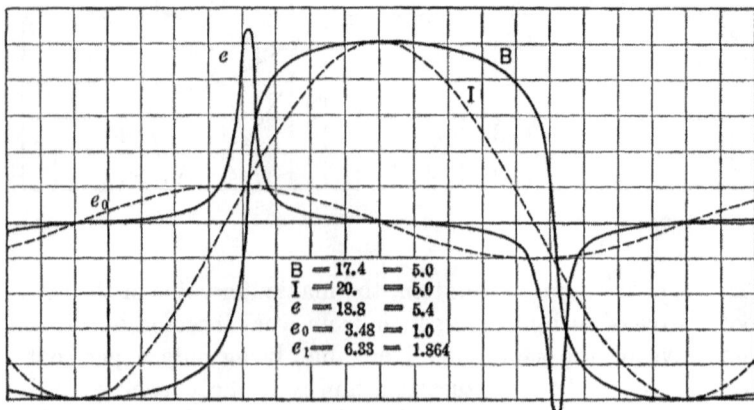

B = 17.4 = 5.0
I = 20. = 5.0
e = 18.8 = 5.4
e_0 = 3.48 = 1.0
e_1 = 6.33 = 1.864

FIG. 61.

saturations are two, five and ten times the maximum current value at beginning saturation, the effective values are only 1.47, 3.1 and 4.47 times higher. Thus, with increasing magnetic saturation, the effective value of current rises much less than the maximum value, and when calculating the exciting current of a saturated magnetic circuit, as an overexcited transformer, from the magnetic characteristic derived by direct current, under the as-

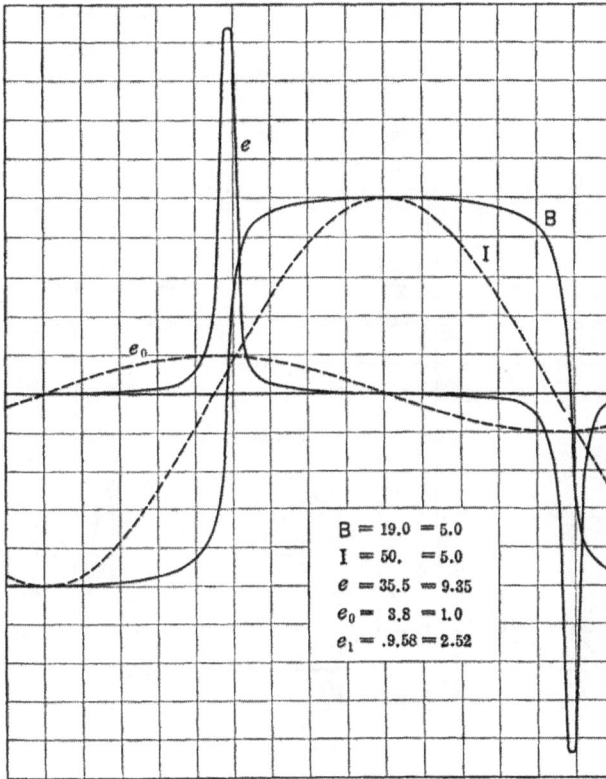

$B = 19.0 = 5.0$

$I = 50. = 5.0$

$e = 35.5 = 9.35$

$e_0 = 3.8 = 1.0$

$e_1 = .9.58 = 2.52$

FIG. 62.

sumption of a sine wave, the calculated exciting current may be more than twice as large as the actual exciting current.

66. Figs. 60 to 63 show, for a sine wave of current, I, traversing a closed magnetic circuit, and the same four magnetic cycles given in Fig. 55, the waves of magnetic flux density, B, of induced voltage, e, the sine wave of voltage, e_0, which would be induced if the

9

magnetic density, B, were a sine wave of the same maximum value, and Fig. 63 also shows the equivalent sine wave, e_1, of the (distorted) induced voltage wave, e.

As seen, already at beginning saturation, Fig. 60, the voltage peak is more than twice as high as it would be with a sine wave,

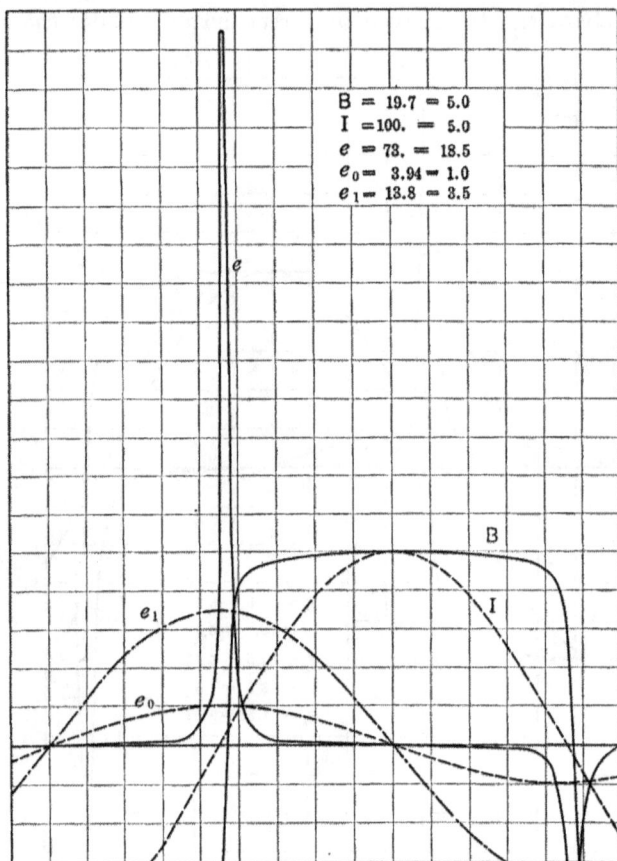

$$
\begin{aligned}
B &= 19.7 = 5.0\\
I &= 100. = 5.0\\
e &= 73. = 18.5\\
e_0 &= 3.94 = 1.0\\
e_1 &= 13.8 = 3.5
\end{aligned}
$$

Fɪɢ. 63.

and rises at higher saturations to enormous values: 18.5 times the sine wave value in Fig. 63.

The magnetic flux wave, B, becomes more and more flat-topped with increasing saturation, and finally practically rectangular, in Fig. 63.

The curves 60 to 63 are drawn with the same maximum values

of current, I, flux density, B, and sine wave voltage, e_0, for better comparison of their wave shapes.

The numerical values are:

TABLE II

	Beginning saturation, $B = 15.4$	Moderate saturation, $B = 17.4$	High saturation, $B = 19.0$	Very high saturation, $B = 19.7$
Sine wave of current, I, maximum	10.0	20.0	50.0	100.0
Flat-top wave of magnetic density, B, maximum	15.4	17.4	19.0	19.7
Peaked voltage wave e, maximum	7.4	18.8	35.5	73.0
Ratio	1.00	2.56	4.80	9.88
Sine wave of voltage, e_0, maximum, for same maximum flux	3.08	3.48	3.80	3.94
Ratio	1.00	1.13	1.23	1.28
Form factor of voltage wave, $\dfrac{e}{e_0}$	2.40	5.40	9.35	18.50
Equivalent sine wave of voltage, e_1, maximum	3.95	6.33	9.58	13.80
Ratio	1.00	1.60	2.42	3.50
$\dfrac{e_1}{e_0}$ (maxima)	1.282	1.864	2.520	3.500
$\dfrac{e}{e_1}$ (maxima)	1.87	2.97	3.70	5.28

As seen, the wave-shape distortion due to magnetic saturation is very much greater with a sine wave of current traversing the closed magnetic circuit, than it is with a sine wave of voltage impressed upon it.

With increasing magnetic saturation, with a sine wave of current, the effective value of induced voltage increases much more rapidly than the magnetic flux increases, and the maximum value of voltage increases still much more rapidly than the effective value: an increase of flux density, B, by 28 per cent., from beginning to very high saturation, gives an increase of the effective value of induced voltage (as measured by voltmeter) by 250 per cent., or 3.5 times, and an increase of the peak value of voltage (which makes itself felt by disruption of insulation, by danger to life, etc.) by 888 per cent., or nearly ten times.

At very high saturation, the voltage wave practically becomes one single extremely high and very narrow voltage peak, which occurs at the reversal of current.

At the very high saturation, Fig. 63, the effective value, e_1, of the voltage is 3.5 times as high as it would be with a sine wave of magnetic flux; the maximum value, e, is more than five times as high as it would be with a sine wave of the same effective value, e_1, that is, more than five times as high, as would be expected from the voltmeter reading, and it is 18.5 times as high as it would be with a sine wave of magnetic flux.

Thus, an oversaturated closed magnetic circuit reactance, which consumes $e_0 = 50$ volts with a sine wave of voltage, e_0, and thus of magnetic density, B, would, at the same maximum magnetic density, that is, the same saturation, with a sine wave of current—as would be the case if the reactance is connected in series in a constant-current circuit—give an effective value of terminal voltage of $e_1 = 3.5 \times 50 = 175$ volts, and a maximum peak voltage of $e = 18.8 \times 50 \times \sqrt{2} = 1330$ volts.

Thus, while supposed to be a low-voltage reactance, $e_0 = 50$ volts, and even the voltmeter shows a voltage of only $e_1 = 175$, which, while much higher, is still within the limit that does not endanger life, the actual peak voltage $e = 1330$ is beyond the danger limit.

Thus, magnetic saturation may in supposedly low-voltage circuits produce dangerously high-voltage peaks.

A transformer, at open secondary circuit, is a closed magnetic circuit reactance, and in a transformer connected in series into a circuit—such as a current transformer, etc.—at open secondary circuit unexpectedly high voltages may appear by magnetic saturation.

67. From the preceding, it follows that the relation of alternating current to alternating voltage, that is, the reactance of a closed magnetic circuit, within the range of magnetic saturation, is not constant, but varies not only with the magnetic density, B, but for the same magnetic density B, the reactance may have very different values, depending on the conditions of the circuit: whether constant potential, that is, a sine wave of voltage impressed upon the reactance; or constant current, that is, a sine wave of current traversing the circuit; or any intermediate condition, such as brought about by the insertion of various amounts of resistance, or of reactance or capacity, in series to the closed magnetic circuit reactance.

The numerical values in Table III illustrate this.

I gives the magnetic field intensity, and thus the direct current, which produces the magnetic density, *B*—that is, the *B*-H curve of the magnetic material. An alternating current of maximum value, *I*, thus gives an alternating magnetic flux of maximum flux density *B*. If *I* and *B*, were both sine waves, that is, if

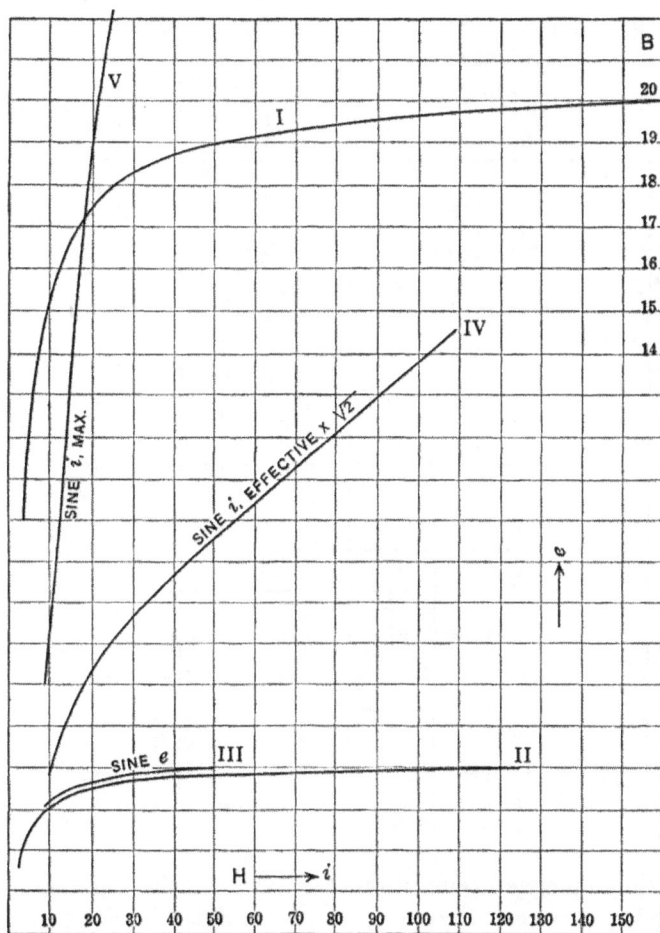

Fɪɢ. 64.

during the cycle current and magnetic flux were proportional to each other, as in an unsaturated open magnetic circuit, e_0, as given in the third column, would be the maximum value of the induced voltage, and $x_0 = \dfrac{e_0}{I}$ the reactance. This reactance varies with the density, and greatly decreases with increasing magnetic saturation, as well known.

However, if e_0 and thus B are sine waves, I can not be a sine wave, but is distorted as shown in Figs. 56 to 59, and the effective value of the current, that is, the current as it would be read by an alternating ammeter, multiplied by $\sqrt{2}$ (that is, the maximum value of the equivalent sine waves of exciting current) is given as i_1. The reactance is then found as $x_p = \dfrac{e_0}{i_1}$. This is the reactance

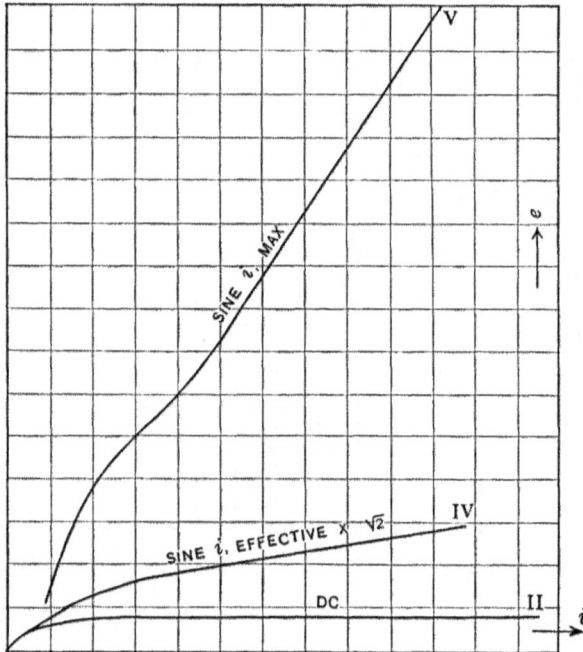

Fig. 65.

of the closed magnetic circuit on constant potential, that is, on a sine wave of impressed voltage, and, as seen, is larger than x_0.

If, however, the current, I, which traverses the reactance, is a sine wave, then the flux density, B, and the induced voltage are not sines, but are distorted as in Figs. 60 to 63, and the effective value of the induced voltage (that is, the voltage as read by alternating voltmeter), multiplied by $\sqrt{2}$ (that is; the maximum of the equivalent sine wave of voltage) is given as e_1 in Table III, and the true maximum value of the induced voltage wave is e.

The reactance, as derived by voltmeter and ammeter readings under these conditions. that is, on a constant-current circuit, or with a sine wave of current traversing the magnetic circuit, is $x_c = \dfrac{e_1}{I}$, thus larger than the constant-potential reactance, x_p.

Much larger still is the reactance derived from the actual maximum values of voltage and current: $x_m = \dfrac{e}{I}$.

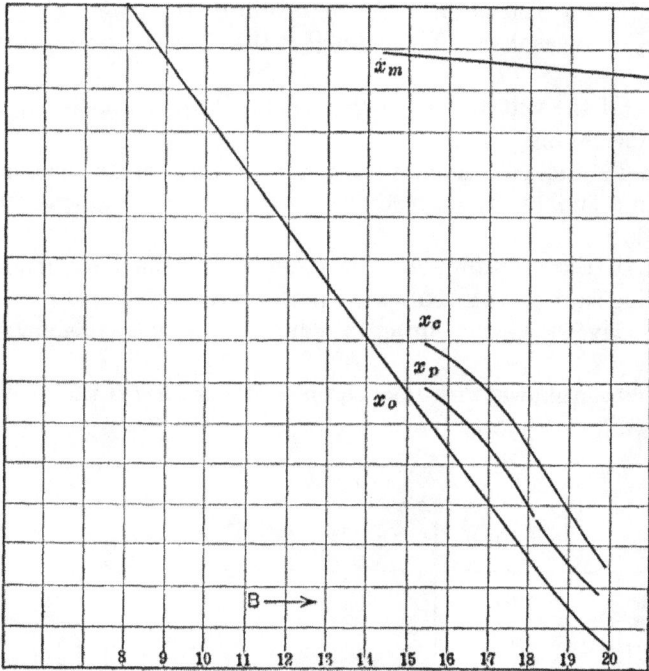

FIG. 66.

It is interesting to note that x_m, the peak reactance, is approximately constant, that is, does not decrease with increasing magnetic saturation. (The higher value at beginning saturation, for $I = 20$, may possibly be due to an inaccuracy in the hysteresis cycle of Fig. 55, a too great steepness near the zero value, rather than being actual.)

It is interesting to realize, that when measuring the reactance of a closed magnetic circuit reactor by voltmeter and ammeter readings, it is not permissible to vary the voltage by series resistance, as this would give values indefinite between x_p and x_c, depending on the relative amount of resistance. To get x_p, the generated supply voltage of a constant-potential source must be varied; to get x_c, the current in a constant-current circuit must be varied. As seen, the differences may amount to several hundred per cent.

As graphical illustration, Fig. 64 shows:

As curve I the magnetic characteristic, as derived with direct current.

Curve II the volt-ampere characteristic of the closed circuit reactance, I, e_0, as it would be if I and B, that is, e_0, both were sine waves.

Curve III the volt-ampere characteristic on constant-potential alternating supply, i_1, e_0.

Curve IV the volt-ampere characteristic on constant-current alternating supply, as derived by voltmeter and ammeter, I, e_1, and as

Curve V the volt-ampere characteristic on constant-current alternating supply, as given by the peak values of I and e.

Fig. 65 gives the same curves in reduced scale, so as to show V completely.

Fig. 66 then shows, with B as abscissæ, the values of the reactances x_0, x_p, x_c, and x_m.

TABLE III

I	B	e_0	$x_0 = \dfrac{e_0}{I}$	i_1	$x_p = \dfrac{e_0}{i_1}$	e_1	$x_c = \dfrac{e_1}{I}$	e	$x_m = \dfrac{e}{I}$	p	p_0
2.0	7.30	.....	0.7300	.....	.....	.....	.....	.....	.....	1.00	
3.0	10.00	.....	0.6670	.....	.....	.....	.....	.....	.....	1.09	
4.0	11.50	.....	0.5750	.....	.....	.....	.....	.....	.....	1.27	
5.0	12.50	.....	0.5000	.....	.....	.....	.....	.....	.....	1.46	
7.5	14.30	.....	0.3810	.....	.....	.....	.....	.....	.....	1.92	
10.0	15.40	3.08	0.3080	9.0	0.342	3.95	0.395	7.4	0.74	2.37	2.40
15.0	16.70	.....	0.2230	.....	.....	.....	.....	.....	.....	3.27	
20.0	17.40	3.48	0.1740	14.1	0.247	6.33	0.316	18.8	0.94	4.20	5.40
30.0	18.30	.....	0.1220	.....	.....	.....	.....	.....	.....	6.00	
40.0	18.70	.....	0.0930	.....	.....	.....	.....	.....	.....	7.85	
50.0	19.00	3.80	0.0760	29.8	0.127	9.58	0.912	35.5	0.71	9.60	9.35
75.0	19.35	.....	0.0520	.....	.....	.....	.....	.....	.....	14.10	
100.0	19.70	3.94	0.0394	43.0	0.092	13.80	0.138	73.0	0.73	18.50	18.5
125.0	19.85	.....	0.0320	.....	.....	.....	.....	.....	.....	22.80	
150.0	19.95	.....	0.0270	.....	.....	.....	.....	.....	.....	27.00	

68. Another way of looking at the phenomenon is this: while with increasing current traversing a closed magnetic circuit, the magnetic flux density is limited by saturation, the induced voltage

peak is not limited by saturation, as it occurs at the current reversal, but it is proportional to the rate of change of the magnetic
flux density at the current reversal, and thus approximately proportional to the current.

Thus, approximately, within the range of magnetic saturation,
with increasing current traversing the closed magnetic circuit
(like that of a series transformer):

The magnetic flux density, and therefore the mean value of induced voltage remains constant;

The peak value of induced voltage increases proportional to the
current, and therefore;

The effective value of induced voltage increases proportional
to the square root of the current.

Thus, if the exciting current of a series transformer is 5 per cent.
of full-load current, and the secondary circuit is opened, while the
primary current remains the same, the effective voltage consumed
by the transformer increases approximately $\sqrt{20} = 4.47$ times,
and the maximum voltage peak 20 times above the full-load
voltage of the transformer.

As the shape of the magnetic flux density and voltage waves are
determined by the current and flux relation of the hysteresis cycles, and the latter are entirely empirical and can not be expressed
mathematically, therefore it is not possible to derive an exact
mathematical equation for these distorted and peaked voltage
waves from their origin. Nevertheless, especially at higher saturation, where the voltage peaks are more pronounced, the equation of the voltage wave can be derived and represented by a
Fourier series with a fair degree of accuracy. By thus deriving
the Fourier series which represents the peaked voltage waves, the
harmonics which make up the wave, and their approximate values can be determined and therefrom their probable effect on the
system, as resonance phenomena, etc., estimated.

The characteristic of the voltage-wave distortion due to magnetic saturation in a closed magnetic circuit traversed by a sine
wave of current is, that the entire voltage wave practically con-
tracts into a single high peak, at, or rather shortly after, the moment of current reversal, as shown in Figs. 63, 62, etc.

With the same maximum value of magnetic density, B, and thus
of flux, Φ, the area of the induced voltage wave, and thus the mean
value of the voltage, is the same, whatever may be the wave of
magnetism and thus of voltage, since $\Phi = \int e\, dt$, and the area of

the peaked voltage wave of the saturated magnetic circuit, e, thus is the same as that of a sine wave of voltage, e_0. Neglecting then the small values of voltage, e, outside of the voltage peak, if this voltage peak of e is p times the maximum value of the sine wave, e_0, its width is $\frac{1}{p}$ of that of the sine wave, and if the sine wave of voltage, e_0, is represented by the equation

$$e_0 \cos \phi \tag{11}$$

the peak of the distorted voltage wave is represented, in first approximation, by assuming it as of sinusoidal shape, by

$$p e_0 \cos p\phi \tag{12}$$

That is, the distorted voltage wave, e, can be considered as represented by $p e_0 \cos p\phi$ within the angle

$$-\frac{\pi}{2\,p} < \phi < \frac{\pi}{2\,p} \tag{13}$$

and by zero outside of this range.

The value of p follows, approximately, from the consideration that the peak reactance, x_m, is independent of the saturation, or constant, since it depends on the rate of change of magnetism with current near the zero value, where there is no saturation, and the ratio $\frac{dB}{dI}$ thus (approximately) constant.

Or, in other words, if below saturation, in the range where the magnetic permeability is a maximum, the current, i, produces the magnetic flux, Φ, and thereby induces the voltage, e', the reactance is

$$x' = \frac{e'}{i} \tag{14}$$

This is the maximum reactance, below saturation, of the magnetic circuit, and can be calculated from the dimensions and the magnetic characteristic, in the usual manner, by assuming sine waves of i and B.

The peak reactance, x_m, of the saturated magnetic circuit is approximately equal to x', and thus can be calculated with reasonable approximation, from the dimensions of the magnetic circuit and the magnetic characteristic.

If now, in the range of magnetic saturation, a sine wave of current, of maximum value I, traverses the closed magnetic circuit, the peak value of the (distorted) induced voltage is

$$e = x_m I \qquad (15)$$

where

$$x_m = x' = \frac{e'}{i} \qquad (16)$$

is the maximum reactance of the magnetic circuit below saturation, derived by the assumption of sine waves, e' and i.

If B is the maximum value of the magnetic density produced by the sine wave of current of maximum value, I, and, e_0, the maximum value of the sine wave of voltage induced by a sinusoidal variation of the magnetic density, B, the "form factor" of the peaked voltage wave of the saturated magnetic circuit is

$$p = \frac{e}{e_0} = \frac{x_m I}{e_0} \qquad (17)$$

thus determined, approximately.

As illustrations are given, in the second last column of Table III, the form factors, p, calculated in this manner, and in the last column are given the actual form factors, p_0, derived from the curves 60 to 63. As seen, the agreement is well within the uncertainty of observation of the shape of the hysteresis cycles, except perhaps at $I = 20$, and there probably the calculated value is more nearly correct.

69. The peaked voltage wave induced by the saturated closed magnetic circuit can, by assuming it as symmetrical and counting the time from the center of the peak, be represented by the Fourier series.

$$\left.\begin{aligned} e &= a_1 \cos \phi + a_3 \cos 3\,\phi + a_5 \cos 5\,\phi + a_7 \cos 7\,\phi + \ldots \\ &= \Sigma\, a_n\, \cos\, n\phi \end{aligned}\right\} \quad (18)$$

where

$$a_n = \frac{4}{\pi} \int_0^{\frac{\pi}{2}} e \cos n\phi \, d\phi \qquad (19)$$

$$= 2\, \mathrm{avg}(e \cos n\phi)_0^{\frac{\pi}{2}} \qquad (20)$$

The slight asymmetry of the peak would introduce some sine terms, which might be evaluated, but are of such small values as to be negligible.

(*a*) For the lower harmonics, where n is small compared to p,

$\cos n\phi$ is practically constant and $= 1$ during the short voltage peak $e = pe_0 \cos p\phi$, and it is, therefore,

$$a_n = 2 \operatorname{avg}(e)_0^{\frac{\pi}{2}}$$

$$= 2 \operatorname{avg}(pe_0 \cos p\phi)_0^{\frac{\pi}{2}}$$

$$= \frac{2}{p} \operatorname{avg}(pe_0 \cos p\phi)_0^{\frac{\pi}{2}p}$$

$$= 2 e_0 \operatorname{avg} \cos = \frac{4}{\pi} e_0.$$

(b) For the harmonic, where $n = p$, it is

$$a_p = 2 \operatorname{avg}(pe_0 \cos^2 p\phi)_0^{\frac{\pi}{2}}$$

$$= \frac{2}{p} \operatorname{avg}(pe_0 \cos^2 p\phi)_0^{\frac{\pi}{2}}$$

$$= 2 e_0 \operatorname{avg} \cos^2 = e_0.$$

(c) For still higher harmonics than $n = p$, $\cos n\phi$ assumes negative values within the range of the voltage peak, and a_n thereby rapidly decreases, finally becomes zero and then negative, at $n = 3 p$, positive again at $n = 5 p$, etc., but is practically negligible.

Thus, the coefficients of the Fourier series decrease gradually, with increasing order, n, from $\frac{4}{\pi} e_0$ as maximum, to e_0 for $n = p$, and then with increasing rapidity fall off to negligible values.

Their exact values can easily be derived by substituting (12) into (19),

$$a_n = \frac{4}{\pi}\int_0^{\frac{\pi}{2p}} pe_0 \cos p\phi \cos n\phi \, d\phi \tag{21}$$

here the integration is extended to $\frac{\pi}{2p}$ only, as beyond this, the voltage, e, is not given by equation (12) any more, but is zero.

(21) integrates by

$$a_n = \frac{2 pe_0}{\pi} \left/ \frac{\sin(p + n)\phi}{p + n} + \frac{\sin(p - n)\phi}{p - n} \right/ _0^{\frac{\pi}{2p}}$$

$$= \frac{2 pe_0}{\pi} \left\{ \frac{\sin \frac{\pi}{2}\left(1 + \frac{n}{p}\right)}{p + n} + \frac{\sin \frac{\pi}{2}\left(1 - \frac{n}{p}\right)}{p - n} \right\},$$

but since $\quad \sin \dfrac{\pi}{2}\left(1 + \dfrac{n}{p}\right) = \sin \dfrac{\pi}{2}\left(1 - \dfrac{n}{p}\right)$, it is

$$a_n = \frac{4\,e_0 \sin \dfrac{\pi}{2}\left(1 - \dfrac{n}{p}\right)}{\pi\left(1 - \dfrac{n^2}{p^2}\right)} \tag{22}$$

and

$$e = \frac{4\,e_0}{\pi} \sum \frac{\sin \dfrac{\pi}{2}\left(1 - \dfrac{n}{p}\right)}{1 - \dfrac{n^2}{p^2}} \cos n\phi \tag{23}$$

as the equations of the voltage wave distorted by magnetic saturation.

70. These coefficients, a_n, are very easily calculated, and as instances are given in Table IV, the coefficients of the distorted voltage wave of Fig. 62, which has the form factor $p = 9.35$.

TABLE IV

$p = 9.35$				$a_n = \dfrac{4\,e_0}{\pi}\ \dfrac{\sin\dfrac{\pi}{2}\left(1 - \dfrac{n}{p}\right)}{1 - \dfrac{n^2}{p^2}}$					
$n = 1$	3	5	7	9	11	13	15	17	19
$\dfrac{a_n}{e_0} = 1.270$	1.242	1.188	1.114	1.018	0.906	0.786	0.658	0.528	0.406
$n = 21$	23	25	27	29	31	33			
$\dfrac{a_n}{e_0} = 0.292$	0.189	0.101	0.031	−0.023	−0.060	−0.082			

As seen, after $n = 9$, the values of a_n rapidly decrease, and become negative, though of negligible value, after $n = 27$.

In Fig. 67 the successive values of $\dfrac{a_n}{e_0}$ are shown as curve.

In reality, the peaked voltage wave of magnetic saturation, as shown in Figs. 61 to 63, is not half a sine wave, but is rounded off at the ends, toward the zero values. Physically, the meaning of the successive harmonics is, that they raise the peak and cut off the values outside of the peak. It is the high harmonics, which sharpen the edge of the peak, and the rounded edge of the peak in the actual wave thus means that the highest harmonics, which give very small or negative values of a_n, are lower than given by equations (23), or rather are absent.

Thus, by omitting the highest harmonics, the wave is rounded off and brought nearer to its actual shape. Thus, instead of following the curve, a_n, as calculated and given in Fig. 67, we cut it off before the zero value of a_n, about at $n = 23$, and follow the curve line, a'_n, which is drawn so that $\Sigma \dfrac{a'_n}{e_0} = 9.35$, that is, that the voltage peak has the actual value.

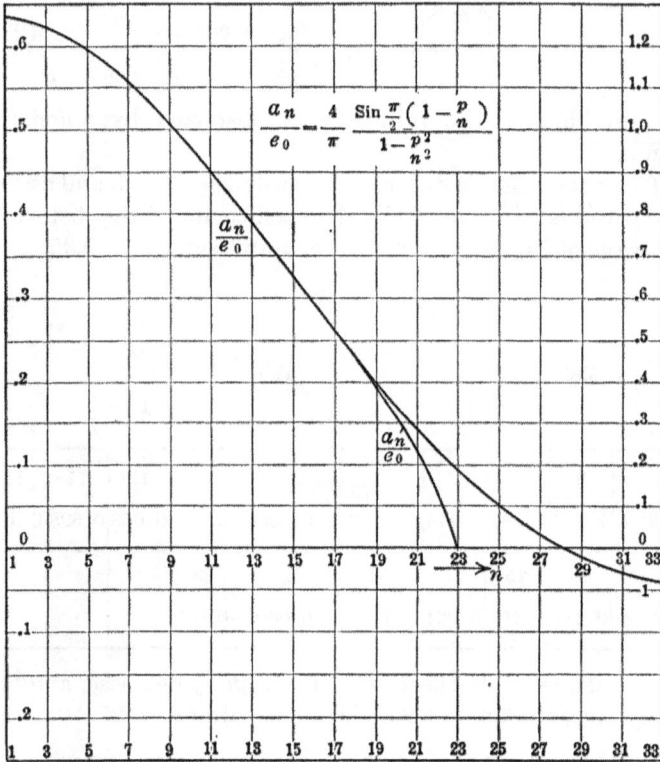

$$\frac{a_n}{e_0} = \frac{4}{\pi} \frac{\mathrm{Sin}\ \frac{\pi}{2}\left(1 - \frac{p}{n}\right)}{1 - \frac{p^2}{n^2}}$$

Fig. 67.

The equation of the peaked voltage in Fig. 62 then becomes

$e = e_0 \{1.270 \cos \phi + 1.242 \cos 3\phi + 1.188 \cos 5\phi + 1.114 \cos 7\phi$
$\quad + 1.018 \cos 9\phi + 0.906 \cos 11\phi + 0.786 \cos 13\phi + 0.658 \cos 15\phi$
$\quad + 0.529 \cos 17\phi + 0.400 \cos 19\phi + 0.240 \cos 21\phi\}.$

Or, in symbolic writing,

$e = e_0\{1.270_1 + 1.242_3 + 1.188_5 + 1.114_7 + 1.018_9 + 0.906_{11}$
$\quad + 0.786_{13} + 0.658_{15} + 0.529_{17} + 0.400_{19} + 0.240_{21}\}$

$$= 1.270 \, e_0 \, \{1_1 + 0.978_3 + 0.953_5 + 0.877_7 + 0.800_9 + 0.713_{11}$$
$$+ \, 0.617_{13} + 0.517_{15} + 0.416_{17} + 0.315_{19} + 0.189_{21}\}.$$

It is of interest to note how extended a series of powerful harmonics is produced. It is easily seen that in the presence of capacity, these large and very high harmonics may be of considerable danger. In any reactance, which is intended for use in series to a high-voltage circuit, the use of a closed magnetic circuit thus constitutes a possible menace from excessive voltage peaks if saturation occurs.

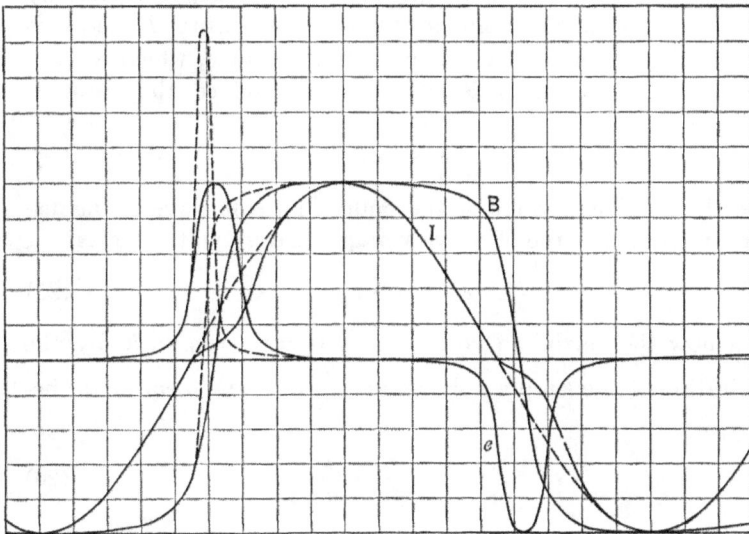

Fig. 68.

71. Such high-voltage peaks by magnetic saturation in a closed magnetic circuit traversed by a sine wave of current can occur only if the available supply voltage is sufficiently high. If the total supply voltage of the circuit is less than the voltage peak produced by magnetic saturation, obviously this voltage peak must be reduced to a value below the voltage available in the supply circuit, and in this case simply the current wave can not remain a sine, but is flattened at the zero values, and with it the wave of magnetic density.

Thus, if in Fig. 62 the maximum supply voltage is $E = 19.0$, the maximum peak voltage can not rise to $e = 35.5$, but stops at

$e \leqq E$, and when this value is reached, the rate of change of flux density, B, and thus of current, I, decreases, as shown in Fig. 68, in drawn lines. In dotted lines are added the curves corresponding to unlimited supply voltage. The voltage peak is thereby reduced, correspondingly broadened, and retarded, and the current is flattened at and after its zero value, the more, the lower the maximum supply voltage.

The reactance is reduced hereby also, from $x_c = 0.192$, in Fig. 62, to $x_c = 0.140$.

In other words, if p is the form factor of the distorted voltage wave, which would, with unlimited supply voltage, be induced by the saturated magnetic circuit of maximum density, B, and e_0 is the maximum value of the sine wave of voltage, which a sinusoidal flux of maximum density, B, would induce, the distorted voltage peak is

$$e = pe_0 \tag{24}$$

and the maximum value of the equivalent sine wave of the distorted voltage, or the effective voltage read by voltmeter, is

$$e_1 = \sqrt{p}\, e_0 \tag{25}$$

If now the maximum voltage peak is cut down to E, by the limitation of the supply voltage, and $\dfrac{e}{E} = q$, the form factor becomes

$$p' = \frac{E}{e_0} = \frac{p}{q}, \tag{26}$$

and the effective value of the distorted voltage, times $\sqrt{2}$, that is, the maximum of the equivalent sine wave, is

$$e'_1 = \sqrt{p'}\, e_0 = \frac{e_1}{\sqrt{q}} \tag{27}$$
$$= \sqrt{e_0 E},$$

thus varies with the supply voltage, E.

The reactance then is

$$x'_c = \frac{e'_1}{I} = \frac{x_c}{\sqrt{q}} \tag{28}$$

Thus, for $e = 35.5$, $E = 19.0$, it is

$$q = 1.87,$$

and as $e_0 = 3.80$; $p = 9.35$, it is

$$p' = \frac{p}{q} = 5.0,$$

$$e'_1 = \frac{e_1}{\sqrt{q}} = \frac{9.58}{1.37} = 7.0,$$

$$x'_c = \frac{x_c}{\sqrt{q}} = \frac{0.192}{1.37} = 1.40.$$

These values, however, are only fair approximations, as they are based on the assumption of sinusoidal shape of the peaks.

72. In the preceding, the assumption has been made, that the magnetic flux passes entirely within the closed magnetic circuit, that is, that there is no magnetic leakage flux, or flux which closes through non-magnetic space outside of the iron conduit.

If there is a magnetic leakage flux—and there must always be some—it somewhat reduces the voltage peak, the more, the greater is the proportion of the leakage flux to the main flux. The leakage flux, in open magnetic circuit, is practically proportional to the current, and that part of the voltage, which is induced by the leakage flux, therefore, is a sine wave, with a sine wave of current, hence does not contribute to the voltage peak.

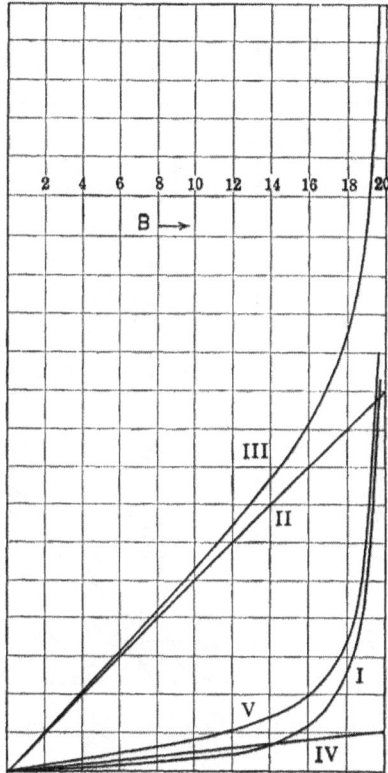

Fig. 69.

Such high magnetic saturation peaks occur only in a closed magnetic circuit. If the magnetic circuit is not closed, but contains an air-gap, even a very small one, the voltage peak, with a sine wave of current, is very greatly reduced, since in the air-gap magnetic flux and magnetizing current are proportional.

10

Thus, below saturation and even at beginning saturation, an air-gap in the magnetic circuit, of one-hundredth of its length, makes the voltage wave practically a sine wave, with a sine wave of current, as discussed in "Theory and Calculation of Alternating-current Phenomena."

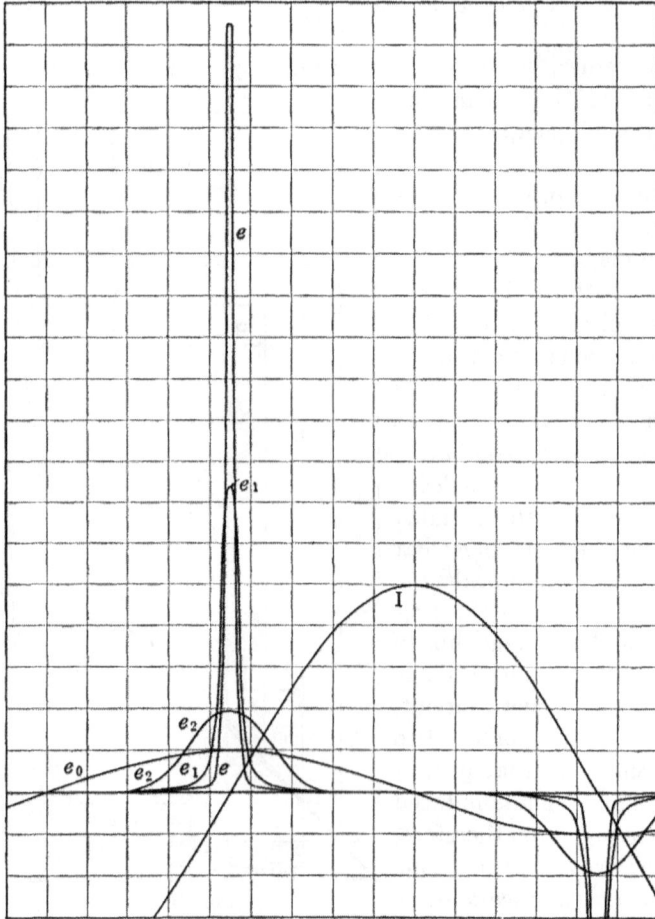

FIG. 70.

The enormous reduction of the voltage peak by an air-gap of 1 per cent. of the length of the magnetic circuit is shown in Figs. 69 and 70.

In Fig. 69, with the magnetic flux density, B, as abscissæ, the m.m.f. of the iron part of the magnetic circuit is shown as curve I. This would be the magnetizing current if the magnetic circuit were closed. Curve II show the m.m.f. consumed in an air-gap of 1 per cent. of the length of the magnetic circuit of curve I, and curve III, therefore, shows the total m.m.f. of the magnetizing current of the magnetic circuit with 1 per cent. air-gap.

Choosing as instance the very high saturation $B = 19.7$, the same as illustrated in Fig. 63, and neglecting the hysteresis—which is permissible, as the hysteresis does not much contribute to the wave-shape distortion—the corresponding voltage waves are plotted in Fig. 70, in the same scale as Figs. 56 to 63: for a sine wave of current, curves Fig. 69 give the corresponding values of magnetic flux, and from the magnetic flux wave is derived, as $\dfrac{dB}{d\phi}$, the voltage wave. The waves of magnetism are not plotted. e_0 is the sine wave of voltage, which would be induced by a sinusoidal variation of magnetic flux; e is the peaked voltage wave induced in a closed magnetic circuit of the same maximum values of magnetism, of form factor $p = 18.5$ (the same as Fig. 63), and e_2 is the voltage wave induced in a magnetic circuit having an air-gap of 1 per cent. of its length. As seen, the excessive peak of e has vanished, and e_2 has a moderate peak only, of form factor $p = 1.9$.

Even a much smaller air-gap has a pronounced effect in reducing the voltage peak. Thus curves IV and V show the m.m.fs. of the air-gap and of the total magnetic circuit, respectively, when containing an air-gap of one-thousandth of the length of magnetic circuit. e_1 in Fig. 70 then shows the voltage wave corresponding to V in Fig. 69: of form factor $p = 7.4$.

Thus, while excessive voltage peaks are produced in a highly saturated closed magnetic circuit, even an extremely small air-gap, such as given by some butt-joints, materially reduces the peak: from form factor $p = 18.5$ to 7.4 at one-thousandth gap length, and with an air-gap of 1 per cent. length, only a moderate peakedness remains at the highest saturation, while at lower saturation the voltage wave is practically a sine.

73. Even a small air-gap in the magnetic circuit of a reactor greatly reduces the wave-shape distortion, that is, makes the voltage wave more sinusoidal, and cuts off the saturation peak. The latter, however, is the case only with a complete air-gap. A partial air-gap or bridged gap, while it makes the wave shape more sinusoidal elsewhere, does not reduce but greatly increases the voltage peak, and produces excessive peaks even below saturation, with a sine wave of current, and such bridged gaps are, therefore, objectionable with series reactors in high-voltage circuits. In shunt reactors, or reactors having a constant sine wave of impressed voltage, the bridged gap merely produces a short flat zero of the current wave, thus is harmless, and for these purposes the bridged gap reactance—shown diagrammatically in Fig. 71 —is extensively used, due to its constructive advantages: greater

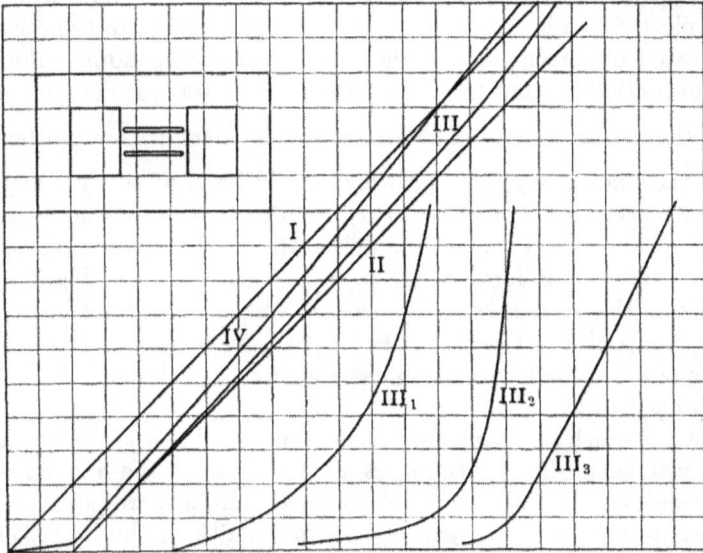

Fig. 71.

rigidity or structure and, therefore, absence of noise, and reduced magnetic stray fields and eddy-current losses resulting therefrom.

Assuming that one-tenth of the gap is bridged, and that the length of the gap is one one-hundredth that of the entire magnetic circuit, as shown diagrammatically in Fig. 71. With such a bridged gap, with all but the lowest m.m.fs. the narrow iron bridges of the gap are saturated, thus carry the flux density $S + H$, where S = metallic saturation density, = 20 kilolines per cm.^{2} in these figures, and H the magnetizing force in the gap.

For one-tenth of the gap, the flux density thus is $H + S$, for the other nine-tenths, it is H, and the average flux density in the gap thus is

$$B = H + 0.1\,S = H + 2,$$

or, if p = bridged fraction of gap,

$$B = H + pS.$$

Curve II in Fig. 71 shows, with the average flux density, B, as abscissæ, the m.m.f. required by the gap,

$$H = B - 0.1\,S$$
$$= B - 2,$$

while curve I shows the m.m.f. which an unbridged gap would require.

Adding to the ordinates of II the values of the m.m.f. required for the iron part of the magnetic circuit, or the other 99 per cent., gives as curve III the total m.m.f. of the reactor.

The lower part of curve III is once more shown, with five times the abscissæ B, and 1000, 100 and 10 times, respectively, the ordinates H, as III_1. III_2, III_3.

74. From $B = 2$ upward, curve III is practically a straight line, and plotting herefrom for a sine wave of current, I and thus m.m.f., H, the wave of magnetism, B, and of voltage, e, these curves become within this range similar to a sine wave as shown as B and e in Fig. 72. Below $B = 2$, however, the slope of the B-H curve and with this their wave shapes change enormously. The B wave becomes practically vertical, that is, B abruptly reverses, and corresponding thereto, the voltage abruptly rises to an excessive peak value, that is, a high and very narrow voltage peak appears on top of the otherwise approximately sine-shaped voltage wave, e.

Choosing the same value as in Fig. 60, $B = 15.4$ or beginning saturation, as the maximum value of flux density: at this, in an entirely closed magnetic circuit the voltage peak is still moderate. On the B-H curve III of Fig. 71, the flux density, $B = 15.4$, requires the m.m.f., $H = 14.4$ If then B and H would vary sinusoidally, giving a sine wave of voltage, e_0, the average value of this voltage wave, e_0, would be proportional to the average rate of magnetic change, or to $\dfrac{B}{H} = \dfrac{15.4}{14.4} = 1.07$, and the maximum value of the sine wave of voltage would be $\dfrac{\pi}{2}$ as high, or,

$$e_0 = \frac{\pi}{2}\frac{B}{H} = \frac{1.07\,\pi}{2} = 1.68.$$

The maximum value of the actual voltage curve, e, occurs at the moment where B passes through zero, and is, from curve III_1,

$$e = \left[\frac{B}{H}\right]_0 = \frac{290}{5} = 580.$$

This, then, is the peak voltage of the actual wave, while, if it were a sine wave, with the same maximum magnetic flux, the maximum voltage would be $e_0 = 1.68$.

The voltage peak produced by the bridged gap and the form factor thus is

$$p = \frac{e}{e_0} = \frac{580}{1.68} = 345,$$

that is, 345 times higher than it would be with a sine wave.

Obviously, such peak can hardly ever occur, as it is usually beyond the limit of the available supply voltage. It thus means, that during the very short moment of time, when during the current reversal the flux density in the iron bridge of the gap changes from saturation to saturation in the reverse direction, a voltage peak rises up to the limits of voltage given by the supply system. This peak is so narrow that even the oscillograph usually does not completely show it.

However, such practically unlimited peaks occur only in a perfectly closed magnetic circuit, containing a bridged gap. If, in addition to the bridged gap of 1 per cent., an unbridged gap of 0.1 per cent.—such as one or several butt-joints—is present, giving the B–H curve IV of Fig. 71, the voltage peak is greatly reduced. It is

$$e_0 = \frac{\pi}{2}\frac{B}{H} = \frac{\pi}{2}\frac{15.4}{15.95} = 1.51,$$

$$e = \left[\frac{B}{H}\right]_0 = \frac{1000}{100} = 10,$$

hence, the relative voltage peak, or form factor,

$$p = \frac{e}{e_0} = 6.6.$$

That is, by this additional gap of one one-thousandth of the length of magnetic circuit, the peak voltage is reduced from 345 times that of the sine wave, to only 6.6 times, or to less than 2 per cent. of its previous value.

As seen from the reasoning in paragraph and Fig. 67, the

peaked wave of Fig. 72 contains very pronounced harmonics up to about the 701th, which at 60 cycles of fundamental frequency, gives frequencies up to 42,000, or well within the range of the danger frequencies of high-voltage power transformers, that is,

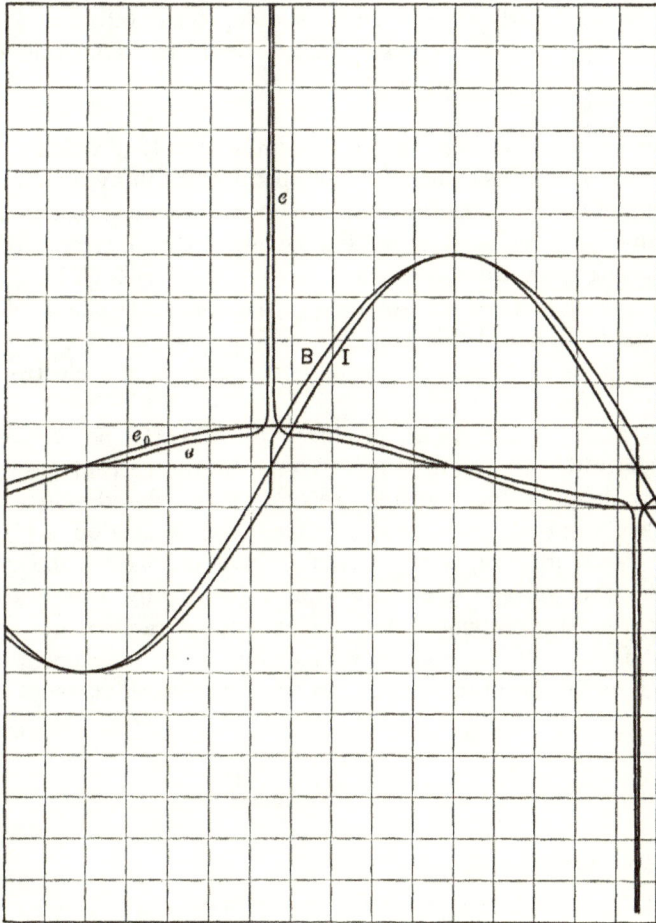

Fig. 72.

frequencies with which the high-voltage coils of transformers, as circuits of distributed capacity, can resonate.

75. Magnetic saturation, and closed or partly closed magnetic circuits thus are a likely source of wave-shape distortion, resulting in high voltage peaks, and where they are liable to occur, as in current transformers, series transformers at open secondary circuit, autotransformers or reactors, etc., they may be guarded against by using a small air-gap in the magnetic circuit, or by providing the extra insulation required to stand the voltage, and the secondary circuit, even if of an effective voltage which is not dangerous to life when a sine wave, should be carefully handled as the voltage peak may reach values which are dangerous to life, without the voltmeter—which reads the effective value—indicating this.

Inversely, such voltage peaks are intentionally provided in some series autotransformers for the operation of individual arcs of the type, in which slagging and consequent failures to start may occur, due to a high-resistance slag covering the electrode tips. By designing the autotransformer so as to give a very high voltage peak at open circuit—and providing in the apparatus the insulation capable to stand this voltage—reliability of starting is secured by puncturing any non-conducting slag on the electrode tips, by the voltage peak.

These high voltage peaks, produced by magnetic saturation, etc., greatly decrease and vanish if considerable current is produced by them. Thus, when the secondary of a closed magnetic circuit series transformer is open, at magnetic saturation, a high voltage peak appears; with increasing load on the secondary, however, the voltage peak drops and practically disappears already at relatively small load. Thus such arrangements are suitable for producing voltage peaks only when no current is required, as for disruptive effects, or only very small currents.

CHAPTER IX

WAVE SCREENS. EVEN HARMONICS

76. The elimination of voltage and current distortion, and production of sine waves from any kind of supply wave, that is, the reverse procedure from that discussed in the preceding chapter, is accomplished by what has been called "wave screens."

Series reactance alone acts to a considerable extent as wave screen, by consuming voltage proportional to the frequency and the current, and thereby reducing the harmonics of voltage in the rest of the circuit the more, the higher their order.

Let the voltage impressed upon the circuit be denoted symbolically by

$$e = e_1 + e_3 + e_5 + e_7 + \ldots$$

$$= \Sigma \, e_n. \tag{29}$$

where n denotes the order of the harmonic of absolute numerical value e_n.

If, then, the reactance x (at fundamental frequency) is inserted into the circuit of resistance, r, the impedance is

$$z_1 = \sqrt{r^2 + x^2} \text{ for the fundamental frequency, and}$$

$$z_n = \sqrt{r^2 + n^2 x^2} \text{ for the } n\text{th harmonic,} \tag{30}$$

and the current thus is

$$i = \frac{e}{z} = \Sigma \frac{e}{\sqrt{r^2 + n^2 x^2}}, \tag{31}$$

or, denoting

$$\frac{r}{x} = c, \tag{32}$$

it is

$$i = \frac{1}{x} \Sigma \frac{e_n}{\sqrt{n^2 + c^2}} = \frac{e_1}{x\sqrt{1 + c^2}} + \frac{e_3}{x\sqrt{9 + c^2}} + \frac{e_5}{x\sqrt{25 + c^2}} + \ldots \tag{33}$$

if r is small compared with x, c^2 is negligible compared with 1, 9, 25, etc., and it is

$$i = \frac{1}{x} \left\{ e_1 + \frac{e_3}{3} + \frac{e_5}{5} + \frac{e_7}{7} + \dots \right\},$$

that is, the current, i, and thus the voltage across the resistance, r, shows the harmonics of the supply voltage, e, reduced in proportion to their order, n.

Even if r is large compared with x, and thus $c^2 > 1$, finally c^2 becomes negligible with n^2, and the harmonics decrease with their order.

77. The screening effect of the series reactance is increased by shunting a capacity, C, beyond the inductance, L, that is, across the resistance, r, as shown in Fig. 73. By consuming current

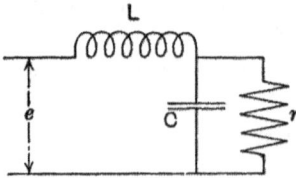

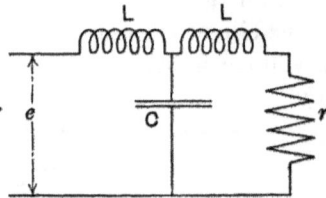

Fig. 73.   Fig. 74.

proportional to frequency and voltage, the condenser shunts the more of the current passing through the reactance, the higher the frequency, and thereby still further reduces the higher harmonics of current in the resistance, r, and thus of voltage across this resistance. Its effect is limited, however, by the decreasing voltage distortion at r and thus at the condenser, C.

Thus the screening effect is still further increased by inserting a second inductance, L, beyond the condenser, C, in series to the resistance, r, as shown in Fig. 74. By making the second inductance equal to the first one, and making the condenser, C, of the same reactance, for the fundamental wave, as each of the two inductances, we get what probably is the most effective wave screen. This T-connection or resonating circuit will be discussed more fully in Chapter XIV, in its feature of constant-potential constant-current transformation.

Under the condition, that the two inductive reactances and the capacity reactance are equal, the equation of the current in the resistance, r, is (page 291), for the nth harmonic,

$$I = \frac{je_0}{xn(n^2 - 2) - jr(n^2 - 1)}. \tag{34}$$

or, absolute,

$$i = \frac{e_0}{x} \times \frac{1}{\sqrt{n^2(n^2 - 2)^2 + c^2(n^2 - 1)^2}} \tag{35}$$

where

$$c = \frac{r}{x} \tag{36}$$

If c is small, that is, r small compared with x, the current becomes

$$i = \frac{e_0}{xn(n^2 - 2)} \tag{37}$$

or, for higher values of n,

$$i = \frac{e_0}{xn^3}, \tag{38}$$

that is, it decreases with increasing order of harmonic, and proportional to the *cube* of the order n, thus shows an extremely rapid decrease.

If c is not negligible, the denominator in (35) is larger, and i, therefore, still smaller.

As illustration may be shown the current, i_0, and thus the voltage, e_0, across a resistance, r, under the very greatly distorted and peaked voltage of Fig. 62:

(a) for a series reactance, x, equal to r, that is, $c = 1$;

(b) for the complete wave screen of two inductances and one capacity.

It is impressed voltage,

$$e = 1.27 \ e_0 \ \{ \ 1_1 + 0.978_3 + 0.935_5 + 0.877_7 + 0.800_9 + 0.713_{11} \\ + 0.617_{13} + 0.517_{15} + 0.416_{17} + 0.315_{19} + 0.189_{21}\}.$$

(a) Reduction factor of the nth harmonic,

$$\frac{1}{\sqrt{n^2 + c^2}} = \frac{1}{\sqrt{n^2 + 1}},$$

hence,

$$e_1 = \frac{1.27}{\sqrt{2}} e_0 \Big\{ 1_1 + 0.442_3 + 0.258_5 + 0.175_7 + 0.125_9 + 0.091_{11} \\ + 0.067_{13} + 0.049_{15} + 0.034_{17} + 0.023_{19} + 0.013_{21}\}.$$

(b) Reduction factor of the nth harmonic,

$$\frac{1}{n(n^2 - 2)},$$

hence,

$e_2 = 1.27\, e_0\, \{1_1 + 0.047_3 + 0.008_5 + 0.003_7 + 0.001_9 + 0.001_{11}\}.$

That is, the third harmonic is reduced to less than 5 per cent., the fifth to less than 1 per cent., and the higher ones are practically entirely absent.

While in the supply voltage wave, e, the voltage peak (by adding the numerical values of all the harmonics: $1 + 0.978 + 0.935 + \ldots$) is 7.36 times that of the fundamental wave, it is reduced by series reactance to less than 2.28 times the maximum of the fundamental wave, that is, very greatly reduced, and by the complete wave screen to less than 1.06 times the maximum of the fundamental. That is, in the last case the voltage is practically a perfect sine wave.

78. By "wave screens" the separation of pulsating currents into their alternating and their continuous component, or the separation of complex alternating currents—and thus voltages—into their constituent harmonics can be accomplished, and inversely,

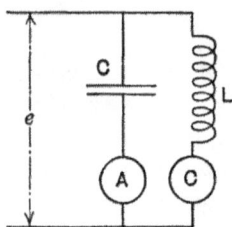

Fig. 75.

the combination of alternating and continuous currents or voltages into resultant complex alternating or pulsating currents.

The simplest arrangement of such a wave screen for separating, or combining alternating and continuous currents into pulsating ones, is the combination, in shunt with each other, of a capacity, C, and an inductance, L, as shown in Fig. 75. If, then, a pulsating voltage, e, is impressed upon the system, the pulsating current, i, produced by it divides, as the continuous component can not pass through the condenser, C, and the alternating component is barred by the inductance, L, the more completely, the higher this inductance. Thus the current, i_1, in the apparatus, A, is a true alternating current, while the current, i_0, in the apparatus, C, is a slightly pulsating direct current.

Inversely, by placing a source of alternating voltage, such as an alternator or the secondary of a transformer, at A, and a source of continuous voltage, such as a storage battery or direct-current generator, at C, in the external circuit a pulsating voltage, e, and pulsating current, i, result.

If the capacity, C, is so large as to practically short-circuit the alternating voltage, and the inductance, L, so high as to practically open-circuit the alternating voltage, the separation—of combination—is practically complete, and independent of the frequency of the alternating wave.

Wave screens based on resonance for a definite frequency by series connection of capacity and inductance, can be used to separate the current of this frequency from a complex current or voltage wave, such as those given in Figs. 56 to 63, and thus can be used for separation of complex waves into their components, by "harmonic analysis."

Thus in Fig. 76, if the successive capacities and inductances are chosen such that

$$2\,\pi f L_1 = \frac{1}{2\,\pi f C_1},$$

$$6\,\pi f L_3 = \frac{1}{6\,\pi f C_3},$$

$$10\,\pi f L_5 = \frac{1}{10\,\pi f C_5},$$

$$2n\,\pi f L_n = \frac{1}{2\,\pi f n C_n} \quad (39)$$

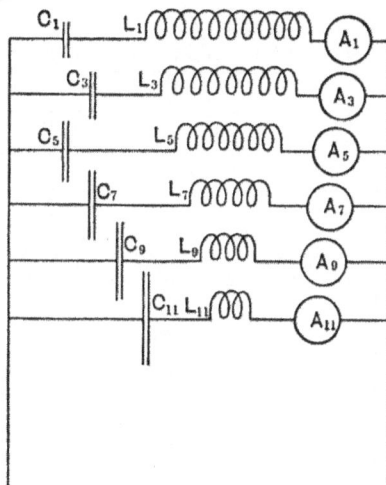

FIG. 76.

where f = frequency of the fundamental wave.

Then, through any of the branch circuits C_n, L_n, only the nth harmonic, i_n, can pass to an appreciable extent.

Such resonant wave screen, however, has the serious disadvantage to require very high constancy of f, since the resonance condition between C_n and L_n depends on the square of f,

$$\frac{1}{C_n} = 4\,\pi^2 f^2 L_n.$$

79. Even harmonics are produced in a closed magnetic circuit by the superposition of a continuous current upon the alternating wave. With an alternating sine wave impressed upon an iron magnetic circuit, saturation, or in general the lack of proportionality between magnetic flux and m.m.f., produces a wave-shape distortion, that is, higher harmonics, of voltage with a sine wave of current, of current with a sine wave of impressed voltage. The constant term of a wave, however, is the first even harmonic, and thus, if the impressed wave comprises a fundamental sine and a

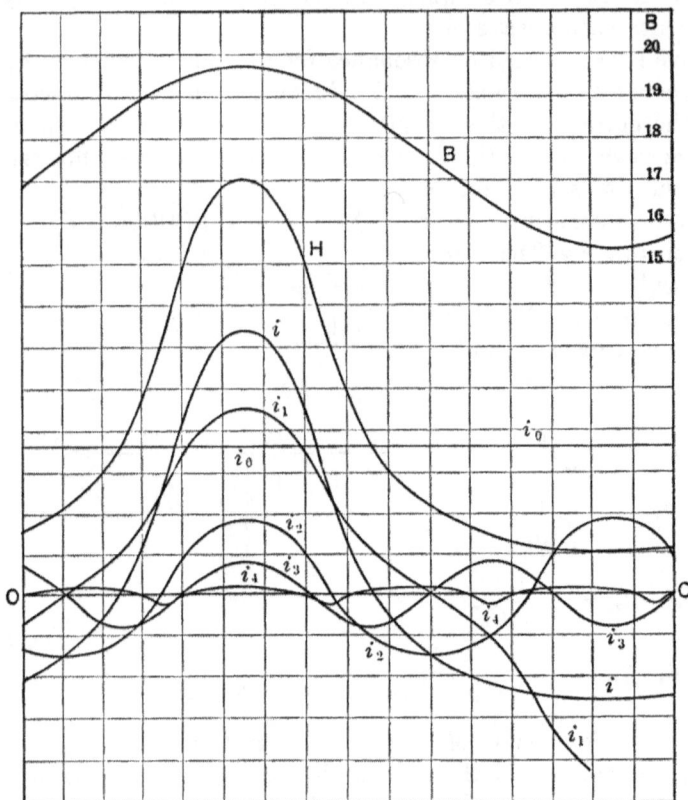

FIG. 77.

constant term, the former gives rise to the odd harmonics, the latter to the even harmonics.

Let, then, on the alternating sine wave of impressed voltage a continuous current by superimposed. The magnetic flux then oscillates sinusoidally, not between equal and opposite values, but between two unequal values, which may be of the same, or of opposite signs. That is, it performs an unsymmetrical magnetic cycle. Neglecting again hysteresis, that is, assuming the rising and the decreasing magnetization curve as coincident—which is permissible as approximation, since the hysteresis contributes little to distortion—and choosing the same magnetization curve as in the preceding, curve I in Fig. 64, we may as an instance consider a sinusoidal magnetic pulsation between the limits $+15.4$ and $+19.7$, corresponding to a variation of the m.m.f. between $H = +10$ and $H = +100$.

Fig. 77 then gives, as curve B, the sinusoidally pulsating magnetic flux density. Taking from curve I, Fig. 64, the values of H corresponding to the values in curve B, Fig. 77, gives curve H. This, resolved ("Engineering Mathematics," paragraph 92) gives the constant term $i_0 = 36$, and the alternating current, i. The latter is unsymmetrical, having one short half-wave of a peak value 64, and one long half-wave of maximum value 26. It thus resolves into the odd harmonics, i_1, alternating between ± 45, and the even harmonics, mainly the second harmonic, alternating between maximum values $+18$ and -15. i_1 is peaked with flat zero, thus showing a third harmonic, which is separated as i_3, and i_2 is unsymmetrical, showing further even harmonics, which are separated as i_4, but are rather small.

Thus the pulsating exciting current of the sinusoidally varying unidirectional magnetic flux

$$B = 17.55 + 2.15 \cos \phi$$

is given by

$$H = 36 + 37 \cos \phi + 16.5 \cos 2 \phi + 8 \cos 3 \phi + 2 \cos 4 \phi + \ . \ .$$

Instead of superimposing a direct current upon an alternating wave, as by connecting in series an alternator and a direct-current generator or storage battery, two separate coils can be used on the magnetic circuit, one energized by an alternating impressed voltage, the other by a direct current. A high inductive reactance would then be connected in the latter circuit, to eliminate the current pulsation which would be caused by the alternating voltage induced in this coil.

Connecting two such magnetic circuits with their direct-current magnetizing coils in series, but in opposition (without the use of a series reactance) eliminates the induced fundamental wave, but leaves the second harmonic in the direct-current circuit, which thus can be separated. Numerous arrangements can then be devised by two magnet cores energized by separate alternating-current exciting coils and saturated by one common direct-current exciting coil, surrounding both cores, or their common return, etc.

80. The preceding may illustrate some of the numerous wave-shape distortions which are met in electrical engineering, their characteristics, origin, effects, use and danger.   Numerous other wave distortions, such as those produced by arcs, by unidirectional conductors, by dielectric effects such as corona, by Y connection of transformers for reactors, by electrolytic polarization, by pulsating resistance or reactance, etc., are discussed in other chapters or may be studied in a similar manner.

CHAPTER X

INSTABILITY OF CIRCUITS: THE ARC

A. General

81. During the earlier days of electrical engineering practically all theoretical investigations were limited to circuits in stable or stationary condition, and where phenomena of instability occurred, and made themselves felt as disturbances or troubles in electric circuits, they either remained ununderstood or the theoretical study was limited to the specific phenomenon, as in the case of lightning, dropping out of step of induction motors, hunting of synchronous machines, etc., or, as in the design of arc lamps and arc-lighting machinery, the opinion prevailed that theoretical calculations are impossible and only design by trying, based on practical experience, feasible.

The first class of unstable phenomena, which was systematically investigated, were the transients, and even today it is questionable whether a systematic theoretical classification and investigation of the conditions of instability in electric circuits is yet feasible. Only a preliminary classification and discussion of such phenomena shall be attempted in the following.

Three main types of instability in electric systems may be distinguished:

I. The transients of readjustment to changed circuit conditions.

II. Unstable electrical equilibrium, that is, the condition in which the effect of a cause increases the cause.

III. Permanent instability resulting from a combination of circuit constants which can not coexist.

I. TRANSIENTS

82. Transients are the phenomena by which, at the change of circuit conditions, current, voltage, etc., readjust themselves from the values corresponding to the previous condition to the values corresponding to the new condition of the circuit. For instance, if a switch is closed, and thereby a load put on the circuit, the current can not instantly increase to the value corresponding to the increased load, but some time elapses, during which the increase of the stored magnetic energy corresponding to the increased current, is brought about. Or, if a motor switch is closed, a period of acceleration intervenes before the flow of current becomes stationary, etc.

The characteristic of transients therefore is, as implied in the term, that they are of limited, usually very short duration, intervening between two periods of stationary conditions.

Considerable theoretical work has been done, more or less systematically, on transients, and a great mass of information is thus available in the literature. These transients are more extensively treated in "Theory and Calculation of Transient Electric Phenomena and Oscillations," and in "Electric Discharges, Waves and Impulses," and therefore will be omitted in the following. However, to some extent, the transients of our theoretical literature, still are those of the "phantom circuit," that is, a circuit in which the constants r, L, C, g, are assumed as constant. The effect of the variation of constants, as found more or less in actual circuits: the change of L with the current in circuits containing iron; the change of C and g with the voltage (corona, etc.); the change of r and g with the frequency, etc., has been studied to a limited extent only, and in specific cases.

In the application of the theory of transients to actual electric circuits, considerable judgment thus is often necessary to allow and correct for these "secondary" phenomena which are not included in the theoretical equations.

Especially deficient is our knowledge of the conditions under which the attenuation constant of the transient becomes zero or negative, and the transient thereby becomes permanent, or becomes a cumulative surge, and the phenomenon thereby one of unstable equilibrium.

II. Unstable Electrical Equilibrium

83. If the effect brought about by a cause is such as to oppose or reduce the cause, the effect must limit itself and stability be finally reached. If, however, the effect brought about by a cause increases the cause, the effect continues with increasing intensity, that is, instability results.

This applies not to electrical phenomena alone, but equally to all other phenomena.

Instability of an electric circuit may assume three different forms:

1. Instability leading up to stable conditions.

For instance, in a pyroelectric conductor of the volt-ampere characteristic given in Fig. 78, at the impressed voltage, e_0, three different values of current are possible: i_1, i_2 and i_3. i_1 and i_3 are stable, i_2 unstable. That is, at current, i_2, passing through the conductor under the constant impressed voltage, e_0, a momentary increase of current would give an excess voltage beyond that required by the conductor, thereby increase the current still

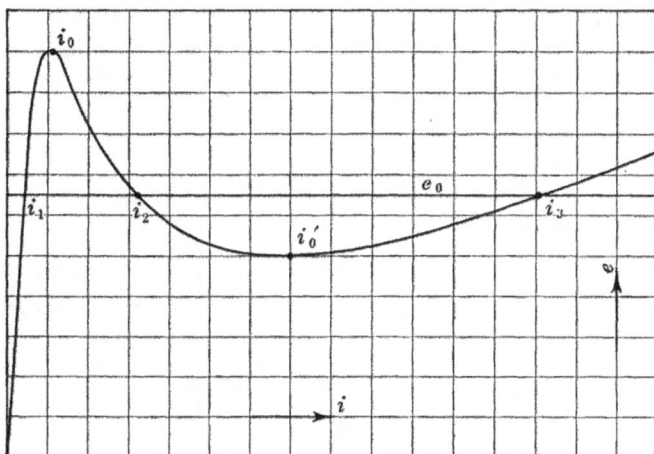

Fig. 78.

further, and with increasing rapidity the current would rise, until it becomes stable at the value, i_3. Or, a momentary decrease of current, by requiring a higher voltage than available, would further decrease the current, and with increasing rapidity the current would decrease to the stable value, i_1.

2. Instability putting the circuit out of service.

An instance is the arc on constant-potential supply. With the volt-ampere characteristic of the arc shown as A, in Fig. 79, a current of 4 amp. would require 80 volts across the arc terminals. At a constant impressed voltage of 80, the current could not remain at 4 amp., but the current would either decrease with increasing rapidity, until the arc goes out, or the current would increase with increasing rapidity, up to short-circuit, that is, until the supply source limits the current.

3. Instability leading again to instability, and thus periodically repeating the phenomena.

For instance, if an arc of the volt-ampere characteristic, A, in Fig. 79 is operated in a constant-current circuit of sufficiently high direct voltage to restart the arc when it goes out, and the arc

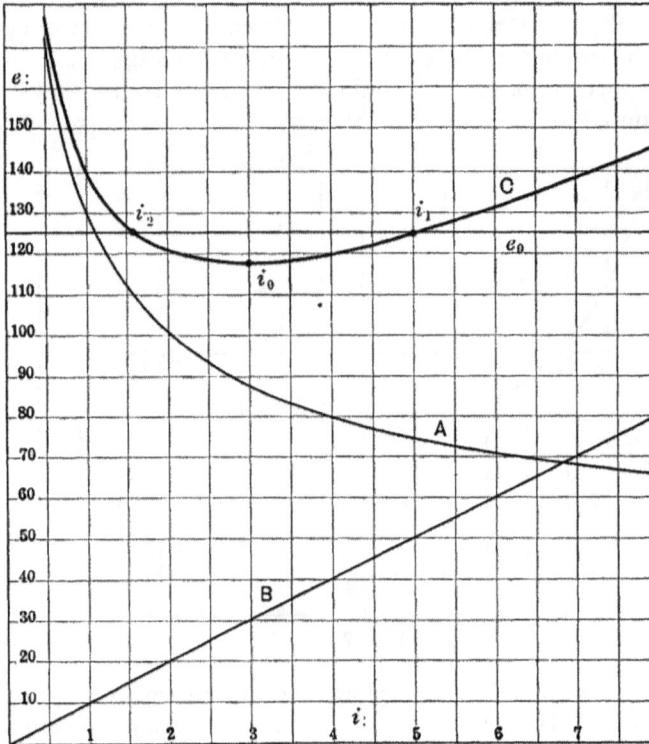

Fig. 79.

is shunted by a condenser, the condenser makes the arc unstable and puts it out; the available supply voltage, however, starts it again, and so periodically the arc starts and extinguishes, as an "oscillating arc."

84. There are certain circuit elements which tend to produce instability, such as arcs, pyroelectric conductors, condensers, induction and synchronous motors, etc., and their recognition therefore is of great importance to the engineer, in guarding

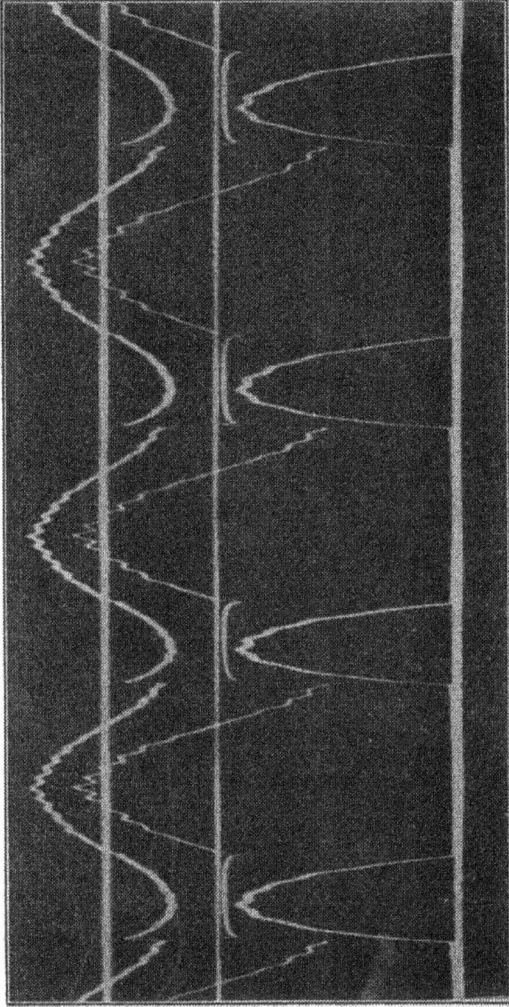

(CD28555.)

FIG. 80.

(*Facing p.* 164.)

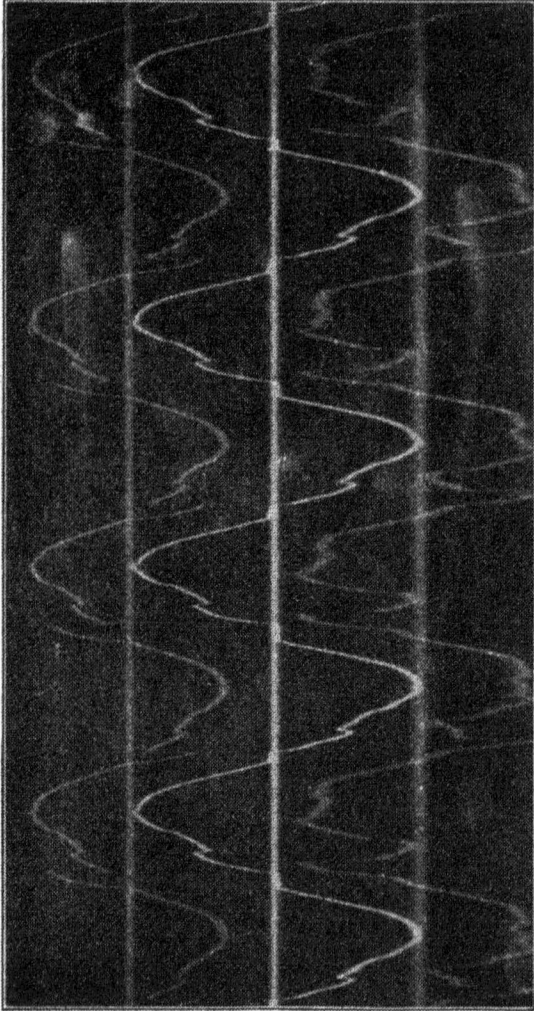

Fig. 81.

(CD15114.)

against instability. Whether instability results, and what form it assumes, depends, however, not only on the "exciting element," as we may call the cause of the instability, but on all the elements of the circuit. Thus an arc is unstable, form (2), on constant-voltage supply at its terminals; it is stable on constant-current supply. But when shunted by a condenser, it becomes unstable on constant current, and the instability may be form (2) or form (3), depending on the available voltage. With a resistance, r, of volt-ampere characteristic ir shown as B, in Fig. 79, the arc is stable on constant-voltage supply for currents above i_0 = 3 amp., unstable below 3 amp., and therefore, with a constant-supply voltage, e_0, two current values, i_1 and i_2, exist, of which the former one is stable, the latter one unstable. That is, current, i_2, can not persist, but the current either runs up to i_1 and the arc then gets stable (form 1), or the current decreases and the arc goes out, instability form (2).

Thus it is not feasible to separately discuss the different forms of instability, but usually all three may occur, under different circuit conditions.

The electric arc is the most frequent and most serious cause of instability of electric circuits, and therefore should first be suspected, especially if the instability assumes the form of high-frequency disturbances or abrupt changes of current or voltage, such as is shown for instance in the oscillograms, Figs. 80 and 81.

Somewhat similar effects of instability are produced by pyro-electric conductors.

Induction motors and synchronous motors may show instability of speed: dropping out of step, etc.

III. Permanent instability

85. If the constants of an electric circuit, as resistance, inductance, capacity, disruptive strength, voltage, speed, etc., have values, which can not coexist, the circuit is unstable, and remains so as long as these constants remain unchanged.

Case (3) of II, unstable equilibrium, to some extent may be considered as belonging in this class.

The most interesting class in this group of unstable electric systems are the oscillations resulting sometimes from a change of circuit conditions (switching, change of load, etc.), which continue indefinitely with constant intensity, or which steadily increase in intensity, and may thus be called permanent and cumulative surges, hunting, etc. They may be considered as transients in which the attenuation constant is zero or negative.

In the transient resulting from a change of circuit conditions, the energy which represents the difference of stored energy of the circuit before and after the change of circuit condition, is dissipated by the energy loss in the circuit. As energy losses always occur, the intensity of a true transient thus must always be a maximum at the beginning, and steadily decrease to zero or permanent condition. An oscillation of constant intensity, or of increasing intensity, thus is possible only by an energy supply to the oscillating system brought about by the oscillation. If this energy supply is equal to the energy dissipation, constancy of the phenomenon results. If the energy supply is greater than the energy dissipation, the oscillation is cumulative, and steadily increases until self-destruction of the system results, or the increasing energy loss becomes equal to the energy supply, and a stationary condition of oscillation results. The mechanism of this energy supply to an oscillating system from a source of energy differing in frequency from that of the oscillation is still practically unknown, and very little investigating work has been done to clear up the phenomenon. It is not even generally realized that the phenomenon of a permanent or cumulative line surge involves an energy supply or energy transformation of a frequency equal to that of the oscillation.

Possibly the oldest and best-known instance of such cumulative oscillation is the hunting of synchronous machines.

Cumulative oscillations between electromagnetic and electrostatic energy have been observed by their destructive effects in high-voltage electric circuits on transformers and other apparatus, and have been, in a number of instances where their frequency was sufficiently low, recorded by the oscillograph. They obviously are the most dangerous phenomena in high-voltage electric circuits. Relatively little exact knowledge exists of their origin. Usually—if not always—an arc somewhere in the system is instrumental in the energy supply which maintains the oscillation. In some instances, as in wireless telegraphy, they have found industrial application. A systematic theoretical investigation of these cumulative electrical oscillations probably is one of the most important problems before the electrical engineer today.

The general nature of these permanent and cumulative oscillations and their origin by oscillating energy supply from the transient of a change of circuit condition, is best illustrated by the instance of the hunting of synchronous machines, and this will, therefore, be investigated somewhat more in detail.

B. The Arc as Unstable Conductor

86. The instability of the arc is the result of its dropping volt-ampere characteristic, as discussed in paragraphs 18 to 27 of the

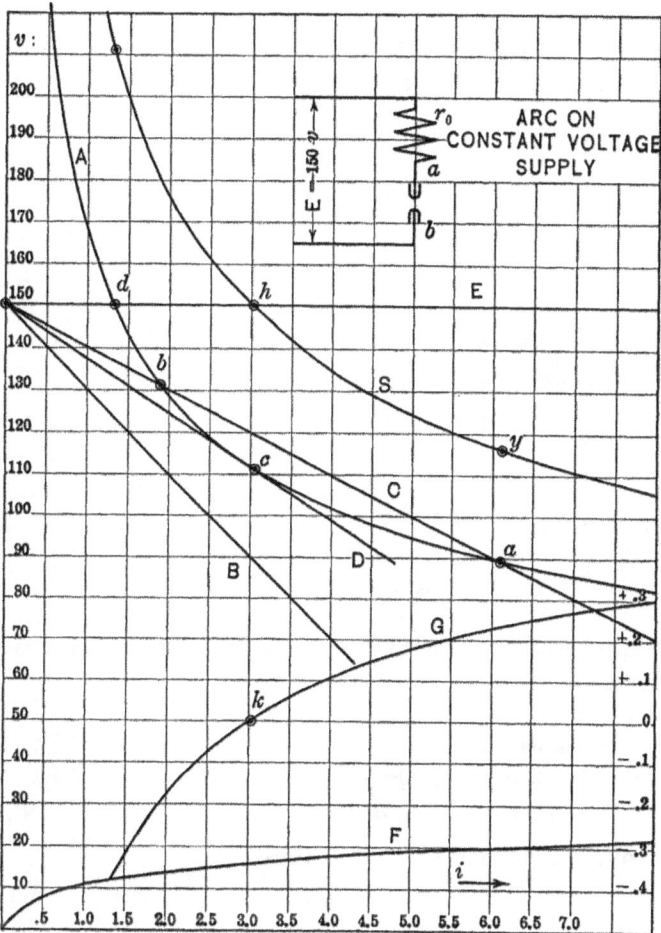

Fig. 82.

chapter on "Electric Conductors." As shown there, the arc is always unstable on constant voltage impressed upon it. Series resistance or reactance produces stability for currents above a certain critical value of current, i_0. Such curves, giving the voltage consumed by the arc and its series resistance as function of the current, thus may be termed stability curves of the arc. Their minimum values, that is, the stability limits corresponding to the different resistances, give the stability characteristic of the arc. The equations of the arc, and of its stability curves and stability characteristic, are given in paragraphs 22 and 23 of the chapter on "Electric Conductors."

Let, in Fig. 82, A present the volt-ampere characteristic of an arc, given approximately by the equation

$$\left. \begin{aligned} e &= a + \frac{c(l + \delta)}{\sqrt{i}} \\ &= a + \frac{b}{\sqrt{i}} \end{aligned} \right\} \tag{1}$$

where

$$e' = \frac{b}{\sqrt{i}} \tag{2}$$

is the stream voltage, that is, voltage consumed by the arc stream.

Fig. 82 is drawn with the constants,

$$a = 35,$$
$$c = 51,$$
$$l = 1.8,$$
$$\delta = 0.8,$$

hence,

$$e = 35 + \frac{133}{\sqrt{i}}.$$

Assuming this arc is operated from a circuit of constant-voltage supply,

$$E = 150 \text{ volts,}$$

through a resistance, r_0

The voltage consumed by the resistance, r_0, then is

$$e_2 = r_0 i, \tag{3}$$

and the voltage available for the arc thus

$$e_1 = E - r_0 i \tag{4}$$

Lines B, C and D of Fig. 82 give e_1, for the values of resistance,

$$r_0 = 20 \text{ ohms } (B)$$
$$= 10 \text{ ohms } (C)$$
$$= 13 \text{ ohms } (D).$$

As seen, line B does not intersect the volt-ampere characteristic, A, of the arc, that is, with 20 ohms resistance in series, this $l = 2.5$ cm. arc can not be operated from $E = 150$ volt supply.

Line C intersects A at a and b, $i = 6.1$ and 1.9 amp. respectively.

At a, $i = 6.1$ amp., the arc is stable;

At b, $i = 1.9$ amp., the arc is unstable;

for the reasons discussed before: an increase of current decreases the voltage consumed by the circuit, $e + e_2$, and thus still further increases the current, and inversely. Thus the arc either goes out, or the current runs up to $i = 6.1$ amp., where the arc gets stable.

Line D is drawn tangent to A, and the contact point, c, thus gives the minimum current, $i = 3.05$ amp., of operation of the arc on $E = 150$ volts, that is, the value of current or of series resistance, at which the arc ceases to be stable: a point of the stability characteristic, S, of the arc.

This stability characteristic is determined by the condition

$$\frac{de_0}{di} = 0, \tag{5}$$

where

$$e_0 = e + r_0 i \tag{6}$$

$$= a + \frac{b}{\sqrt{i}} + r_0 i,$$

this gives

$$r_0 = \frac{b}{2\,i\sqrt{i}} = \frac{e_1}{2\,i} \tag{7}$$

and

$$\left. \begin{aligned} e_0 &= a + \frac{1.5\,b}{\sqrt{i}} \\ &= a + 1.5\,e_1 \end{aligned} \right\} \tag{8}$$

as the equation of the stability characteristic of the arc on a constant-voltage circuit.

87. In general, the condition of stability of a circuit operated on constant-voltage supply, is

$$\frac{de}{di} > 0 \tag{9}$$

where e is the voltage consumed by the current, i, in the circuit.

The ratio of the change of voltage, de, as fraction of the total voltage, e, brought about by a change of current, di, as fraction of the total current, i, thus may be called the *stability coefficient* of the circuit,

$$\delta = \frac{\dfrac{de}{e}}{\dfrac{di}{i}}$$

$$= \frac{\dfrac{de}{di}}{\dfrac{e}{i}}$$

\qquad (10)

In a circuit of constant resistance, r, it is

$$\frac{e}{i} = r,$$

$$\frac{de}{di} = r,$$

hence,

$$\delta = 1,$$

that is, the stability coefficient of a circuit of constant resistance, r, is unity.

In general, if the effective resistance, r, is not constant, but varies with the current, i, it is

$$e = ri,$$

$$\frac{de}{di} = r + i\frac{dr}{di};$$

hence, the stability coefficient

$$\delta = 1 + \frac{\dfrac{dr}{di}}{\dfrac{r}{i}}$$

\qquad (11)

thus in a circuit, in which the resistance increases with the current, the stability coefficient is greater than 1. Such is that of a conductor with positive temperature coefficient of resistance, in which the temperature rise due to the increase of current increases the resistance. A conductor with negative temperature coefficient of resistance gives a stability coefficient less than 1, but as long as δ is still positive, that is, the decrease of resistance slower than the increase of current, the circuit is stable.

$$\delta > 0$$

\qquad (12)

is the condition of stability of a circuit on constant-voltage supply, and

$$\delta < 0 \tag{13}$$

is the condition of instability, and

$$\delta = 0 \tag{14}$$

thus gives the stability characteristic of the circuit.

In the arc,

$$e = a + \frac{b}{\sqrt{i}},$$

the stability coefficient is, by (10),

$$\delta = -\frac{b}{2\,e\sqrt{i}} = -\frac{e'}{2\,e} \tag{15}$$

that is, equals half the stream voltage, $\frac{e'}{2}$, divided by the arc voltage, e.

Or, substituting for e in (15), and rearranging,

$$\delta = -\frac{1}{2\left(1 + \frac{a}{b}\sqrt{i}\right)} \tag{16}$$

$$= -\frac{1}{2(1 + 0.2625\,\sqrt{i})}$$

in Fig. 82.

For $i = 0$, it is $\delta = -0.5$;

$i = \infty$, it is $\delta = 0$.

The stability coefficient of the arc having the volt-ampere characteristic, A, in Fig. 82 is shown as F in Fig. 82.

88. On constant-voltage supply, $E = 150$ volts, the arc having the characteristic, A, Fig. 82, can not be operated at less than 3.05 amperes. At $i = 3.05$ is its stability limit, that is, the stability coefficient of arc plus series resistance, r_0, required to give 150 volts, changes from negative for lower currents, to positive for higher currents.

The stability coefficient of such arcs, operated on constant-voltage supply through various amounts of series resistance, r_0, then would be given by

$$\delta_0 = \frac{\dfrac{de_0}{di}}{\dfrac{e_0}{i}},$$

where

$$e_0 = a + \frac{b}{\sqrt{i}} + r_0 i \tag{17}$$

and the resistance r_0 chosen so as to give

$$e_1 = 150 \text{ volts,}$$

from (17) follows,

$$\delta_0 = \frac{-\dfrac{b}{2\sqrt{i}} + r_0 i}{e_0},$$

and, substituting from (17),

$$i r_0 = e_0 - a - \frac{b}{\sqrt{i}}$$

gives

$$\delta_0 = 1 - \frac{a + \dfrac{1.5\,b}{\sqrt{i}}}{e_0} \tag{18}$$

or,

$$\delta_0 = 1 - \frac{e_0'}{e_0} \tag{19}$$

where e_0 is the supply voltage, e_0' the voltage given by the stability characteristic, S.

δ_0, the stability characteristic of the arc, A, on $E = 150$ volt constant-potential supply, is given as curve, G, in Fig. 82. As seen, it passes from negative—instability—to positive—stability —at the point, k, corresponding to c and h on the other curves.

89. On a constant-current supply, an arc is inherently stable. Instability, however, may result by shunting the arc by a resistance, r_1. Thus in Fig. 83, let $I = 5$ amp. be the constant supply current. The volt-ampere characteristic of the arc is given by A, and shows that on this 5-amp. circuit, the arc consumes 94 volts, point d.

Let now the arc be shunted by resistance, r_1. If $e =$ voltage consumed by the arc, the current shunted by the resistance, r_1, is

$$i_1 = \frac{e}{r_1} \tag{20}$$

and the current available for the arc thus is

$$i = I - i_1 \tag{21}$$

$$= I - \frac{e}{r_1}$$

or

$$e = r_1(I - i). \tag{22}$$

Curves B, C and D of Fig. 83 show the values of equation (22) for

$$r_1 = 32 \quad \text{ohms: line } B$$
$$= 48 \quad \text{ohms: line } C$$
$$= 40.8 \quad \text{ohms: line } D.$$

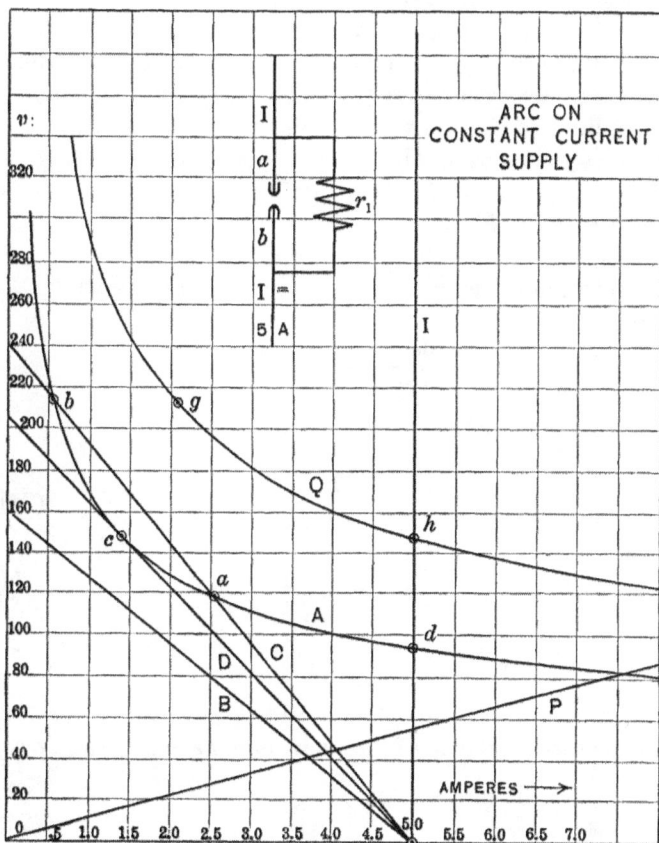

Fig. 83.

Line B does not intersect the arc characteristic, A, that is, with a resistance as low as $r_1 = 32$, no arc can be maintained on the 5-amp. constant-current circuit.

Line C intersects A at two points:

(a) $i = 2.55$ amp., $e = 118$ volts, stable condition;

(b) $i = 0.55$ amp., $e = 214$ volts, unstable condition.

Line D is drawn tangent to A, touches at c: $i = 1.4$ amp.,

$e = 148$ volts, the limit of stability. At $I = 5$ amp., the point h, at $e = 148$ volts, thus gives the voltage consumed by an arc when by shunting it with a resistance the stability limit is reached.

Drawing then from the different points of the abscissæ, i, tangents on A, and transferring their contact points, c, b, to the abscissæ, from which the tangent is drawn, gives the points h, g, of the constant-current stability characteristic of the arc, that is, the curve of arc voltages in a constant-current circuit, I, when by shunting the arc with a resistance, r_1, consuming current, i_1, the stability limit of the arc with current $i = I - i_1$ is reached.

P then gives the curve of the arc currents, i, corresponding to the arc voltage, e, of curve Q, for the different values of the constant-circuit current, I.

The equations of Q and P are derived as follows:

The stability limit, point c, corresponding to circuit current, I, as given by

$$\frac{de}{di} = -r_1,$$

where e = arc voltage, and i = arc current.

Or,

$$r_1 = \frac{b}{2\,i\sqrt{i}}. \tag{23}$$

It is, however,

and

$$\begin{cases} e = a + \dfrac{b}{\sqrt{i}} \\[2ex] \dfrac{e}{I - i} = r_1. \end{cases}$$

From these three equations follows, by eliminating r_1 and i or e, Q,

$$I = \frac{b^2(3\,e - a)}{(e - a)^3} \tag{24}$$

P,

$$I = \frac{i(3\,b + 2\,a\sqrt{i})}{b}. \tag{25}$$

These curves are of lesser interest than the constant-voltage stability curve of the arc, S in Fig. 82.

It is interesting to note, that the resistance, r_1 (23), which makes an arc unstable as shunting resistance in a constant-current circuit, has the same value as the resistance, r_0, (7), which as series resistance makes it unstable in a constant-voltage supply circuit.

90. Due to the dropping volt-ampere characteristic, two arcs can not be operated in parallel, unless at least one of them has a sufficiently high resistance in series.

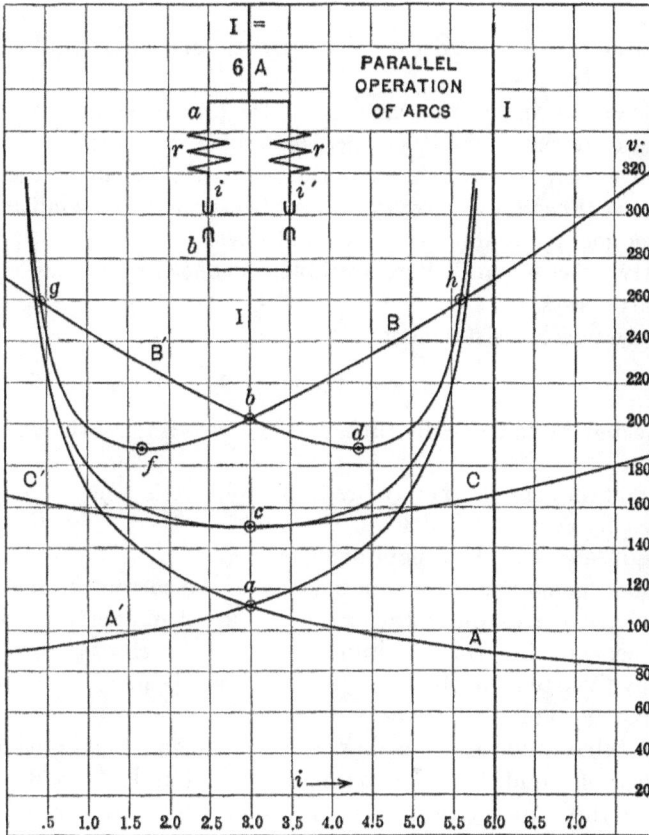

Fig. 84.

Let, as shown in Fig. 84, two arcs be connected in parallel into the circuit of a constant current

$$I = 6 \text{ amp.}$$

Assume at first both arcs of the same length and same electrode material, that is, the same volt-ampere characteristic.

Let i = current in the first arc, thus $i' = I - i$ = current in the second arc.

The volt-ampere characteristic of the first arc, then, is given by A in Fig. 84, that of the second arc by A'.

As the two parallel arcs must have the same voltage, the operating point is the point, a, of the intersection of A and A' in Fig. 84.

The arcs thus would divide the current, each operating at 3 amp.

However, the operation is unstable: if the first arc should take a little more current, its voltage decreases, on curve A, that of the second arc increases, on A', due to the decrease of its current, and the first arc thus takes still more current, thus robs the second arc, the latter goes out and only one arc continues.

Thus two arcs in parallel are unstable, and one of them goes out, only one persists.

Suppose now a resistance of

$$r = 30 \text{ ohms}$$

is connected in series with each of the two arcs, as shown in Fig. 84.

The volt-ampere characteristics of arc plus resistance, r, then, are given by curves B and B'.

These intersect in three points: b, g and h.

Of these, point b is stable: an increase of the current in one of the arcs, and corresponding decrease in the other, increases the voltage consumed by the circuit of the former, decreases that consumed by the circuit of the latter, and thus checks itself.

The points g and h, however, are unstable.

At b, stable condition, the characteristics, B and B', are rising; at a, unstable condition, the characteristics, A and A', are dropping, and the stability limit is at that value of resistance, r, at which the circuit characteristics plus resistance, are horizontal, the point c, where the characteristics, C and C', touch each other.

c is the stability limit of C or C', thus a point of the stability characteristic of either arc, or given by the equation

$$e = a + \frac{1.5\,b}{\sqrt{i}}.$$

Fig. 85 shows the case of two parallel arcs, which are not equal and do not have equal resistances, r, in series, one being a long arc, having no resistance in series, the other a short arc with a resistance $r = 40$ ohms in series.

The volt-ampere characteristic of the long arc is given by A, that of the short arc by B, and that of the short arc plus resistance, r, by C.

A and C intersect at three points, a, b and c. Of these, only the point a is stable, as any change of current from this point limits

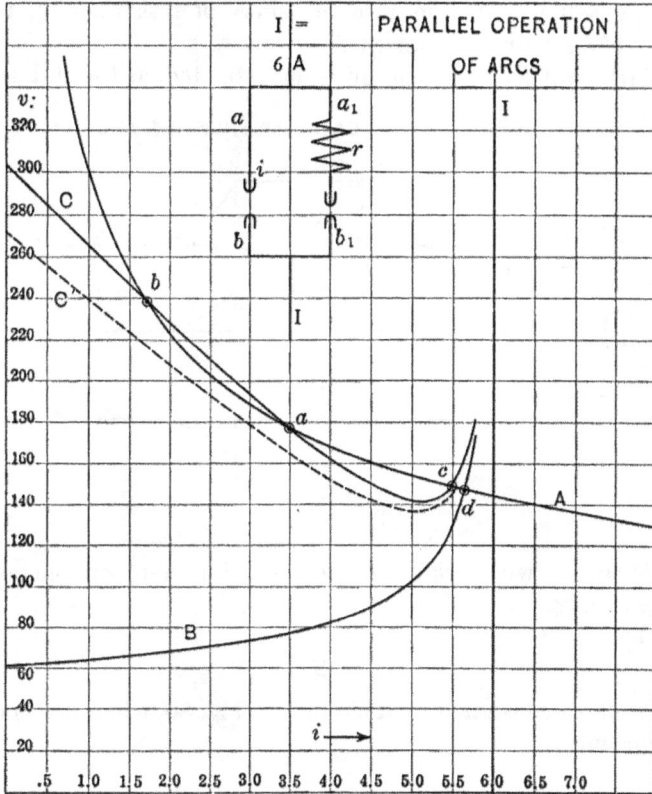

Fig. 85.

itself; b and c, however, are unstable. Thus, at the latter points, the arcs can not run, but the current changes until either one arc has gone out and one only persists, or both run at point a.

However, the angle under which the two curves, A and C, intersect at a is so small, that even at a the two arcs are not very stable.

12

Furthermore, a small change in either of the two curves, A or C, results in the two points of intersection a and b vanishing. Thus, if r is reduced from 40 ohms to 35 ohms, the curve C changes to C', shown dotted in Fig. 85, and as the latter does not intersect A except at the unstable point c, parallel operation is not possible.

That is, two such arcs can be operated in parallel only over a limited range of conditions, and even then the parallel operation is not very stable.

The preceding may illustrate the effect of resistance on the stability of operation of arcs.

Similarly, other conditions can be investigated, as the stability

CAPACITY SHUNTING ARC

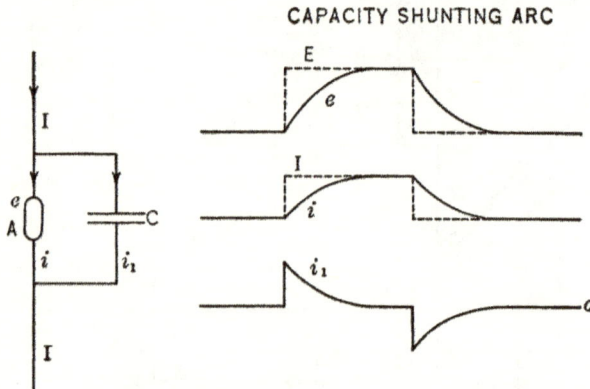

Fig. 86.

condition of arcs with resistance in series and in shunt, on constant, voltage supply, etc.

91. Let

$$e = E$$

be the voltage consumed by a circuit, A, Fig. 86, when traversed by a current

$$i = I.$$

If, then, in this circuit the current changes by δI, to

$$i = I + \delta I,$$

the voltage consumed by the circuit changes by δE, to

$$e = E \pm \delta E,$$

and the change of voltage is of the same sign as that of the current producing it, if A is a resistance or other circuit in which the voltage rises with the current, or is of opposite sign, if the circuit, A, has a dropping volt-ampere characteristic, as an arc.

Suppose now the circuit, A, is shunted by a condenser, C. As long as current, i, and voltage, e, in the circuit, A, are constant, no current passes through the condenser, C. If, however, the voltage of A changes, a current, i_1, passes through the condenser, given by the equation

$$i_1 = C \frac{de}{dt}. \tag{26}$$

If, then, the supply current, I, suddenly changes by δI, from I to $I + \delta I$, and the circuit, A, is a dead resistance, r, without the condenser, C, the voltage of A would just as suddenly change, from E to $E + \delta E$. By (26) this would, however, give an infinite current, i_1, in the condenser. However, the current in the condenser can not exceed δI, as with

$$i_1 = \delta I$$

at the moment of supply current change, the total excess current would in the first moment flow through the condenser, and the circuit, A, thus in this moment not change in current or voltage.

A finite current in the condenser, C, requires a finite rate of change of e in the circuit, A, starting from the previous value, E, at the starting moment, the time, $t = 0$.

Thus, if i = current, e = voltage of circuit, A, at time, t, after the increase of the supply current, I, by δI, it is
current in condenser,

$$i_1 = C \frac{de}{dt}.$$

current in circuit, A,

$$i = I + \delta I - i_1 \tag{27}$$

thus, voltage of circuit, A, of resistance, r,

$$e = ri$$
$$= rI + r\delta I - ri_1 \tag{28}$$

substituting (26) into (28), gives

$$e = r(I + \delta I) - rC \frac{de}{dt},$$

or,

$$\frac{de}{r(I + \delta I) - e} = \frac{1}{rC} dt \tag{29}$$

integrated by

$$e = rI - r\delta I \left(1 - \epsilon^{-\frac{t}{rC}}\right)$$
$$= E - \delta E \left(1 - \epsilon^{-\frac{t}{rC}}\right) \tag{30}$$

since $e = E$ for $t = 0$ is the terminal condition which determines the integration constant.

With a sudden change of the supply current, I, by δI, as shown by the dotted lines, I, in Fig. 86, the voltage, e, and current, i, in the circuit, A, and the current, i_1, in the condenser, C, thus change by the exponential transients shown in Fig. 86 as e, i and i_1.

92. Suppose now, however, that the circuit, A, has a dropping volt-ampere characteristic, is an arc.

A sudden decrease of the supply current, I by δI, to $I - \delta I$, would by the arc characteristic, $e = a + \dfrac{b}{\sqrt{i}}$, cause an increase of the voltage of circuit, A, from E to $E + \delta E$. Such a sudden increase of E would send an infinite current through C, that is, all the supply current would momentarily go through the condenser, C, none through the arc, A, and the latter would thus go out, and that, no matter how small the condenser capacity, C. Thus, with the condenser in shunt to the circuit, A, the voltage, A, can not vary instantly, but at a decrease of the supply current, I, by δI, the voltage of A at the first moment must remain the same, E, and the current in A thus must remain also, and as the supply current has decreased by δI, the condenser, C, thus must feed the current, δI, back into the arc, A. This, however, requires a decreasing voltage rating of A, at decreasing supply current, and this is not the case with an arc.

Inversely, a sudden increase of I, by δI, decreases the voltage of A, thus causes the condenser, C, to discharge into A, still further decreases its voltage, and the condenser momentarily short-circuits through the arc, A; but as soon as it has discharged and the arc voltage again rises with the decreasing current, the condenser, C, robs the arc, A, and puts it out.

Thus, even a small condenser in shunt to an arc makes it unstable and puts it out.

If a resistance, r_0, is inserted in series to the arc in the circuit, A, stability results if the resistance is sufficient to give a rising volt-ampere characteristic, as discussed previously.

Resistance in series to the condenser, C, also produces stability, if sufficiently large: with a sudden change of voltage in the arc circuit, A, the condenser acts as a short-circuit in the first moment, passing the current without voltage drop, and the voltage thus has to be taken up by the shunt resistance, r_1, giving the same condition of stability as with an arc in a constant-current circuit, shunted by a resistance, paragraph 89.

If, in addition to the capacity, C, an inductance, L, and some resistance, r, are shunted across the circuit, A, of a rising volt-ampere characteristic, as shown in Fig. 87, the readjustment occurring at a sudden change of the supply current, I, is not exponential, as in Fig. 86, but oscillatory, as in Fig. 87. As in the circuit, A, assuming it consists of a resistance, r, current and voltage vary simultaneously or in phase, current and voltage in the condenser branch circuit also must be in phase with each other, that is, the

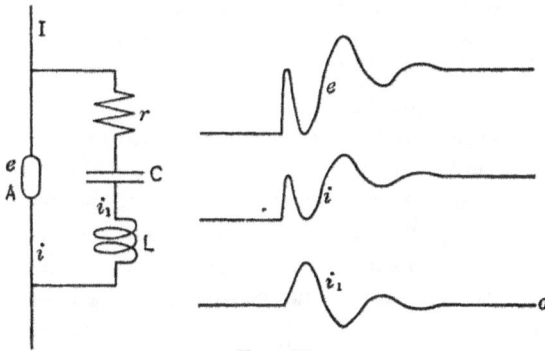

Fig. 87.

frequency of the oscillation in Fig. 87 is that at which capacity, C, and inductance, L, balance, or is the resonance frequency.

If circuit, A, in Fig. 87 is an arc circuit, and the resistance, r, in the shunt circuit small, instability again results, in the same manner as discussed before.

93. Another way of looking at the phenomena resulting from a condenser, C, shunting a circuit, A, is:

Suppose in Fig. 86 at constant-supply current, I, the current in the circuit, A, should begin to decrease, for some reason or another. Assuming as simplest case, a uniform decrease of current.

The current in the circuit, A, then can be represented by

$$i = I\left(1 - \frac{t}{t_0}\right) \tag{31}$$

where t_0 is the time which would be required for a uniform decrease down to nothing.

At constant-supply current, I, the condenser thus must absorb the decrease of current in A, that is, the condenser current is

$$i_1 = I \frac{t}{t_0}. \tag{32}$$

With decrease of current, i, if A is a circuit with rising characteristic, for instance, an ohmic resistance, the voltage of A decreases. The voltage at the condenser increases by the increasing charging current, i_1, thus the condenser voltage tends to rise over the circuit voltage of A, and thus checks the decrease of the voltage and thus of the current in A. Thus, the conditions are stable.

Suppose, however, A is an arc.

A decrease of the current in A then causes an increase of the voltage consumed by A, the arc voltage, e_0.

The same decrease of the current in A, by deflecting the current into the condenser, causes an increase of the voltage consumed by C, the condenser voltage, e_1.

If, now, at a decrease of the arc current, i, the arc voltage, e_0, rises faster than the condenser voltage, e_1, the increase of e_0 over e_1 deflects still more current from A into C, that is, the arc current decreases and the condenser current increases at increasing rate, until the arc current has decreased to zero, that is, the arc has been put out. In this case, the condenser thus produces instability of the arc.

If, however, e_0 increases slower than e_1, that is, the condenser voltage increases faster than the arc voltage, the condenser, C, shifts current over into the arc circuit, A, that is, the decrease of current in the arc circuit checks itself, and the condition becomes stable.

The voltage rise at the condenser is given by

$$\frac{de}{dt} = \frac{1}{C} i_1;$$

hence, by (32),

$$\frac{de}{dt} = \frac{tI}{t_0 C} \tag{33}$$

from the volt-ampere characteristic of the arc,

$$e = a + \frac{b}{\sqrt{i}} \tag{34}$$

follows,

the voltage rise at the arc terminals,

$$\frac{de}{dt} = -\frac{b}{2i\sqrt{i}}\frac{di}{dt} \tag{35}$$

and, by (31),

$$\frac{di}{dt} = -\frac{I}{t_0};$$

hence, substituted into (34),

$$\frac{de}{dt} = \frac{'bI}{2\,t_0 i\sqrt{i}}. \tag{36}$$

The condition of stability is, that the voltage rise at the condenser, (33), is greater than that at the arc, (36), thus,

$$\frac{tI}{t_0C} > \frac{bI}{2\,t_0 i\sqrt{i}},$$

or,

$$\frac{2\,ti\sqrt{i}}{bC} > 1 \tag{37}$$

or, substituting for t from equation (31), gives

$$\frac{2\,t_0 i\sqrt{i}\,(I-i)}{bC} > 1 \tag{38}$$

as the condition of stability, and

$$\frac{2\,t_0 i\sqrt{i}\,(I-i)}{bC} = 1 \tag{39}$$

thus is the stability limit.

94. Integrating (33) and substituting the terminal condition: $t = 0; e = E$, gives

$$e_1 = E + \frac{t^2 I}{2\,t_0 C} \tag{40}$$

as the equation of the voltage at the condenser terminals.

Substitute (31) into (34) gives

$$e_0 = a + \frac{b}{\sqrt{I}\sqrt{1 - \dfrac{t}{t_0}}} \tag{41}$$

as the equation of the arc voltage.

For,

$$a = 35,$$
$$b = 200,$$
$$I = 3,$$

hence,

$$E = 151,$$

and

$$t_0 = 10^{-4} \text{ sec.,}$$

and, for the three values of capacity,

$$C = 10^{-6} \qquad (e_1)$$
$$0.75 \times 10^{-6} \qquad (e_2)$$
$$0.5 \times 10^{-6} \qquad (e_3)$$

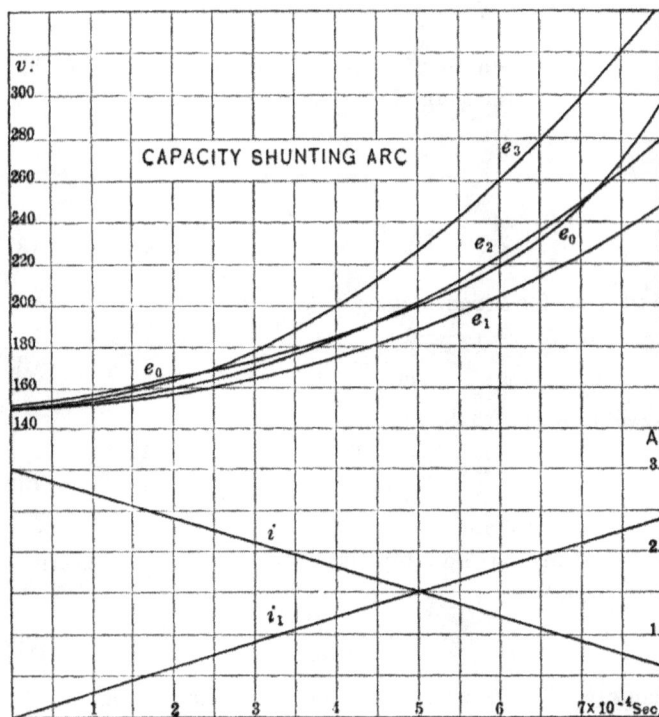

Fig. 88.

the curves of the arc voltage, e_0,
and of the condenser voltage, e_1, e_2, e_3,
are shown on Fig. 88,
together with the values of i and i_1.

As seen, e_1 is below e_0 over the entire range. That is, 1 mf. makes the arc unstable over the entire range. 0.5 mf., e_3, gives instability up to about $t = 0.25 \times 10^{-4}$ sec., then stability results. With 0.75 mf., e_2, there is a narrow range of stability, between

$4\frac{3}{4}$ and $7\frac{1}{4} \times 10^{-4}$ sec., before and after this instability exists.

From equation (37), the condition of stability, it follows that for small values of t, that is, small current fluctuations, the conditions are always unstable. That is, no matter how small a condenser is, it always has an effect in increasing the current fluctuations in the arc, the more so, the higher the capacity, until conditions become entirely unstable.

From equations (40) and (41) follows as the stability limit

$$e_0 = e_1,$$

$$a + \frac{b}{\sqrt{I}\sqrt{1 - \dfrac{t}{t_0}}} = E + \frac{t^2 I}{2\,t_0 C},$$

or, expanded into a series,

$$a + \frac{b}{\sqrt{I}}\left\{ 1 + \frac{t}{2\,t_0} + \ldots \right\} = E + \frac{t^2 I}{2\,t_0 C},$$

cancelling $E = a + \dfrac{b}{\sqrt{I}}$ and rearranging, gives

$$t_1 = \frac{bC}{I\sqrt{I}} \tag{42}$$

thus, at the time,

$$t_1 = \frac{bC}{I\sqrt{I}},$$

the condition changes from unstable to stable.

As t_1 must be smaller than t_0, the total time of change, it follows:

$$t_0 > \frac{bC}{I\sqrt{I}} \tag{43}$$

or,

$$C < \frac{t_0 I\sqrt{I}}{b} \tag{44}$$

are expressions of the (approximate) stability limit of an arc with condenser shunt.

As seen from (44),

the larger t_0 is, that is, the slower the arc changes, the larger is the permissible shunted capacity, and inversely.

As an instance, let

$$b = 200,$$
$$I = \quad 3,$$

and

$$(a) \quad t_0 = 10^{-3},$$

which is probably the approximate magnitude in the carbon arc.
This gives

$$C < 26 \text{ mf.}$$

Let:

$$(b) \quad t_0 = 10^{-5},$$

which is probably the approximate magnitude in the mercury arc.
This gives

$$C < 0.26 \text{ mf.}$$

95. Consider the case of a circuit, A, Fig. 87, supplied by a
constant current, I, but shunted by a capacity, C, inductance, L,
and resistance, r, in series.

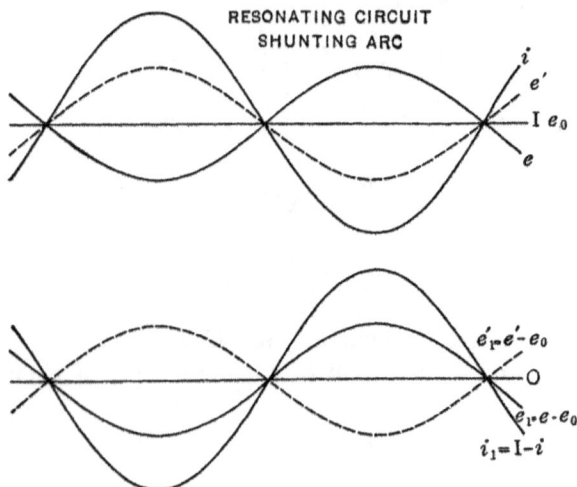

FIG. 89.

As long as the current in the circuit, A—whether resistance or
arc—is steady, no current passes the condenser circuit, and the
current and voltage in A thus are constant, $i = I$, $e = e_0$.

Suppose now a pulsation of the current, i, should be produced
in circuit, A, as shown as i in Fig. 89. Then, with constant-supply current, I, an alternating current,

$$i_1 = I - i,$$

would traverse the condenser circuit, C, since the continuous component of current can not traverse the condenser, C.

Due to the pulsation of current, i in A, the voltage, e, of circuit, A, would pulsate also. These voltage pulsations are in the same direction as the current pulsation, if A is a resistance, in opposite direction, if A is an arc; in either case, however, they are in phase with the current pulsation, and the alternating voltage on the condenser,

$$e_1 = e_0 - e,$$

thus is in phase with the alternating current, i_1, that is, capacity, C, and inductance, L, neutralize.

Thus, the only pulsation of current and voltage, which could occur in a circuit, A, shunted by capacity and inductance, is that of the resonance frequency of capacity and inductance.

Suppose the circuit, A, is a dead resistance. The voltage pulsation produced by a current pulsation, i, in this circuit then would be in the same direction as i, that is, would be as shown in dotted line by e' in Fig. 89. In the condenser circuit, C, the alternating component of voltage thus would be

$$e'_1 = e' - e_0,$$

thus would be in opposition to the alternating current, i_1, as shown in Fig. 89 in dotted line. That is, it would require a supply of power to maintain such pulsation.

Thus, with a dead resistance as circuit, A, or in general with A as a circuit of rising volt-ampere characteristic, the maintenance of a resonance pulsation of current and voltage between A and C, at constant current, I, requires a supply of alternating-current power in the condenser circuit, and without such power supply the pulsation could not exist, hence, if started, would rapidly die out, as oscillation, as shown in Fig. 87.

96. Suppose, however, A is an arc. A current pulsation, i, then gives a voltage pulsation in opposite direction, as shown by e in Fig. 89, and the alternating current, $i_1 = I - i$, and the alternating voltage, $e_1 = e - e_0$, in the condenser circuit, thus would be in phase with each other, as shown by i_1 and e_1 in Fig. 89. That is, they would represent power generation, or rather transformation of power from the constant direct-current supply, I, into the alternating-current resonating condenser circuit, C.

Thus, such a local pulsation of the arc current, i, and corresponding alternating current, i_1, in the condenser circuit, if once started, would maintain itself without external power supply, and would even be able to supply the power represented by voltage, e_1, with current, i_1, into an external circuit, as the resistance, r, shown in Fig. 87, or through a transformer into a wireless sending circuit, etc.

Thus, due to the dropping arc characteristic, an arc shunted by capacity and inductance, on a constant-current supply, becomes a generator of alternating-current power, of the frequency set by the resonance of C and L.

If the resistance, r, or in general, the load on the oscillating circuit, C, is greater than $r_1 = \dfrac{e_1}{i_1}$, that is, if a higher voltage would be required to send the current, i_1, through the resistance, r, than the voltage, e_1, generated by the oscillating arc, A, the pulsations die out as oscillations.

If r is less than $\dfrac{e_1}{i_1}$, the pulsations increase in amplitude, that is, current, i_1, and voltage, e_1, increase, until either, by the internal reaction in the arc, the ratio, $\dfrac{e_1}{i_1}$, drops to equality with the effective resistance of the load, r, and stability of oscillation is reached, or, if $\dfrac{e_1}{i_1}$ never falls to equality with r—for instance, if $r = 0$, the oscillations increase up to the destruction of the circuit: the extinction of the arc.

If, in the latter case, the voltage back of the supply current, I, is sufficiently high to restart the arc, A, the phenomena repeats, and we have a series of successive arc oscillations, each rising until it puts the arc out, and then the arc restarts.

We thus have here the mechanism which produces a *cumulative oscillation*, that is, a transient, which does not die out, but increases in amplitude, until the increasing energy losses limit its further increase, or until it destroys the circuit, and in the latter case, it may become recurrent.

It is very important to realize in electrical engineering, that any electric circuit with dropping volt-ampere characteristic is capable of transforming power into a cumulative oscillation, and thereby is able under favorable conditions to produce cumulative oscillations, such as hunting, etc.

Where the arc oscillations limit themselves, and the alternating current and voltage in the condenser circuit thus reach a constant value, the arc often is called a *"singing arc,"* due to the musical note given by the alternating wave. Where the arc oscillations

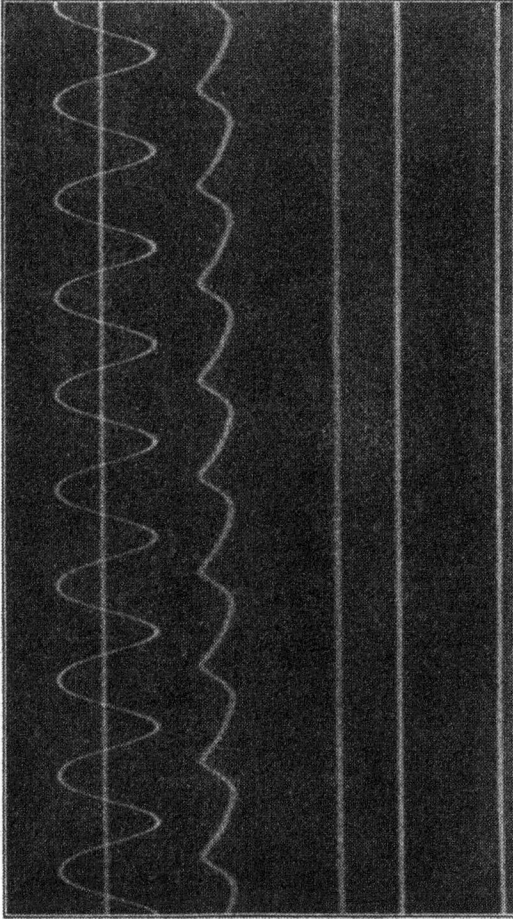

Fig. 90.

(*Facing p.* 188.)

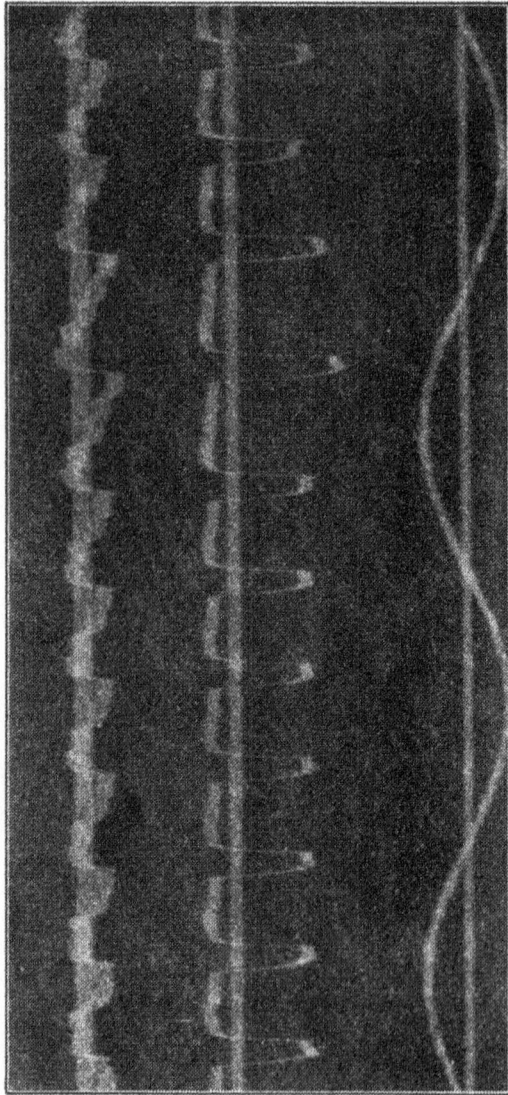

Fig. 91.

(CD15024.)

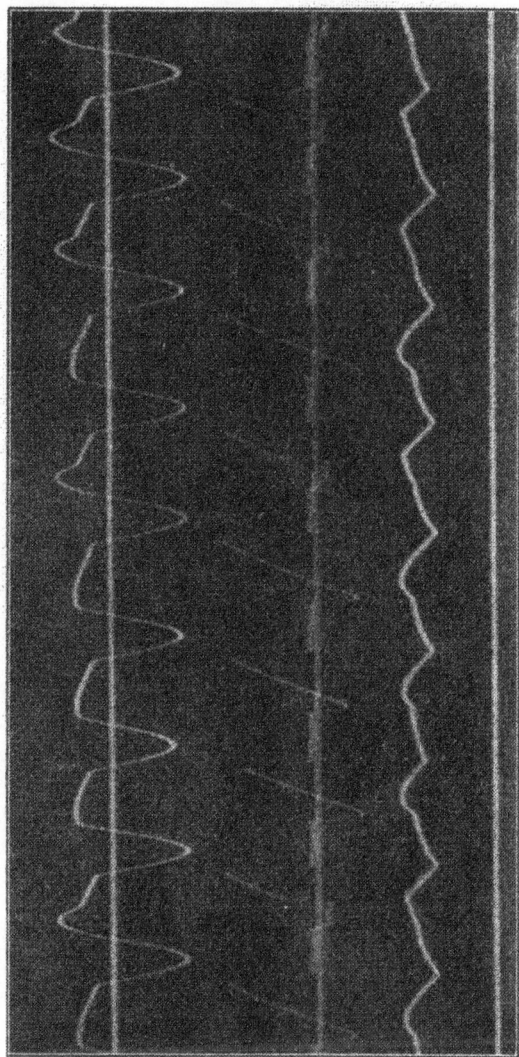

Fig. 92.

(CD26439.)

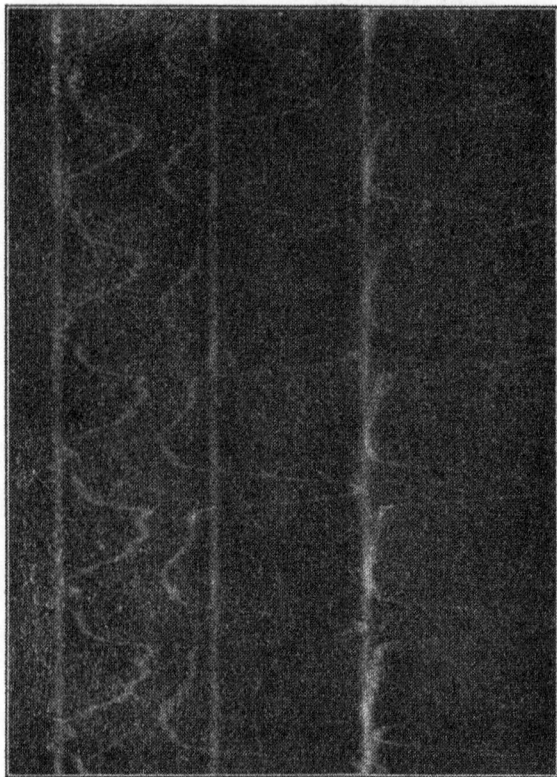

Fig. 93.

rise cumulatively to interruption, and the arc then restarts by the supply voltage and repeats the same phenomenon, it may be called a *"rasping arc,"* by the harsh noise produced by the interrupted cumulative oscillation.

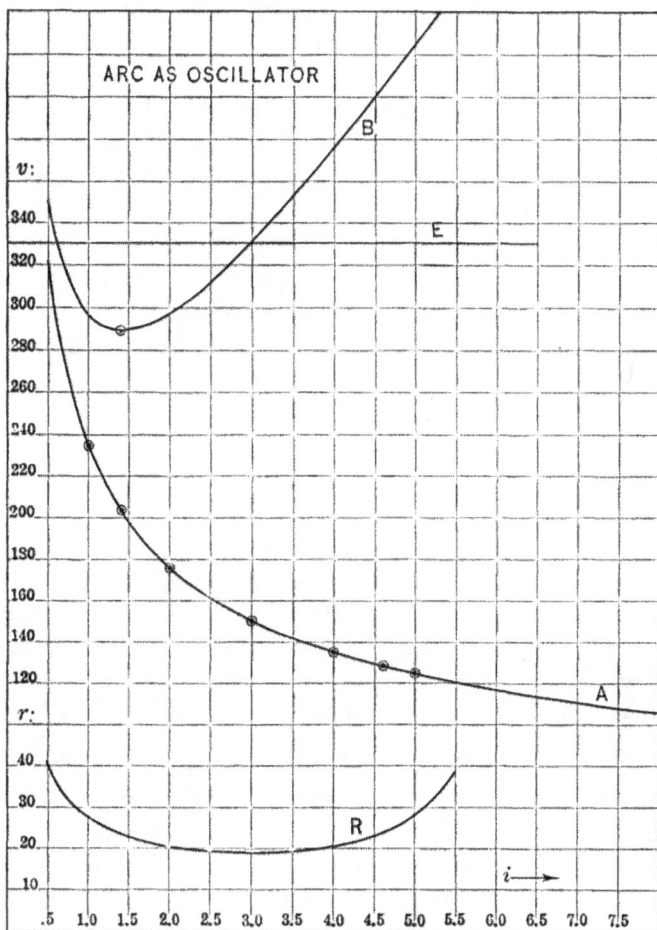

Fɪɢ. 94.

Figs. 90 and 91 give oscillograms of singing arcs; Figs. 92 and 93, of rasping arcs, 90 and 92 in circuits with massed constants, 91 and 93 in transmission lines.

97. As an illustration, let curve, A, in Fig. 94 represent the volt-ampere characteristic of an arc, and assume that this arc is operating steadily at I amp., consuming e_0 volts.

Suppose this arc is shunted by capacity, C, inductance, L, and resistance, r, as shown in Fig. 87.

For a small pulsation of the arc current around its average value $i = I$, the corresponding voltage pulsation is given by

$$\frac{de}{di} = -\frac{b}{2\,i\sqrt{i}}.$$

Or, in general, for any pulsation of current, i, by δi, between i' and i'', around the mean value I, the corresponding voltage pulsation δe, between e' and e'', is given by the volt-ampere characteristic of the arc, A, as

$$\frac{\delta e}{\delta i} = -\frac{e'' - e'}{i'' - i'}.$$

$e_1 = \delta e$ thus is the voltage, made available for the condenser circuit, by the arc pulsation, and in phase with the current, $i_1 = -\delta i$ in the condenser circuit, and

$$R = -\frac{\delta e}{\delta i} = \frac{e'' - e'}{i'' - i'},$$

thus is the permissible effective resistance in the condenser circuit, that is, the maximum value of resistance, through which the pulsating arc can maintain its alternating power supply: with a larger resistance, the oscillations die out; with a smaller resistance, they increase.

From the arc characteristic, A, thus can be derived a curve of effective resistances, R, as the values of $\frac{\delta e}{\delta i}$, for pulsations between $i + \delta i$ and $i - \delta i$, and such a curve is shown as R in Fig. 94.

We may say, that the arc, when shunted by an oscillating circuit, has an effective negative resistance,

$$\frac{\delta e}{\delta i},$$

and thereby generates alternating power, from the consumed direct-current power, and is able to supply alternating power through an effective resistance of the oscillating circuit, of

$$R = -\frac{\delta e}{\delta i}.$$

The arc characteristic in Fig. 94 is drawn with the equation

$$e = 35 + \frac{200}{\sqrt{i}}$$

and for

$$i = I = 3 \text{ amp. as mean value,}$$

the values of the effective resistance, R, increase from

$$R = -\frac{de}{di} = 18.5 \text{ ohms}$$

for very small oscillations, to

$$R = 20.3 \text{ ohms}$$

for oscillations of 1 amp., between $i = 2$ and $i = 4$, to

$$R = 27.5 \text{ ohms}$$

for oscillations of 2 amp., between $i = 1$ and $i = 5$, etc.

Thus, if with this oscillating arc, Figs. 87 and 94, a load resistance $r < 18.5$ ohms is used, oscillation starts immediately, and cumulatively increases.

If the resistance, r, is greater than 18.5 ohms, for instance, is

$$r = 22.5 \text{ ohms,}$$

then no oscillation starts spontaneously, but the arc runs steady, and no appreciable current passes through the condenser circuit. But if once the current in the arc is brought below 1.5 amp., or above 4.5 amp., the oscillation begins and cumulatively increases, since for oscillations of an amplitude greater than between 1.5 and 4.5 amp., the effective resistance, R, is greater than 22.5 ohms.

In either case, however, as soon as an oscillation starts, it cumulatively increases, since the effective resistance, R, steadily increases with increase of the amplitude of oscillation. That is, stability of oscillation, or a "singing arc" can not be reached, but an oscillation, once started, proceeds to the extinction of the arc, and only a "rasping arc" could be produced.

98. However, the arc characteristic, A, of Fig. 94 is the stationary characteristic, that is, the volt-ampere relation at constant current, i, and voltage, e.

If current, i, and thus voltage, e, rapidly fluctuate, the arc characteristic, A, changes, and more or less flattens out. That is, for any value of the current, i, the volume of the arc stream and the temperature of the arc terminals, still partly correspond to previous values of current, thus are lower for rising, higher for decreasing current, and as the result, the arc voltage, e, which de-

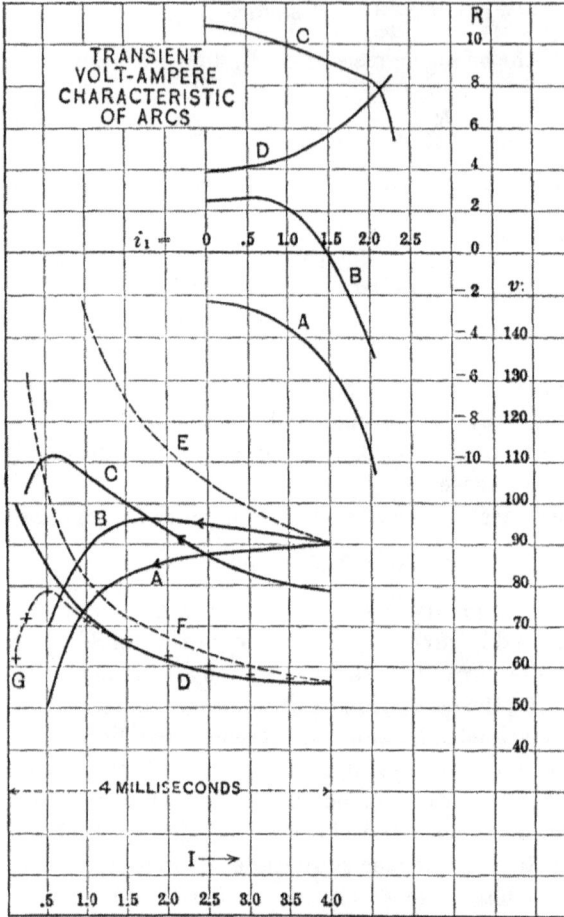

Fig. 95.

pends on the resistance of the arc stream and the potential drop of the terminals, is different, the variation of voltage, for the same variation of current, is less, and the effective negative arc resistance thereby is lowered, or may entirely vanish.

Fig. 95 shows a number of such transient arc characteristics, estimated from oscillographic tests of alternating arcs, and their corresponding effective resistances, R.

They are:

(*A*) Carbon.

(*B*) Hard carbon.

(*C*) Acheson graphite.

(*D*) Titanium carbid.

(*E*) Hard carbon, stationary characteristic.

(*F*) Titanium carbid, stationary characteristic.

As seen from the curves of R in the upper part of Fig. 95, the effective resistances, R, which represent the alternating power generated by the oscillating arc, are much lower with the transient arc characteristic, than would be with the permanent arc characteristic in Fig. 94.

Curve D, titanium carbide, gives under these conditions an unstable or "rasping" arc. That is, with a resistance in the condenser circuit of less than $R = 3.8$ ohms, the oscillation starts spontaneously and cumulatively increases to the extinction of the arc; with a resistance of more than 3.8 ohms, the oscillation does not spontaneously start, but if once started with an amplitude which brings the value of R from curve, D, above that of the resistance in the condenser circuit, cumulative oscillation occurs.

With the carbon arc, A, no oscillations can occur under any condition, the effective resistance, R, is negative, and the arc characteristic rising.

With the hard carbon arc, B, an oscillation starts with a resistance less than 2.4 ohms, cumulatively increases, but its amplitude finally limits itself, to 1.45 amp. if the resistance in the oscillating circuit is zero, to 1.05 amp. with 2 ohms resistance, etc., as seen from the curve, B, in the upper part of Fig. 95. Even with more than 2.4 ohms resistance, up to 2.6 ohms resistance, an oscillation can exist, if once started, as the curve of R, starting from $R = 2.4$ ohms at $i_1 = 0$, rises to $R = 2.6$ ohms at $i_1 = 0.75$, and then drops to zero at $i_1 = 1.45$ ohms, and beyond this becomes negative.

The curve, C, of Acheson graphite, starts with a resistance $R = 10.8$ ohms, but the resistance, R, steadily drops with increasing oscillating current, i_1, down to zero at $i_1 = 2.4$ amp. Thus, with a resistance in the condenser circuit, of 10 ohms, the oscillations would have an amplitude of $i_1 = 0.9$ amp.; with 8 ohms resistance an amplitude of 2.1 amp., etc.

13

From these curves of R, Fig. 95, the regulation curves of the alternating-current generation could now be constructed.

It is interesting to note, that in many of these transient arc characteristics, Fig. 95, the voltage does not indefinitely rise with decreasing current, but reaches a maximum and then decreases again, in B and C, and the oscillation resistance, that is, the resistance through which an alternating current can be maintained by the oscillating arc, thus decreases with increasing amplitude of the oscillation. Thus, if the resistance in the oscillating condenser circuit is less than the permissible maximum, an oscillation starts, cumulatively increases, but finally limits itself in amplitude.

The decrease of the arc voltage with decreasing current, for low values of current in a rapidly fluctuating arc, is due to the time lag of the arc voltage behind the current.

99. The arc voltage, e, consists of the arc terminal drop, a, and the arc stream voltage, e_1:

$$e = a + e_1.$$

The stream voltage, e_1, is the voltage consumed in the effective resistance of the arc stream; but as the arc stream is produced by the current, the volume of the arc stream and its resistance thus depends on the current, i, in the arc, that is, the stream voltage is

$$e_1 = \frac{b}{\sqrt{i}}$$

and the resistance of the arc stream thus

$$r_1 = \frac{e_1}{i} = \frac{b}{i\sqrt{i}}.$$

Thus, if,

$$a = 35$$
$$b = 200,$$

for

$$i = 2 \text{ amp., it is}$$
$$r_1 = 70.7 \text{ ohms,}$$
$$e_1 = 141.4 \text{ volts,}$$
$$e = 176.4 \text{ volts.}$$

But, if the arc current rapidly varies, for instance decreases, then, when the current in the arc is i_1, the volume of the arc stream and thus its resistance is still that corresponding to the previous current, i'_1.

If thus, at the moment where the current in the arc has become

$$i_1 = 2 \text{ amp.},$$

the arc stream still has the volume and thus the resistance corresponding to the previous current,

$$i'_1 = 3 \text{ amp.},$$

this resistance is

$$r'_1 = \frac{200}{3\sqrt{3}} = 38.5 \text{ ohms,}$$

and the stream voltage, at the current

$$i_1 = 2 \text{ amp.},$$

but with the stream resistance, r'_1, corresponding to the previous current, $i'_1 = 3$ amp., thus is

$$e'_1 = r'_1 i_1$$
$$= 77 \text{ volts,}$$

instead of $e_1 = 141.4$ volts, as it would be under stationary conditions.

That is, the stream voltage and thus the total arc voltage at rapidly decreasing current is lower, at rapidly increasing current higher than at stationary current.

With a periodically pulsating current, it follows herefrom, that at the extreme values of current—maximum and minimum— the voltage has not yet reached the extreme values corresponding to these currents, that is, the amplitude of voltage pulsation is reduced. This means the transient volt-ampere characteristic of the arc is flattened out, compared with the permanent characteristic, and caused to bend downward at low currents, as shown by C and B in Fig. 95.

Assuming a sinusoidal pulsation of the current in the arc and assuming the arc stream resistance to lag behind the current by a suitable distance, we then get, from the stationary volt-ampere characteristic of the arc, the transient characteristics.

Thus in Fig. 96, from the stationary arc characteristic, S, the transient arc characteristic, T, is derived. In this figure is shown as S and T the effective resistance corresponding to the stationary characteristic, S, respectively the transient characteristic, T.

As seen, the stationary characteristic, S, gives an arc oscillation which is cumulative and self-destructive, that is, the effective resistance, R, rises indefinitely with increasing amplitude of pulsation. The transient characteristic, however, gives an effective resistance, R, which with increasing amplitude of pulsation

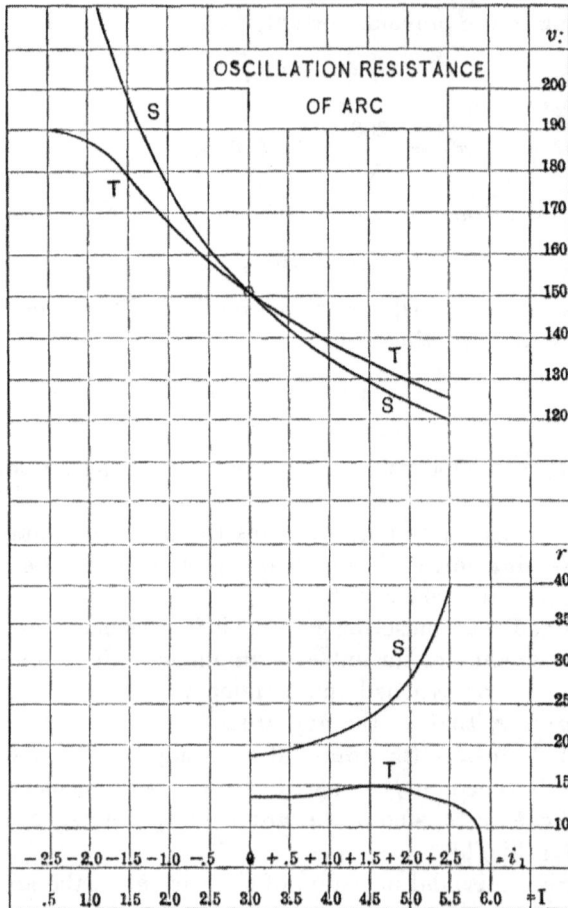

FIG. 96.

first increases, but then decreases again, down to zero, so that the cumulative oscillations produced by this arc are self-limiting, increase in amplitude only up to the value, where the effective resistance, R, has fallen to the value corresponding to the load on the oscillating circuit.

As further illustration, from the stationary volt-ampere characteristic of the titanium arc, shown as F in Fig. 95, values of the transient characteristic have been calculated and are shown in Fig. 95 by crosses. As seen, they fairly well coincide with the transient volt-ampere characteristic, D, of the titanium arc, at least for the larger currents.

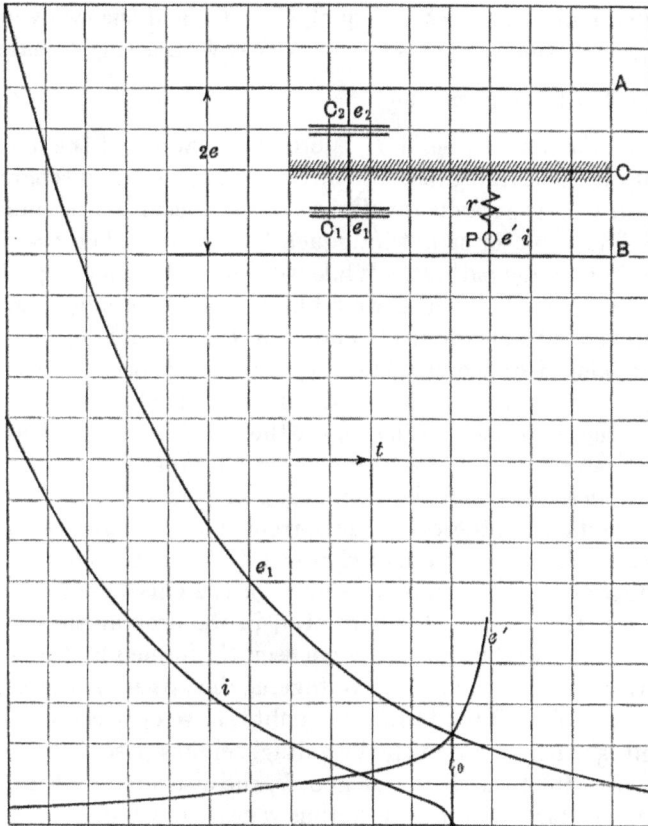

Fig. 97.

In the electric arc we thus have an electric circuit with dropping volt-ampere characteristic. Such a circuit is unstable under various conditions which may occur in industrial circuits, and thereby may be, and frequently is, the source of instability of electric circuits, and of cumulative oscillations appearing in such circuits.

100. For instance, let, in Fig. 97, A and B be two conductors of an ungrounded high-potential transmission line, and $2e$ the voltage impressed between these two conductors. Let C represent the ground.

The capacity of the conductors, A and B, against ground, then, may be represented diagrammatically by two condensers, C_1 and C_2, and the voltages from the lines to ground by e_1 and e_2. In general, the two line capacities are equal, $C_1 = C_2$, and the two voltages to ground thus equal also, $e_1 = e_2 = e$, with a single-phase; $= \dfrac{2e}{\sqrt{3}}$ with a three-phase line.

Assume now that a ground, P, is brought near one of the lines, A, to within the striking distance of the voltage, e. A discharge then occurs over the conductor, P. Such may occur by the puncture of a line insulator as not infrequently the case. Let $r = $ resistance of discharge path, P. While without this discharge path, the voltage between A and C would be $e_1 = e$ (assuming single-phase circuit) with a grounded conductor, P, approaching line A within striking distance of voltage, e, a discharge occurs over P forming an arc, and the circuit of the impressed voltage, $2e$, now comprises the condenser, C_2, in series to the multiple circuit of condenser, C_1, and arc, P, and the condenser, C_1, rapidly discharges, voltage, e_1, decreases, and the voltage, e_2, increases. With a decrease of voltage, e_1, the discharge current, i, also decreases, and the voltage consumed by the discharge arc, e', increases until the two voltages, e_1 and e', cross, as shown in the curve diagram of Fig. 97. At this moment the current, i, in the arc vanishes, the arc ceases, and the shunt of the condenser, C_1, formed by the discharge over P thus ceases. The voltage, e_1, then rises, e_2 decreases and the two voltages tend toward equality, $e_1 = e_2 = e$. Before this point is reached, however, the voltage, e_1, has passed the disruptive strength of the discharge gap, P, the discharge by the arc over P again starts, and the cycle thus repeats indefinitely.

In Fig. 97 are diagrammatically sketched voltage, e_1, of condenser, C_1, the voltage, e', consumed by the discharge arc over P, and the current, i, of this arc, under the assumption that r is sufficiently high to make the discharge non-oscillatory. If r is small, each of these successive discharges is an oscillation.

Such an unstable circuit gives a continuous series of successive discharges, which are single impulses, as in Fig. 97, or more commonly are oscillations.

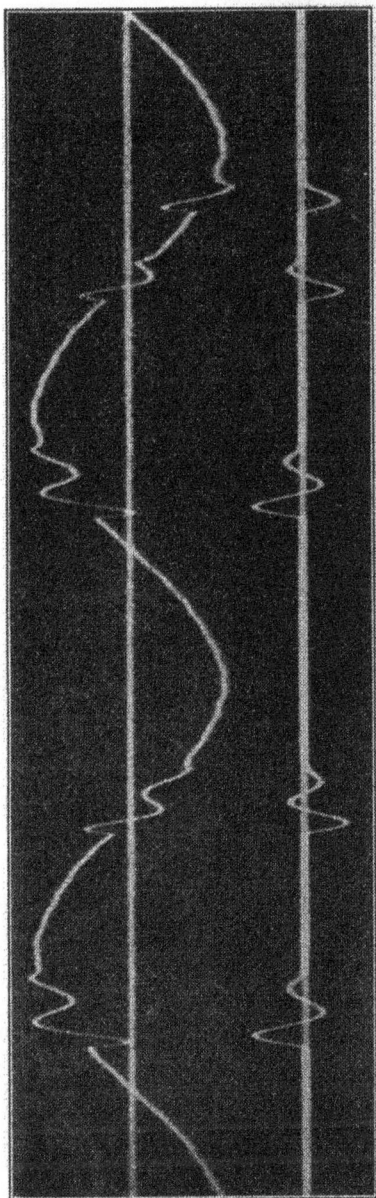

FIG. 98.

(CD23445.)

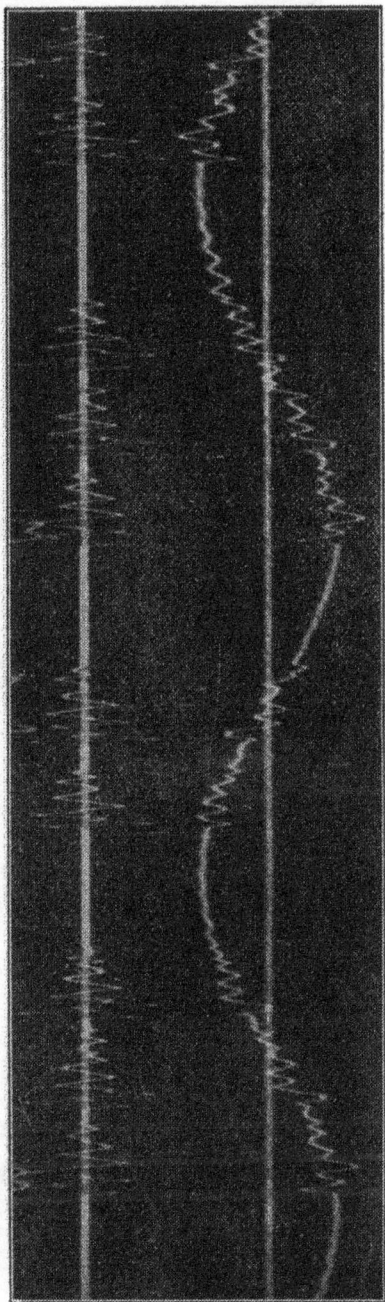

FIG. 99.

(*Facing p.* 198.)

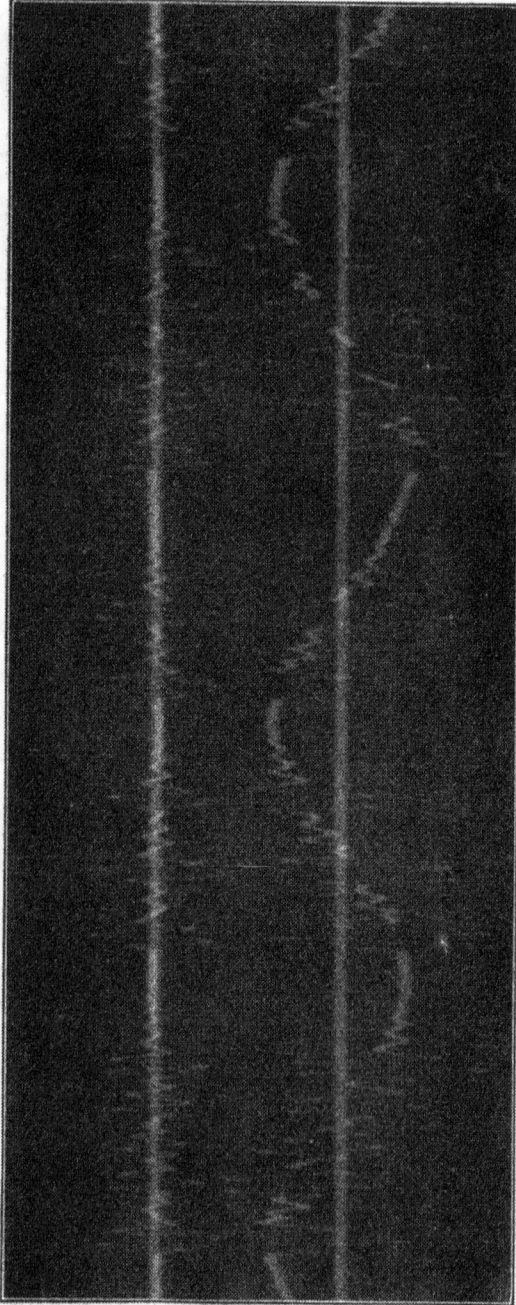

Fig. 100.

(CD23431.)

If the line conductors, A and B, in Fig. 97 have appreciable inductance, as is the case with transmission lines, in the charge of the condenser, C_1, after it has been discharged by the arc over P, the voltage, e_1, would rise beyond e, approaching $2\,e$, and the discharge would thus start over P, even if the disruptive strength of this gap is higher than e, provided that it is still below the voltage momentarily reached by the oscillatory charge of the line condenser, P_1.

This combination of two transmission line conductors and the ground conductor, P, approaching near line, A, to a distance giving a striking voltage above e, but below· the momentary charging voltage, of C_1, then constitutes a circuit which has two permanent conditions, one of stability and one of instability. If the voltage is gradually applied, $e_1 = e_2 = e$, the condition is stable, as no discharge occurs over P. If, however, by some means, as a momentarily overvoltage, a discharge is once produced over the spark-gap, P, the unstable condition of the circuit persists in the form of successive and recurrent discharges.

101. Usually, the resistance, r, of the discharge path is, or after a number of recurrent discharges, becomes sufficiently low to make the discharge oscillatory, and a series of recurrent oscillations then result, a so-called "arcing ground." Oscillograms of such an arcing grounds on a 30-mile 30-kv. transmission line are shown in Figs. 98, 99 and 100.

If, however, the resistance of the discharge path is very low, a sustained or cumulative oscillation results, as discussed in the preceding, that is, the arcing ground becomes a stationary oscillation of constant-resonance frequency, increasing cumulatively in current and voltage amplitude until limited by increasing losses or by destruction of apparatus.

In transmission lines, usually the resistance is too high to produce a cumulative oscillation; in underground cables, usually the inductance is too low and thus no cumulative oscillation results, except perhaps sometimes in single-conductor cables, etc. In the high-potential windings of large high-voltage power transformers, however, as circuits of distributed capacity, inductance and resistance, the resistance commonly is below the value through which a cumulative oscillation can be produced and maintained, and in high-potential transformers, destruction by high voltages resulting from the cumulative oscillation of some arc in the system, and building up to high stationary waves, have frequently been observed.

The "arcing ground" as recurrent single impulses, the "arcing ground oscillation" as more or less rapidly damped recurrent oscillations in transmission lines—of frequencies from a few hundred to a few thousand cycles—and the "stationary oscillations" causing destruction in high-potential transformer windings, at frequencies of 10,000 to 100,000 cycles, thus are the same phenomena of the dropping arc characteristic, causing permanent instability of the electric circuit, and differ from each other merely by the relative amount of resistance in the discharge path.

CHAPTER XI

INSTABILITY OF CIRCUITS: INDUCTION AND SYNCHRONOUS MOTORS

C. Instability of Induction Motors

102. Instability of electric circuits may result from causes which are not electrical: thus, mechanical relations between the torque given by a motor and the torque required by its load, may lead to instability.

Let

D = torque given by a motor at speed, S, and

D' = torque required by the load at speed, S.

The motor, then, could theoretically operate, that is, run at constant speed, at that speed, S, where

$$D = D' \tag{1}$$

However, at this speed and load, the operation may be stable, that is, the motor continue to run indefinitely at constant speed, or the condition may be unstable, that is, the speed change with increasing rapidity, until stability is reached at some other speed, or the motor comes to a standstill, or it destroys itself.

In general, the motor torque, D, and the load torque, D', change with the speed, S.

If, then,

$$\frac{dD'}{dS} > \frac{dD}{dS} \tag{2}$$

the conditions are stable, that is, any change of speed, S, changes the motor torque less than the load torque, and inversely, and thus checks itself.

If, however,

$$\frac{dD'}{dS} < \frac{dD}{dS} \tag{3}$$

the operation is unstable, as a change of speed, S, changes the motor torque, D, more than the load torque, D', and thereby further increases the change of speed, etc.

$$\frac{dD'}{dS} = \frac{dD}{dS} \tag{4}$$

thus is the expression of the stability limit.

For instance, assuming a load requiring a constant torque at all speeds. The load torque thus is given by a horizontal line

$$D' = \text{const.} \qquad (5)$$

in Fig. 101.

Let then the speed-torque curve of the motor be represented by the curve, D, in Fig. 101. D approximately represents the torque curve of a series motor. At the constant-load torque, D', the motor runs at the speed, $S = 0.6$, point a of Fig. 101, and the speed is stable, as any tendency to change of speed, checks itself. If

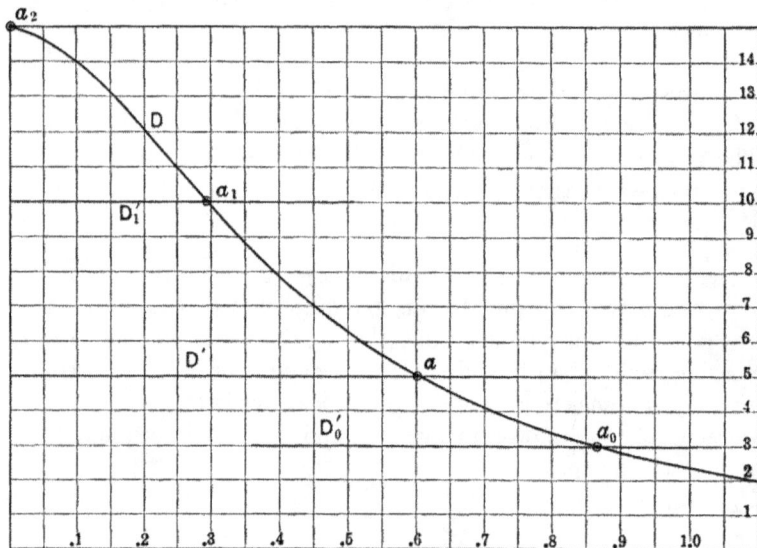

Fig. 101.

the load torque decreases to D'_0, the speed rises to $S = 0.865$, point a_0; if the load torque increases to D'_1, the speed drops to $S = 0.29$, point a_1, but the conditions are always stable, until finally with increasing load torque, D', and decreasing speed, standstill is reached at point a_2.

Let now the speed-torque curve of a motor be represented by D in Fig. 102: the curve of a squirrel-cage induction motor with moderately high resistance secondary. The horizontal line, D', corresponding to a load torque of $D' = 10$, intersects D at two points, a and b.

At a, $S = 0.905$, the speed is stable.   At b, however, $S = 0.35$, the conditions are unstable, and the motor thus can not run at b, but either—if the speed should drop or the load rise ever so little —the motor begins to slow down, thereby, on curve, D, its torque falls below that of the load, D', thus it slows down still more, and so, with increasing rapidity the motor comes to a standstill.   Or, if the motor speed should be a little higher, or the load momentarily a little lower, the motor speed rises, until stability is reached at point a.

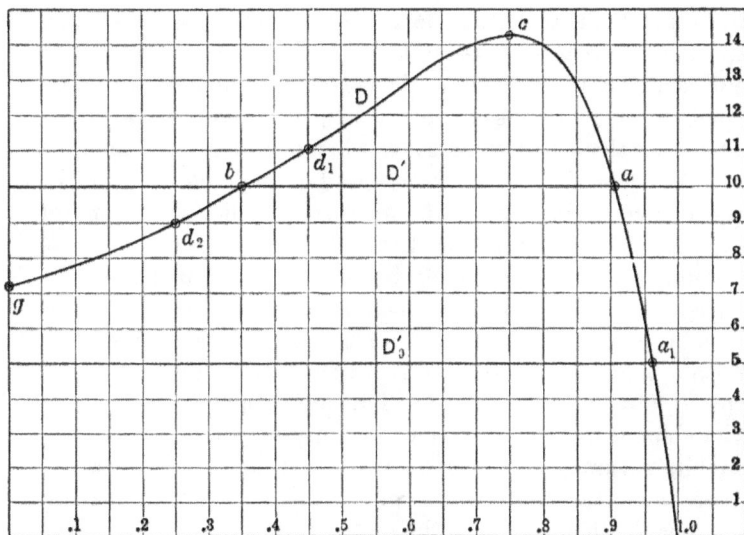

FIG. 102.

With increasing load torque, D', the speed gradually drops, from $S = 0.905$ at $D' = 10$, point a, down to point c, at $S = 0.75$, $D' = 14.3$; from there, however, the speed suddenly drops to standstill, that is, it is not possible to operate the motor at speeds less than $S = 0.75$, at constant load-torque, and the branch of the motor characteristic from the starting point, g, up to the maximum torque point, c, is unstable on a load requiring constant torque.

At load torque, $D' = 10$, the motor can not start the load, can not carry it below b, $S = 0.35$; at speeds from b to a, $S = 0.35$ to 0.905, the motor speeds up; at speeds above a, $S = 0.905$, the motor slows down, and drops into stable condition at a.

With a load torque, $D'_0 = 5$, the motor starts and runs up to speed a_1, $S = 0.96$.

$D' = 7.2$, point g, thus, is the maximum load torque which the motor can start.

103. Suppose now, while running in stable condition, at point a, with the load torque, $D' = 10$, the load torque is momentarily increased. If this increase leaves D' lower than the maximum motor torque, $D_0 = 14.3$, the motor speed slows down, but remains above c, and thus when the increase of load is taken off, the motor again speeds up to a.

If, however, the temporary increase of load torque exceeds the maximum motor torque, $D_0 = 14.3$—for instance by starting a line of shafting or other mass of considerable momentum—then the motor speed continues to drop as long as the excess load exists, and whether the motor will recover when the excess load is taken off, or not, depends on the loss of speed of the motor during the period of overload: if, when the overload is relieved, the motor has dropped to point d_1 in Fig. 102, its speed thus is still above b, the motor recovers; if, however, its speed has dropped to d_2, below the speed b, $S = 0.35$, at which the motor torque drops below the load torque, then the motor does not recover, but stops.

With a lighter load torque, D'_0, which is less than the starting torque, g, obviously the motor will always recover in speed

The amount, by which the motor drops in speed at temporary overload, naturally depends on the duration of the overload, and on the momentum of the motor and its moving masses: the higher the momentum of the motor and of the masses driven by it at the moment of overload, the slower is the drop of speed of the motor, and the higher thus the speed retained by it at the moment when the overload is relieved.

Thus a motor of low starting torque, that is, high speed regulation, may be thrown out of step by picking up a load of high momentum rapidly, while by adding a flywheel to the motor, it would be enabled to pick up this load. Or, it may be troublesome to pick up the first load of high momentum, while the second load of this character may give no trouble, as, due to the momentum of the load already picked up, the speed would drop less.

Thus a motor carrying no load, may be thrown out of step by a load which the same motor, already partly loaded (with a load of considerable momentum), would find no difficulty to pick up.

The ability of an induction motor, to carry for a short time without dropping out of step a temporary excessive overload, naturally also depends on the excess of the maximum motor torque (at c in Fig. 102) over the normal load torque of the motor. A motor, in which the maximum torque is very much higher—several hundred per cent.—than the rated torque, thus could momentarily carry overloads which a motor could not carry, in which the maximum torque exceeds the rated torque only by 50 per cent., as was the case with the early motors. However, very high maximum torque means low internal reactance and thus high exciting current, that is, low power-factor at partial loads, and of the two types of motors:

(a) High overload torque, but poor power-factor and efficiency at partial loads;

(b) Moderate overload torque, but good power-factor and efficiency at partial loads;

the type (b) gives far better average operating conditions, except in those rare cases of operation at constant full-load, and is therefore preferable, though a greater care is necessary to avoid momentary excessive overloads.

Gradually the type (a) had more and more come into use, as the customers selected the motor, and the power supply company neglected to pay much attention to power-factor, and it is only in the last few years, that a realization of the harmful effects of low power-factors on the economy of operation of the systems is again directing attention to the need of good power-factors at partial loads, and the industry thus is returning to type (b), especially in view of the increasing tendency toward maximum output rating of apparatus.

In distributing transformers, the corresponding situation had been realized by the central stations since the early days, and good partial load efficiencies and power-factors secured.

104. The induction motor speed-torque curve thus has on a constant-torque load a stable branch, from the maximum torque point, c, Fig. 102, to synchronism; and an unstable branch, from standstill to the maximum torque point.

However, it would be incorrect to ascribe the stability or instability to the induction motor-speed curve; but it is the character of the load, the requirement of constant torque, which makes a part of the speed curve unstable, and on other kinds of load no instability may exist, or a different form of instability.

Thus, considering a load requiring a torque proportional to

the speed, such as would be given, approximately, by an electric generator at constant field excitation and constant resistance as load.

The load-torque curves, then, would be straight lines going through the origin, as shown by D'_1, D'_2, D'_3, etc., for increasingly larger values of load, in Fig. 103. The motor-torque curve, D, is the same as in Fig. 102. As seen, all the lines, D', intersect D at points, a_1, a_2, a_3 . . ., at which the speed is stable, since

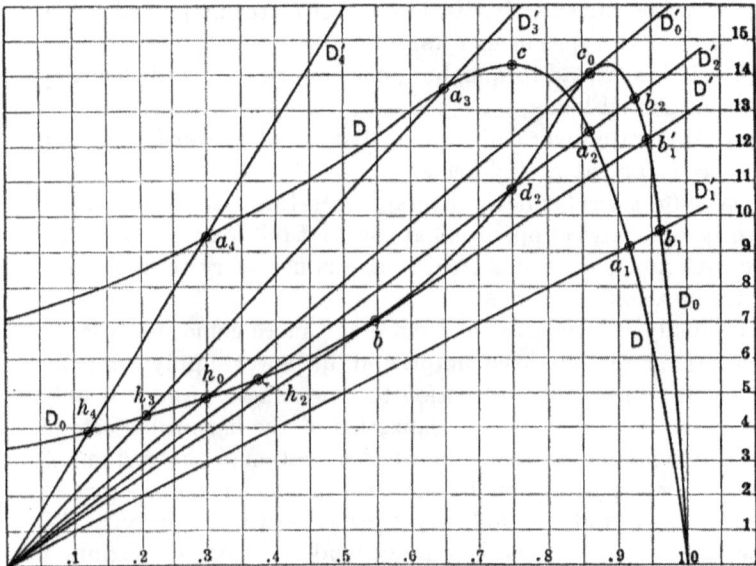

FIG. 103.

$$\frac{dD'}{dS} > \frac{dD}{dS}.$$

Thus, with this character of load, a torque required proportional to the speed, and the motor-torque curve, D, no instability exists, but conditions are stable from standstill to synchronism, just as in Fig. 101. That is, with increasing load, the speed decreases and increases again with decreasing load.

If, however, the motor curve is as shown by D_0 in Fig. 103, that is, low starting torque and a maximum torque point close to synchronism, as corresponds to an induction motor with low resistance secondary, then for a certain range of load, between

D' and D'_0, the load-torque line, D'_2, intersects the motor curve, D_0, in three points b_2, d_2, h_2.

At b_2, $S = 0.925$, and at h_2, $S = 0.375$, conditions are stable; at d_2, $S = 0.75$, instability exists.

Thus with this load, D'_2, the motor can run at two different speeds in stable conditions: a high speed, above c_0, and a low speed, below b; while there is a third, theoretical speed, d_2, which is unstable. In the range below h_2, the motor speeds up to h_2; in the range between h_2 and d_2, the motor slows down to h_2; in the range between d_2 and b_2, the motor speeds up to b_2, and in the range above b_2, the motor slows down to b_2.

There is thus a (fairly narrow) range of loads between D' and D'_0, in which an unstable branch of the induction motor-torque curve exists, at intermediate speeds; at low speed as well as at high speed conditions are stable.

For loads less than D', conditions are stable over the entire range of speed; for loads above D'_0, the motor can run only at low speeds, h_3, h_4, but not at high speeds; but there is no load at which the motor would not start and run up to some speed.

Obviously, at the lower speeds, the current consumed by the motor is so large, that the operation would be very inefficient.

It is interesting to note, that with this kind of load, the "maximum torque point," c, is no characteristic point of the motor-torque curve, but two points, c_0 and b, exist, between which the operation of the motor is unstable, and the speed either drops down below b, or rises above c_0.

105. With a load requiring a torque proportional to the square of the speed, such as a fan, or a ship propeller, conditions are almost always stable over the entire range of speed, from standstill to synchronism, and an unstable range of speed may occur only in motors of very low secondary resistance, in which the drop of torque below the maximum torque point, c, of the motor characteristic is very rapid, that is, the torque of the motor decreases more rapidly than with the square of the speed. This may occur with very large motors, such as used on ship propellers, if the secondary resistance is made too low.

More frequently instability with such fan or propeller load or other load of similar character may occur with single-phase motors, as in these the drop of the torque curve below maximum torque is much more rapid, and often a drop of torque with increasing speed occurs, especially with the very simple and cheap starting devices economically required on very small motors, such as fan motors.

Instability and dropping out of step of induction motors also may be the result of the voltage drop in the supply lines, and furthermore may result from the regulation of the generator voltage being too slow. Regarding hereto, however, see "Theory and Calculation of Electrical Apparatus," in the chapter on "Stability of Induction Machines."

D. Hunting of Synchronous Machines

106. In induction-motor circuits, instability almost always assumes the form of a steady change, with increasing rapidity, from the unstable condition to a stable condition or to standstill, etc.

Oscillatory instability in induction-motor circuits, as the result of the relation of load to speed and electric supply, is rare. It has been observed, especially in single-phase motors, in cases of considerable oversaturation of the magnetic circuit.

Oscillatory instability, however, is typical of the synchronous machine, and the hunting of synchronous machines has probably been the first serious problem of cumulative oscillations in electric circuits, and for a long time has limited the industrial use of synchronous machines, in its different forms:

(*a*) Difficulty and failure of alternating-current generators to operate in parallel.

(*b*) Hunting of synchronous converters.

(*c*) Hunting of synchronous motors.

While considerable theoretical work has been done, practically all theoretical study of the hunting of synchronous machines has been limited to the calculation of the frequency of the transient oscillation of the synchronous machine, at a change of load, frequency or voltage, at synchronizing, etc. However, this transient oscillation is harmless, and becomes dangerous only if the oscillation ceases to be transient, but becomes permanent and cumulative, and the most important problem in the study of hunting thus is the determination of the cause, which converts the transient oscillation into a cumulative one, that is, the determination of the source of the energy, and the mechanism of its transfer to the oscillating system. To design synchronous machines, so as to have no or very little tendency to hunting, obviously requires a knowledge of those characteristics of design which are instrumental in the energy transfer to the oscillating system, and thereby cause hunting, so as to avoid them and produce the greatest possible inherent stability.

If, in an induction motor running loaded, at constant speed, the load is suddenly decreased, the torque of the motor being in excess of the reduced load causes an acceleration, and the speed increases. As in an induction motor the torque is a function of the speed, the increase of speed decreases the torque, and thereby decreases the increase of speed until that speed is reached at which the motor torque has dropped to equality with the load, and thereby acceleration and further increase of speed ceases, and the motor continues operation at the constant higher speed, that is, the induction motor reacts on a decrease of load by an increase of speed, which is gradual and steady without any oscillation.

If, in a synchronous motor running loaded, the load is suddenly decreased, the beginning of the phenomenon is the same as in the induction motor, the excess of motor torque causes an acceleration, that is, an increase of speed. However, in the synchronous motor the torque is not a function of the speed, but in stationary condition the speed must always be the same, synchronism, and the torque is a function of the relative position of the rotor to the impressed frequency. The increase of speed, due to the excess torque resulting from the decreased load, causes the rotor to run ahead of its previous relative position, and thereby decreases the torque until, by the increased speed, the motor has run ahead from the relative position corresponding to the previous load, to the relative position corresponding to the decreased load. Then the acceleration, and with it the increase of speed, stops. But the speed is higher than in the beginning, that is, is above synchronism, and the rotor continues to run ahead, the torque continues to decrease, is now below that required by the load, and the latter thus exerts a retarding force, decreases the speed and brings it back to synchronism. But when synchronous speed is reached again, the rotor is ahead of its proper position, thus can not carry its load, and begins to slow down, until it is brought back into its proper position. At this position, however, the speed is now below synchronism, the rotor thus continues to drop back, and the motor torque increases beyond the load, thereby accelerates again to synchronous speed, etc., and in this manner conditions of synchronous speed, with the rotor position

14

behind or ahead of the position corresponding to the load, alternate with conditions of proper relative position of the rotor, but below or above synchronous speed, that is, an oscillation results which usually dies down at a rate depending on the energy losses resulting from the oscillation.

107. As seen, the characteristic of the synchronous machine is, that readjustment to a change of load requires a change of relative position of the rotor with regard to the impressed frequency, without any change of speed, while a change of relative position can be accomplished only by a change of speed, and this results in an over-reaching in position and in speed, that is, in an oscillation.

Due to the energy losses caused by the oscillation, the successive swings decrease in amplitude, and the oscillation dies down. If, however, the cause which brings the rotor back from the position ahead or behind its normal position corresponding to the changed load (excess or deficiency of motor torque over the torque required by the load) is greater than the torque which opposes the deviation of the rotor from its normal position, each swing tends to exceed the preceding one in amplitude, and if the energy losses are insufficient, the oscillation thus increases in amplitude and becomes cumulative, that is, hunting.

In Fig. 104 is shown diagrammatically as p, the change of the relative position of the rotor, from p_1 corresponding to the previous load to p_2 the position further forward corresponding to the decreased load.

v then shows the oscillation of speed corresponding to the oscillation of position.

The dotted curve, w_1, then shows the energy losses resulting from the oscillation of speed (hysteresis and eddies in the pole faces, currents in damper windings), that is, the damping power, assumed as proportional to the square of the speed.

If there is no lag of the synchronizing force behind the position displacement, the synchronizing force, that is, the force which tends to bring the rotor back from a position behind or ahead of the position corresponding to the load, would be—or may approximately be assumed as—proportional to the position displacement, p, but with reverse sign, positive for acceleration when p is negative or behind the normal position, negative or retarding when p is ahead. The synchronizing power, that is, the power exerted by the machine to return to the normal position, then is derived by multiplying $-p$ with v, and is shown dotted as w_2 in Fig. 104. As seen, it has a double-frequency alternation with zero as average.

The total resultant power or the resulting damping effect which restores stability, then, is the sum of the synchronizing power w_2 and the damping power w_1, and is shown by the dotted

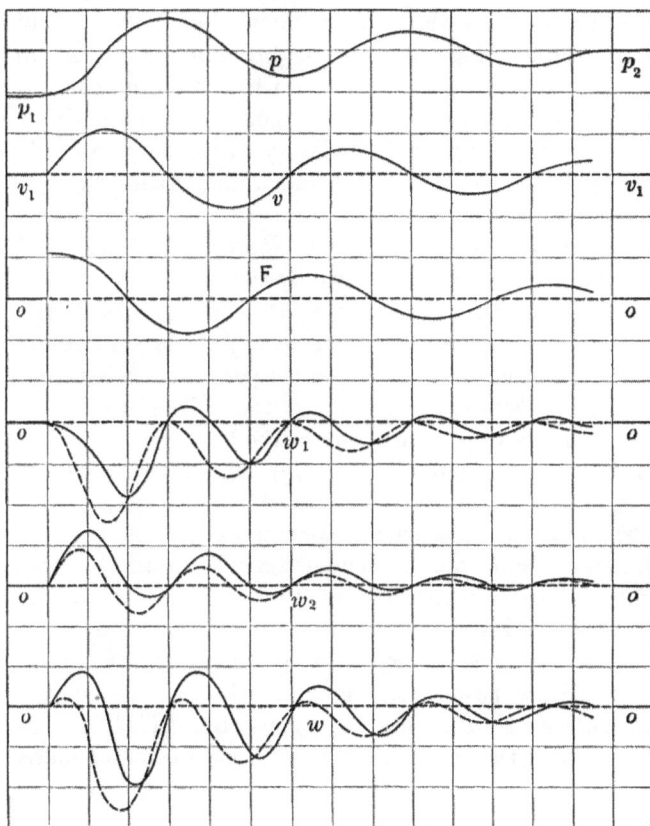

Fɪɢ. 104.

curve w. As seen, under the assumption or Fig. 104, in this case a rapid damping occurs.

If the damping winding, which consumes a part of all the power, w_1, is inductive—and to a slight extent it always is—the current in the damping winding lags behind the e.m.f. induced in it by the oscillation, that is, lags behind the speed, v. The power, w_1, or that part of it which is current times voltage, then ceases to be continuously negative or damping, but contains a positive period, and its average is greatly reduced, as shown by the drawn curve, w_1, in Fig. 104, that is, inductivity of the damper winding is very harmful, and it is essential to design the damper winding as non-inductive as possible to give efficient damping.

With the change of position, p, the current, and thus the armature reaction, and with it the magnetic flux of the machine, changes. A flux change can not be brought about instantly, as it represents energy stored, and as a result the magnetic flux of the machine does not exactly correspond with the position, p, but lags behind it, and with it the synchronizing force, F, as shown in Fig. 104, lags more or less, depending on the design of the machine.

The synchronizing power of the machine, Fv, in the case of a lagging synchronizing force, F, is shown by the drawn curve, w_2. As seen, the positive ranges of the oscillation are greater than the negative ones, that is, the average of the oscillating synchronizing power is positive or supplying energy to the oscillating system, which energy tends to increase the amplitude of the oscillation—in other words, tends to produce cumulative hunting.

The total resulting power, $w = w_1 + w_2$, under these conditions is shown by the drawn curve, w, in Fig. 104. As seen, its average is still negative or energy-consuming, that is, the oscillation still dies out, and stability is finally reached, but the average value of w in this case is so much less than in the case above discussed, that the dying out of the oscillation is much slower.

If now, the damping power, w_1, were still smaller, or the average synchronizing power, w_2, greater, the average w would become positive or supplying energy to the oscillating system. In other words, the oscillation would increase and hunting result.

That is:

If the average synchronizing power resulting from the lag of the synchronizing force behind the position exceeds the average damping power, hunting results. The condition of stability of the synchronous machine is, that the average damping power exceeds the average synchronizing power, and the more this is the case, the more stable is the machine, that is, the more rapidly the transient oscillation of readjustment to changed circuit conditions dies out.

Or, if

a = attenuation constant of the oscillating system,

$a < 0$ gives cumulative oscillation or hunting.

$a > 0$ gives stability.

108. Counting the time, t, from the moment of maximum backward position of the rotor, that is, the moment at which the load on the machine is decreased, and assuming sinusoidal variation, and denoting

$$\phi = 2\pi ft = \omega t \tag{1}$$

where

$$f = \text{frequency of the oscillation} \tag{2}$$

the relative position of the rotor then may be represented by

$$p = -p_0 \epsilon^{a\phi} \cos \phi,$$

where

$p_0 = p_2 - p_1$ = position difference of rotor resulting from change of load, $\tag{3}$

$$a = \text{attenuation constant of oscillation.} \tag{4}$$

The velocity difference from that of uniform rotation then is

$$v = \frac{dp}{dt} = \omega \frac{dp}{d\phi} = \omega p_0 \epsilon^{-a\phi} (\sin \phi + a \cos \phi). \tag{5}$$

Let

$$a = \tan \alpha; \quad 1 + a^2 = A^2 \tag{6}$$

hence,

$$\sin \alpha = \frac{a}{A}; \quad \cos \alpha = \frac{1}{A} \tag{7}$$

it is

$$v = \omega p_0 A \epsilon^{-a\phi} \sin (\phi + \alpha). \tag{8}$$

Let

γ = lag of damping currents behind e.m.f. induced in damper windings $\tag{9}$

the damping power is

$$w_1 = -cvv_\gamma$$
$$= -c\omega^2 p_0^2 A^2 \epsilon^{-2a\phi} \sin (\phi + \alpha) \sin (\phi + \alpha - \gamma) \tag{10}$$

where

$c = \dfrac{w}{v^2}$ = damping power per unit velocity and v_γ is v,

lagged by angle γ. $\tag{11}$

Let

β = lag of synchronizing force behind position displacement p (12)

and

$$\beta = \omega t_0 \qquad (13)$$

where

$$t_0 = \text{time lag of synchronizing force.} \qquad (14)$$

The synchronizing force then is

$$F = b p_0 \epsilon^{-a\phi} \cos (\phi - \beta) \qquad (15)$$

where

$$b = \frac{F_0}{p_0} = \text{ratio of synchronizing force to po-}$$

sition displacement, or specific synchronizing force. (16)

The synchronizing power then is

$$w_2 = Fv = b\omega p_0 A \epsilon^{-2a\phi} \sin (\phi + \alpha) \cos (\phi - \beta). \qquad (17)$$

The oscillating mechanical power is

$$w = \frac{d}{dt} \frac{mv^2}{e} = m\omega v \frac{dv}{d\phi}$$

$$= m\omega\beta p_0^2 A^2 \epsilon^{-2\,a\phi} \sin (\phi + \alpha) \\ \{\cos (\phi + \alpha) - \alpha \sin (\phi + \alpha)\} \qquad (18)$$

where

$$m = \text{moving mass reduced to the radius, on which } p \text{ is measured.} \qquad (19)$$

It is, however,

$$w_1 + w_2 - w = 0 \qquad (20)$$

hence, substituting (10), (17), (18) into (20) and canceling,

$$b \cos (\phi - \beta) - c\omega A \sin (\phi + \alpha - \gamma) - $$

$$m\omega^2 A \cos (\phi + \alpha) + m\omega^2 A a \sin (\phi + a) = 0. \qquad (21)$$

This gives, as the coefficients of $\cos \phi$ and $\sin \phi$ the equations

$$\left. \begin{array}{l} b \cos \beta - c\omega A \sin (a - \gamma) - ma^2 A \cos \alpha + m\omega^2 A a \sin \alpha = 0 \\ b \sin \beta - c\omega A \cos (a - \gamma) + m\omega^2 A \sin a + m\omega^2 A \cos \alpha = 0 \end{array} \right\} \quad (22)$$

Substituting (6) and (7) and approximating from (13), for β as a small quantity,

$$\cos \beta = 1; \quad \sin \beta = \omega t_0 \qquad (23)$$

gives

$$\left. \begin{array}{l} b - c\omega \, (\, a \cos \gamma - \sin \gamma) - m\omega^2 \, (1 - a^2) = 0 \\ b t_0 - c \, (\cos \gamma + a \sin \gamma) + 2m\omega a = 0 \end{array} \right\} \quad (24)$$

This gives the values, neglecting smaller quantities

$$a = \frac{c \cos \gamma - bt_0}{\sqrt{4\,mb - c^2 \cos^2 \gamma + b^2 t_0^2}} \tag{25}$$

$$\omega = \frac{1}{2m}\left\{\sqrt{4\,mb - c^2 \cos 2\gamma + b^2 t_0^2} + c \sin \gamma\right\} \tag{26}$$

$$f = \frac{\omega}{2\,\phi} \tag{27}$$

These equations (25) and (26) apply only for small values of a, but become inaccurate for larger values of a, that is, very rapid damping. However, the latter case is of lesser importance.

$$a = 0$$

gives

$$bt_0 = c \cos \gamma,$$

hence,

$$c > \frac{bt_0}{\cos \gamma} \Bigg\}$$

or,

$$\left. t_0 < \frac{c \cos \gamma}{b} \right\} \tag{28}$$

are the conditions of stability of the synchronous machine.

If

$$t_0 = 0$$

$$\gamma = 0$$

it is

$$a = \frac{c}{\sqrt{4\,mb - c^2}},$$

$$\omega = \frac{\sqrt{4\,mb - c^2}}{2\,m},$$

and, if also,

$$c = 0:$$

it is

$$\omega = \sqrt{\frac{b}{m}}.$$

CHAPTER XII

REACTANCE OF INDUCTION APPARATUS

109. An electric current passing through a conductor is accompanied by a magnetic field surrounding this conductor, and this magnetic field is as integral a part of the phenomenon, as is the energy dissipation by the resistance of the conductor. It is represented by the inductance, L, of the conductor, or the number of magnetic interlinkages with unit current in the conductor. Every circuit thus has a resistance, and an inductance, however small the latter may be in the so-called "non-inductive" circuit. With continuous current in stationary conditions, the inductance, L, has no effect on the energy flow; with alternating current of frequency, f, the inductance, L, consumes a voltage $2\pi fLi$, and is, therefore, represented by the reactance, $x = 2\pi fL$, which is measured in ohms, and differs from the ohmic resistance, r, merely by being wattless or reactive, that is, representing not dissipation of energy, but surging of energy.

Every alternating-current circuit thus has a resistance and a reactance, the latter representing the effect of the magnetic field of the current in the conductor.

When dealing with alternating-current apparatus, especially those having several circuits, it must be realized, however, that the magnetic field of the circuit may have no independent existence, but may merge into and combine with other magnetic fields, so that it may become difficult what part of the magnetic field is to be assigned to each electric circuit, and circuits may exist which apparently have no reactance. In short, in such cases, the magnetic fields of the reactance of the electric circuit may be merely a more or less fictitious component of the resultant magnetic field.

The industrial importance hereof is that many phenomena, such as the loss of power by magnetic hysteresis, the m.m.f. required for field excitation, etc., are related to the resultant magnetic field, thus not equal to the sum of the corresponding effects of the components.

As the transformer is the simplest alternating-current apparatus, the relations are best shown thereon.

Leakage Flux of Alternating-current Transformer

110. The alternating-current transformer consists of a magnetic circuit, interlinked with two electric circuits, the primary
circuit, which receives power from its impressed voltage, and
the secondary circuit, which supplies power to its external circuit.

For convenience, we may assume the secondary circuit as reduced to the primary circuit by the ratio of turns, that is, assume
ratio of turns $1 \div 1$.

Let

$Y_0 = g - jb =$ primary exciting admittance;

$Z_0 = r_0 + jx_0 =$ primary self-inductive impedance;

$Z_1 = r_1 + jx_1 =$ secondary self-inductive impedance (reduced
to the primary).

The transformer thus comprises three magnetic fluxes: the
mutual magnetic flux, Φ, which, being interlinked with primary
and secondary, transforms the power from primary to secondary,
and is due to the resultant m.m.f of primary and secondary circuit; the primary leakage flux, Φ'_0, due to the m.m.f. of the primary
circuit, F_0, and interlinked with the primary circuit only, which is
represented by the self-inductive or leakage reactance, x_0; and the
secondary leakage flux, Φ'_1, due to the m.m.f. of the secondary
circuit, F_1, and interlinked with the secondary circuit only
which is represented by the secondary reactance, x_1.

As seen in Fig. 105o, the mutual flux, Φ—usually—has a closed
iron circuit of low reluctance, ρ, thus low m.m.f., F, and high intensity; the self-inductive flux or leakage reactance flux, Φ'_0 and Φ'_1,
close through the air circuit between the primary and secondary
electric circuits, thus meet with a high reluctance, ρ_0, respectively
ρ_1, usually many hundred times higher than ρ. Their m.m.fs., F_0
and F_1, however, are usually many times greater than F; the latter is the m.m.f. of the exciting current, the former that of full
primary or secondary current.

For instance, if the exciting current is 5 per cent. of full-load
current, the reactance of the transformer 4 per cent., or 2 per cent.
primary and 2 per cent. secondary, then the m.m.f. of the leakage
flux is 20 times that of the mutual flux, and the mutual flux 50
times the leakage flux, hence the reluctance of leakage flux 50
$\times 20 = 1000$ times that of the mutual or main flux: $\rho_1 = 1000\,\rho$.

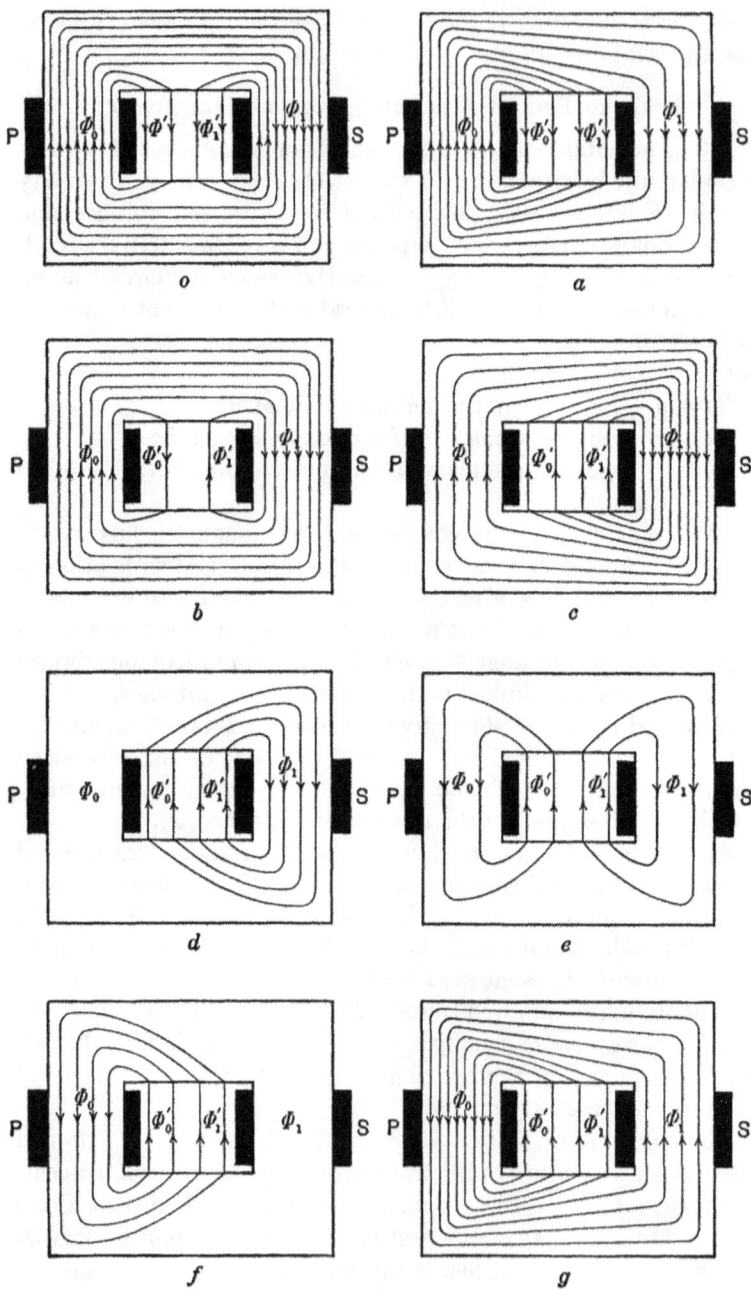

Fig. 105.

111. Usually, as stated, the leakage fluxes are not considered as such, but represented by their reactances, in the transformer diagram. Thus, at non-inductive load, it is, Fig. 106,

$O\Phi$ = mutual, or main magnetic flux, chosen as negative vertical.

OF = m.m.f. required to produce flux, $O\Phi$, and leading it by the angle of hysteretic advance of phase, $FO\Phi$.

OE'_1 = e.m.f. induced in the secondary circuit by the mutual flux, and 90° behind it.

TRANSFORMER DIAGRAM
NON-INDUCTIVE LOAD
SHOWING MAGNETIC FLUXES

$\rho = .32$; $\rho_0 = \rho_1 = 4$
$\Phi = 6.2$; $F = 6$
HENCE
$F = 2$; $\Phi' = 1.5$; $\Phi_1 = 6$
$F_0 = 7.6$; $\Phi'_0 = 1.9$; $\Phi_0 = 7.6$

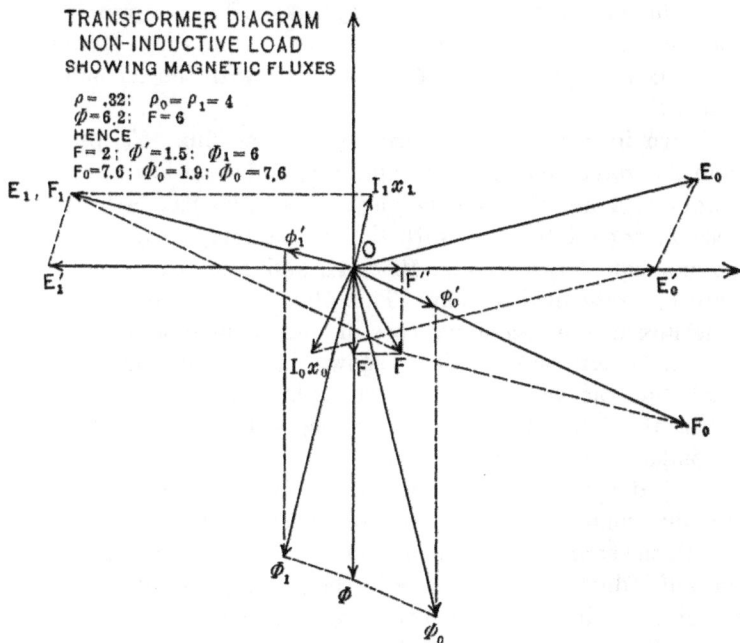

Fig. 106.

$I_1 x_1$ = secondary reactance voltage, 90° behind the secondary current, and combining with OE'_1 to

OE_1 = true secondary induced voltage. From this subtracts the secondary resistance voltage, $I_1 r_1$, leaving the secondary terminal voltage, and, in phase with it at non-inductive load, the secondary current and secondary m.m.f., OF_1.

From component, OF_1, and resultant, OF, follows the other component,

OF_0 = primary m.m.f. and in phase with it the primary current.

OE'_0 = primary voltage consumed by mutual flux, equal and opposite to OE'_1.

$I_0 x_0$ = primary reactance voltage, 90° ahead of the primary current OF_0.

From $I_0 x_0$ as component and E'_0 as resultant follows the other component, $\overline{OE_0}$, and adding thereto the primary resistance voltage, $I_0 r_0$, gives primary supply voltage.

In this diagram, Fig. 106, the primary leakage flux is represented by $O\Phi'_0$, in phase with the primary current, $\overline{OF_0}$, and the secondary leakage flux is represented by $\overline{O\Phi'_1}$, in phase with the secondary current, OF_1.

As shown in Fig. 105o, the primary leakage flux, Φ'_0, passes through the iron core inside of the primary coil, together with the resultant flux, Φ, and the secondary leakage flux, Φ'_1, passes through the secondary core, together with the mutual flux, Φ. However, at the moment shown in Fig. 105o, Φ'_1 and Φ in the secondary core are opposite in direction. This obviously is not possible, and the flux in the secondary core in this moment is $\Phi - \Phi'_1$, that is, the magnetic disposition shown in Fig. 105o is merely nominal, but the actual magnetic distribution is as shown in Fig. 105a; the flux in the primary core, $\Phi_0 = \Phi + \Phi'_0$, the flux in the secondary core, $\Phi_1 = \Phi - \Phi'_1$.

As seen, at the moment shown in Fig. 105o and 105a, all the leakage flux comes from and interlinks with the primary winding, none with the secondary winding, and it thus would appear, that all the self-inductive reactance is in the primary circuit, none in the secondary circuit, or, in other words, that the secondary circuit of the transformer has no reactance.

However, at a later moment of the cycle, shown in Fig. 105c, all the leakage flux comes from and interlinks with the secondary, and this figure thus would give the impression, that all the leakage reactance of the transformer is in the secondary, none in the primary winding.

In other words, the leakage fluxes of the transformer and the mutual or main flux are not independent fluxes, but partly traverse the same magnetic circuit, so that each of them during a part of the cycle is a part of any other of the fluxes. Thus, the reactance voltage and the mutual inductive voltage of the transformer are not separate e.m.fs., but merely mathematical fictions, components of the resultant induced voltage, OE_1 and OE_0, induced by the resultant fluxes, $\overline{O\Phi_0}$ in the primary, and $\overline{O\Phi_1}$ in the secondary core.

112. In Fig. 107 are plotted, in rectangular coördinates, the magnetic fluxes:

The mutual or main magnetic flux, Φ;

The primary leakage flux, Φ'_0;

The resultant primary flux, $\Phi_0 = \Phi + \Phi'_0$;

The secondary leakage flux, Φ'_1;

The resultant secondary flux, $\Phi_1 = \Phi - \Phi'_1$;

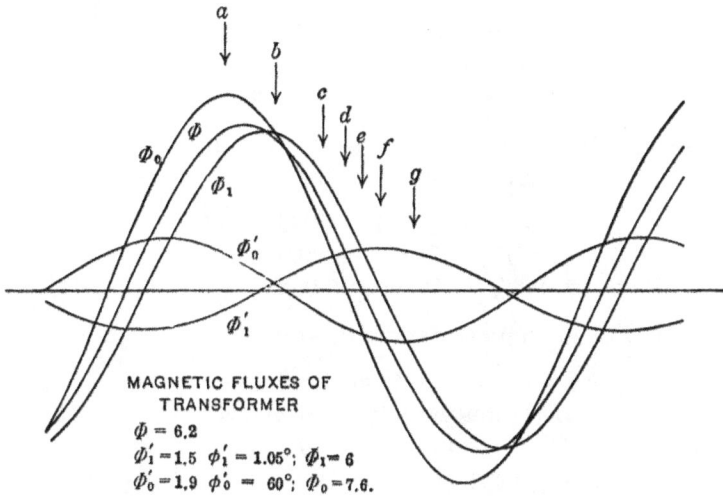

MAGNETIC FLUXES OF
TRANSFORMER

$\Phi = 6.2$

$\Phi'_1 = 1.5$ $\phi'_1 = 1.05°$; $\Phi_1 = 6$

$\Phi'_0 = 1.9$ $\phi'_0 = 60°$; $\Phi_0 = 7.6$.

Fig. 107.

and the magnetic distribution in the transformer, during the moments marked as a, b, c, d, e, f, g, in Fig. 107, is shown in Fig. 105.

In Fig. 105a, the primary flux is larger than the secondary, and all leakage fluxes (x_0 and x_1) come from the primary flux, that is, there is no secondary leakage flux.

In Fig. 105b, primary and secondary flux equal, and primary and secondary leakage flux equal and opposite, though small.

In Fig. 105c, the secondary flux is larger, all leakage flux (x_0 and x_1) comes from the secondary flux, that is, there is no primary leakage flux.

In Fig. 105*d*, there is no primary flux, and all the secondary flux is leakage flux.

In Fig. 105*e*, there is no mutual flux, all primary flux is primary leakage flux, and all secondary flux is secondary leakage flux.

In Fig. 105*f*, there is no secondary flux, and all primary flux is leakage flux.

In Fig. 105*g*, the primary flux is larger than the secondary, and all leakage flux comes from the primary, the same as in 105*a*.

Figs. 105*a* to 105*f*, thus show the complete cycle, corresponding to diagrams, Figs. 106 and 107.

These figures are drawn with the proportions,

$$\rho \div \rho_0 \div \rho_1 = 1 \div 12.5 \div 12.5$$
$$F \div F_0 \div F_1 = 1 \div 3.8 \div 3$$
$$\Phi \div \Phi'_0 \div \Phi'_1 = 1 \div 0.317 \div 0.25.$$

thus are greatly exaggerated, to show the effect more plainly. Actually, the relations are usually of the magnitude,

$$\rho \div \rho_0 \div \rho_1 = 1 \div 1000 \div 1000$$
$$F \div F_0 \div F_1 = 1 \div 20.6 \div 20$$
$$\Phi \div \Phi'_0 \div \Phi'_1 = 1 \div 0.02 \div 0.02$$

113. In symbolic representation, denoting,

Φ = mutual magnetic flux.

E = mutual induced voltage.

Φ_0 = resultant primary flux.

Φ'_0 = primary leakage flux.

E_0 = primary terminal voltage.

I_0 = primary current.

$Z_0 = r_0 + jx_0$ = primary self-inductive impedance.

Φ_1 = resultant secondary flux.

Φ'_1 = secondary leakage flux.

E_1 = secondary terminal voltage.

I_1 = secondary current.

$Z_1 = r_1 + jx_1$ = secondary self-inductive impedance.

and

$$c = 2\pi f n$$

where n = number of turns.

It then is

$$c\Phi'_1 = jx_0 I_0$$
$$c\Phi'_1 = jx_1 I_1$$
$$c\Phi = E = E_0 - Z_0 I_0 = E_1 + Z_1 I_1$$
$$c\Phi_0 = E_0 - r_0 I_0 = E + jx_0 I_0$$
$$c\Phi_1 = E_1 + r_1 I_1 = E - jx_1 I_1$$
$$\Phi'_0 = \Phi_0 - \Phi$$
$$\Phi'_1 = \Phi + \Phi_1,$$

thus, the total leakage flux

$$\Phi' = \Phi'_0 + \Phi'_1 = \Phi_0 - \Phi_1.$$

114. One of the important conclusions from the study of the actual flux distribution of the transformer is that the distinction between primary and secondary leakage flux, Φ'_0 and Φ'_1, is really an arbitrary one. There is no distinct primary and secondary leakage flux, but merely one leakage flux, Φ', which is the flux passing between primary and secondary circuit, and which during a part of the cycle interlinks with the primary, during another part of the cycle interlinks with the secondary circuit Thus the corresponding electrical quantities, the reactances, x_0 and x_1, are not independent quantities, that is, it can not be stated that there is a definite primary reactance, x_0, and a definite secondary react-ance, x_1, but merely that the transformer has a definite reactance, x, which is more or less arbitrarily divided into two parts; $x = x_0 + x_1$, and the one assigned to the primary, the other to the second-ary circuit.

As the result hereof, "mutual magnetic flux" Φ, and the mutual induced voltage, E, are not actual quantities, but rather mathe-matical fictions, and not definite but dependent upon the distri-bution of the total reactance between the primary and the sec-ondary circuit.

This explains why all methods of determining the transformer reactance give the total reactance $x_0 + x_1$.

However, the subdivision of the total transformer reactance into a primary and a secondary reactance is not entirely arbitrary. Assuming we assign all the reactance to the primary, and consider the secondary as having no reactance. Then the mutual mag-netic flux and mutual induced voltage would be

$$c\Phi = E = E_0 - [r_0 + j(x_0 + x_1)] I_0$$

and the hysteresis loss in the transformer would correspond hereto, by the usual assumption in transformer calculations.

Assigning, however, all the reactance to the secondary circuit, and assuming the primary as non-inductive, the mutual flux and mutual induced voltage would be $c\Phi = E = E_0 - r_0 I_0$, hence larger, and the hysteresis loss calculated therefrom larger than under the previous assumption. The first assumption would give too low, and the last too high a calculated hysteresis loss, in most cases.

By the usual transformer theory, the hysteresis loss under load is calculated as that corresponding to the mutual induced voltage, E. The proper subdivision of the total transformer reactance, x, into primary reactance, x_0, and secondary reactance, x_1, would then be that, which gives for a uniform magnetic flux, Φ, corresponding to the mutual induced voltage, E, the same hysteresis loss, as exists with the actual magnetic distribution of $\Phi_0 = \Phi + \Phi'_0$ in the primary, and $\Phi_1 = \Phi - \Phi'_1$ in the secondary core. Thus, if V_0 is the volume of iron carrying the primary flux, Φ_0, at flux density, B_0, V_1 the volume of iron carrying the secondary flux, Φ_1, at flux density, B_1, the flux density of the theoretical mutual magnetic flux would be given by

$$B^{1.6} = \frac{V_0 B_0^{1.6} + V_1 B_1^{1.6}}{V_0 + V_1}$$

from B then follows Φ, E, and thus x_0 and x_1.

This does not include consideration of eddy-current losses. For these, an approximate allowance may be made by using 1.7 as exponent, instead of 1.6.

Where the magnetic stray field under load causes additional losses by eddy currents, these are not included in the loss assigned to the mutual magnetic flux, but appear as an energy component of the leakage reactances, that is, as an increase of the ohmic resistances of the electric circuits, by an effective resistance.

115. Usually, the subdivision of x into x_0 and x_1, by this assumption of assigning the entire core loss to the mutual flux, is sufficiently close to equality, to permit this assumption. That is, the total transformer reactance is equally divided between primary and secondary circuit.

This, however, is not always justified, and in some cases, the one circuit may have a higher reactance than the other. Such, for instance, is the case in some very high voltage transformers, and usually is the case in induction motors and similar apparatus.

It is more commonly the case, where true self-inductive fluxes exist, that is, magnetic fluxes produced by the current in one circuit, and interlinked with this circuit, closing upon themselves in a path which is entirely distinct from that of the mutual magnetic flux, that is, has no part in common with it. Such, for instance, frequently is the self-inductive flux of the end connections of coils in motors, transformers, etc. To illustrate: in the high-voltage shell-type transformer, shown diagrammatically in Fig. 108, with primary coil 1, closely adjacent to the core, and high-voltage secondary coil 2 at considerable distance:

The primary leakage flux consists of the flux in spaces, a, between the yokes of the transformer, closing through the iron core, C, and the flux through the spaces, b, outside of the transformer, which enters the faces, F, of the yokes and closes through the central core, C.

The secondary leakage flux contains the same two components: the flux through the spaces, a, between the yokes closing, however, through the outside shells, S, and the flux through the spaces, b, outside of the transformer, and entering the faces, F, but in this case closing through the shells, S. In addition to these two components, the secondary leakage flux contains a third component, passing through the spaces, b, between the coils, but closing, through outside space, c, in a complete air circuit. This flux has no corresponding component in the primary, and the total secondary leakage reactance in this case thus is larger than the total primary reactance.

Similar conditions apply to magnetic structures as in the induction motor, alternator, etc.

In such a case as represented by Fig. 108, the total reactance of the transformer, with (2) as primary and (1) as secondary, would be greater than with (1) as primary and (2) as secondary.

In this case, when subdividing the total reactance into primary reactance and secondary reactance, it would appear legitimate to divide it in proportion of the total reactances with (1) and (2) as primary, respectively. That is,

if x = total reactance, with coil (1) as primary, and (2) as secondary, and

x' = total reactance, with coil (2) as primary, and (1) as secondary, then it is:

With coil (1) as primary and (2) as secondary,

15

Primary reactance,

$$x_0 = \frac{x}{x + x'}x = \frac{x^2}{x + x'}.$$

Secondary reactance,

$$x_1 = \frac{x'}{x + x'}x = \frac{xx'}{x + x'}.$$

With coil (2) as primary and (1) as secondary,

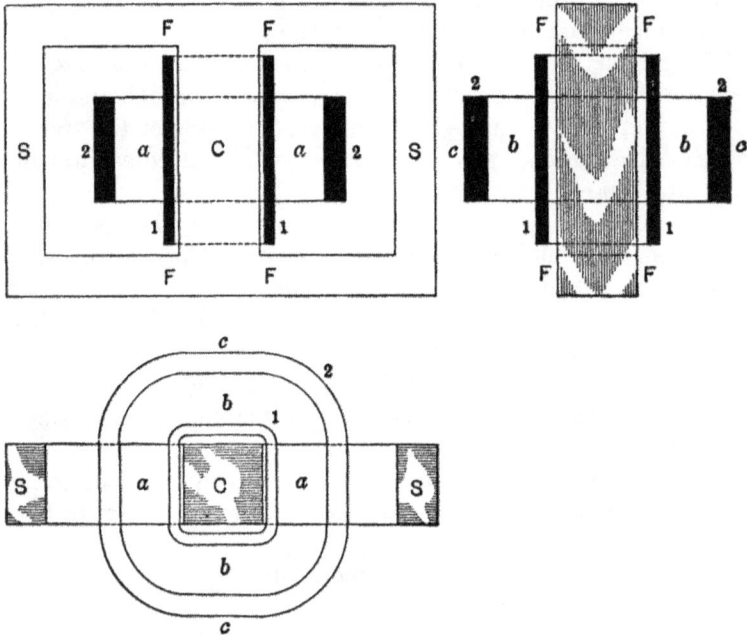

FIG. 108.

Primary reactance,

$$x_0 = \frac{x'}{x + x'}x' = \frac{x'^2}{x + x'}.$$

Secondary reactance,

$$x_1 = \frac{x}{x + x'}x' = \frac{xx'}{x + x'}.$$

116. By test, the two total reactances, x and x', can be derived by considering, that in Fig. 107 at the moments, f and d, the total flux is leakage flux, as more fully shown in Fig. 105f and 105d, and the flux measured from f, gives the reactance, x, measured from d, gives the reactance, d.

Assuming we connect primary coil and secondary coil in series with each other, but in opposition, into an alternating-current circuit, as shown in Fig. 109, and vary the number of primary and secondary turns, until the voltage, e_1, across the secondary coil, s, becomes equal to r_1i. Then no flux passes through the secondary coil, that is, the condition, Fig. 107f, exists, and the voltage, e_0, across the primary coil, p, gives the total reactance, x, for p as primary,

$$e_0{}^2 = i^2 (r_0{}^2 + x^2).$$

Varying now the number of turns so that the voltage across the primary coil equals its resistance drop, $e_0 = r_0i$, then the

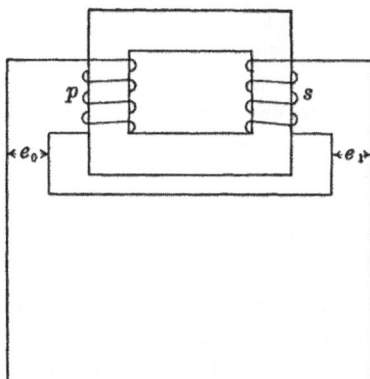

FIG. 109.

voltage across the secondary coil, s, gives the total reactance, x', for s as primary,

$$e_1{}^2 = i^2 (r_1{}^2 + x'^2).$$

It would rarely be possible to vary the turns of the two coils, p and s. However, if we short-circuit s and pass an alternating current through p, then at the very low resultant magnetic flux and thus resultant m.m.f., primary and secondary current are practically in opposition and of the same m.m.f., and the magnetic flux in the secondary coil is that giving the resistance drop r_1i_1, that is, $e'_1 = r_1 i_1$ is the true primary voltage in the secondary, and the voltage across the primary terminals thus is that giving primary resistance drop, r_0i_0, total self-inductive reactance, xi_0, and the secondary induced voltage, r_1i_1. Thus,

$$e_0{}^2 = (r_0i_0 + r_1i_1)^2 + x^2i_0{}^2,$$

or, since i_1 practically equals i_0,

$$e_0{}^2 = i_0{}^2 [(r_0 + r_1)^2 + x^2],$$

and inversely, impressing a voltage upon coil, s, and short-circuiting the coil p, gives the leakage reactance, x', for s as primary,

$$e_1{}^2 = i_1{}^2 [(r_0 + r_1)^2 + x'{}^{\cdot})].$$

Thus, the so-called "impedance test" of the transformer gives the total leakage reactance $x_0 + x_1$, for that coil as primary, which is used as such in the impedance test.

Where an appreciable difference of the total leakage flux is expected when using the one coil as primary, as when using the other coil, the impedance tests should be made with that coil as primary, which is intended as such. Since, however, the two leakage fluxes are usually approximately equal, it is immaterial which coil is used as primary in the impedance test, and generally that coil is used, which gives a more convenient voltage and current for testing.

Magnetic Circuits of Induction Motor

117. In general, when dealing with a closed secondary winding, as an induction-motor squirrel-cage, we consider as the mutual inductive voltage, E, the voltage induced by the mutual magnetic flux, Φ, that is, the magnetic flux due to the resultant of the primary and the secondary m.m.f. This voltage, E, then is consumed in the closed secondary winding by the resistance, $r_1 I_1$, and the reactance, $jx_1 I_1$, thus giving, $E = (r_1 + jx_1) I_1$.

The reactance voltage, $jx_1 I_1$, is consumed by a self-inductive flux, Φ_1', that is, a magnetic leakage flux produced by the secondary current and interlinked with the secondary circuit, and the actual or resultant magnetic flux interlinked with the secondary circuit, that is, the magnetic flux, which passes beyond the secondary conductor through the armature core, thus is the vector difference, $\Phi_1 = \Phi - {}_1\Phi'$, and the actual voltage induced in the secondary circuit by the resultant magnetic flux interlinked with it thus is, $E_1 = E - jx_1 I_1$. This voltage is consumed by the resistance of the secondary circuit, $E_1 = r_1 I_1$, and the voltage consumed by self-induction, $jx_1 I_1$, is no part of E_1, but as stated, is due to the self-inductive flux, Φ_1', which vectorially subtracts from the mutual magnetic flux, Φ, and thereby leaves the flux, Φ_1', which induces E_1.

In other words:

In any closed secondary circuit, as a squirrel-cage of an induction motor, the true induced e.m.f. in the circuit, that is, the e.m.f. induced by the actual magnetic flux interlinked with the circuit, is the resistance drop of the circuit, $E_1 = r_1 I_1$.

This is true whether there is one or any number of closed secondary circuits—or squirrel-cages in an induction motor. In each the current, I_1 is $\dfrac{E_1}{r_1}$, where r_1 is the resistance of the circuit, and E_1 the voltage induced by the flux which passes through the circuit. The E_1 of the different squirrel-cages then would differ from each other by the voltage induced by the leakage flux which passes between them, and which is represented by the self-inductive reactance of the next squirrel-cage:

$$E'_1 = E_1 - jx'_1 I'_1$$

where $I'_1 = \dfrac{E'_1}{r'_1}$ is the current in the inner squirrel-cage of voltage, E'_1, and resistance, r'_1, and $x'_1 I'_1$, is the reactance of the flux between the two squirrel-cages.

The mutual magnetic flux and the mutual induced e.m.f. of the common induction motor theory thus are mathematical fictions and not physical realities.

The advantage of the introduction of the mutual magnetic flux, Φ, and the mutual induced voltage, E, in the induction-motor theory, is the ease and convenience of passing therefrom to the secondary as well as the primary circuit. Where, however, a number of secondary circuits exist, as in a multiple squirrel-cage, it is preferable to start from the innermost magnetic flux, that is, the magnetic flux passing through the innermost squirrel-cage, and the voltage induced by it in the latter, which is the resistance drop of this squirrel-cage.

In the same manner, in a primary circuit, the actual or total magnetic flux interlinked with the circuit, Φ_0, is that due to the impressed voltage, E_0, minus the resistance drop, $r_0 I_0$, $E'_0 = E_0 - r_0 I_0$. Of this magnetic flux, Φ_0, a part, Φ'_0, passes as primary leakage flux between primary and secondary, without reaching the secondary, and is represented by the primary reactance voltage, $jx_0 I_0$, and the remainder—usually the major part—is impressed upon the secondary circuit as mutual magnetic flux, $\Phi = \Phi_0 - \Phi'_0$, corresponding to the mutual inductive voltage, $E = E'_0 - jx_0 I_0$. The mutual magnetic flux, Φ, then is impressed upon the secondary, and as stated above, a part of it, the secondary leakage flux, Φ'_1, is shunted across outside of the secondary circuit, the remainder, $\Phi' = \Phi - \Phi'_1$, passes through the secondary circuit and corresponds to $r_1 I_1$.

118. Applying this to the polyphase induction motor with single squirrel-cage secondary. Let

$$Y_0 = g - jb = \text{primary exciting admittance;}$$
$$Z_0 = r_0 + jx_0 = \text{primary self-inductive impedance;}$$
$$Z_1 = r_1 + jx_1 = \text{secondary self-inductive impedance}$$
at full frequency, reduced to the primary.

Let

$$E_1 = \text{the true induced voltage in the secondary, at full frequency, corresponding to the magnetic flux in the armature core.}$$

The secondary current then is

$$I_1 = \frac{sE_1}{r_1}.$$

The mutual inductive voltage at full frequency,

$$E = E_1 + jx_1 I_1$$
$$= E_1 \left(1 + j\frac{sx_1}{r_1}\right).$$

Thus the exciting current,

$$I_{00} = Y_0 E$$
$$= (g - jb)\left(1 + j\frac{sx_1}{r_1}\right) E_1$$
$$= (q_1 - jq_2) E_1,$$

where

$$q_1 = g + \frac{sbx_1}{r_1}$$

$$q_2 = b - \frac{sgx_1}{r_1},$$

and the total current,

$$I_0 = I_1 + I_{00}$$
$$= E_1 \left\{\frac{s}{r_1} + q_1 - jq_2\right\},$$

hence, the primary impressed voltage,

$$E_0 = E + Z_0 I_0$$
$$= E_1 \left\{1 + j\frac{sx_1}{r_1} + (r_0 + jx_0)\left[\frac{s}{r_1} + q_1 - jq_2\right]\right\},$$
$$= E_1 (c_1 + jc_2),$$

where

$$c_1 = 1 + r_0\left(\frac{s}{r_1} + q_1\right) + x_0 q_2 = 1 + s\frac{r_0}{r_1} + r_0 q_1 + x_0 q_2$$

$$c_2 = \frac{sx_1}{r_1} + x_0\left(\frac{s}{r_1} + q_1\right) - r_0 q_2 = \frac{s(x_1 + x_0)}{r_1} + x_0 q_1 - r_0 q_2,$$

choosing now the impressed voltage as zero vector,

$$\underline{E}_0 = e_0$$

gives

$$\underline{E}_1 = \frac{e_0}{c_1 + jc_2},$$

or, absolute,

$$e_1 = \frac{e_0}{\sqrt{c_1{}^2 + c_2{}^2}}$$

the torque of the motor is

$$D = /\underline{E}_1, \underline{I}_1/ \,^1$$

$$= \frac{se_1{}^2}{r_1} = \frac{se_0{}^2}{r_1(c_1{}^2 + c_2{}^2)},$$

the power,

$$P = \frac{se_1{}^2(1 - s)}{r_1} = \frac{s(1 - s)e_0{}^2}{r_1(c_1{}^2 + c_2{}^2)}$$

the volt-ampere input,

$$Q = e_0 i_0$$

etc.

As seen, this method is if anything, rather less convenient than the conventional method, which starts with the mutual inductive voltage $\underline{E}$.

It becomes materially more advantageous, however, when dealing with double and triple squirrel-cage structures, as it permits starting with the innermost squirrel-cage, and gradually building up toward the primary circuit. See "Multiple Squirrel-cage Induction Motor," "Theory and Calculation of Electrical Apparatus."

CHAPTER XIII

REACTANCE OF SYNCHRONOUS MACHINES

119. The synchronous machine—alternating-current generator, synchronous motor or synchronous condenser—consists of an armature containing one or more electric circuits traversed by alternating currents and synchronously revolving relative to a unidirectional magnetic field, excited by direct current. The armature circuit, like every electric circuit, has a resistance, r, in which power is being dissipated by the current, I, and an inductance, L, or reactance, $x = 2\pi fL$, which represents the magnetic flux produced by the current in the armature circuit, and interlinked with this circuit. Thus, if E_0 = voltage induced in the armature circuit by its rotation through the magnetic field—or, as now more usually the case, the rotation of the magnetic field through the armature circuit—the terminal voltage of the armature circuit is

$$E = E_0 - (r + jx)\, I.$$

In Fig. 110 is shown diagrammatically the path of the field flux, in two different positions, A with an armature slot standing midway between two field poles, B with an armature slot standing opposite the field pole.

In Fig. 111 is shown diagrammatically the magnetic flux of armature reactance, that is, the magnetic flux produced by the current in the armature circuit, and interlinked with this circuit, which is represented by the reactance x, for the same two relative positions of field and armature.

As seen, field flux and armature flux pass through the same iron structures, thus can not have an independent existence, but actual is only their resultant. This resultant flux of armature self-induction and field excitation is shown in Fig. 112, for the same two positions, A and B, derived by superpositions of the fluxes in Figs. 110 and 111.

As seen, in Fig. 112A, all the lines of magnetic forces are interlinked with the field circuit, but there is no line of magnetic flux interlinked with the armature circuit only, that is, there is ap-

232

parently no self-inductive armature flux, and no true self-inductive reactance, x, and the self-inductive armature flux of Fig. 111
thus merely is a mathematical fiction, a theoretical component of
the resultant flux, Fig. 112. The effect of the armature current,

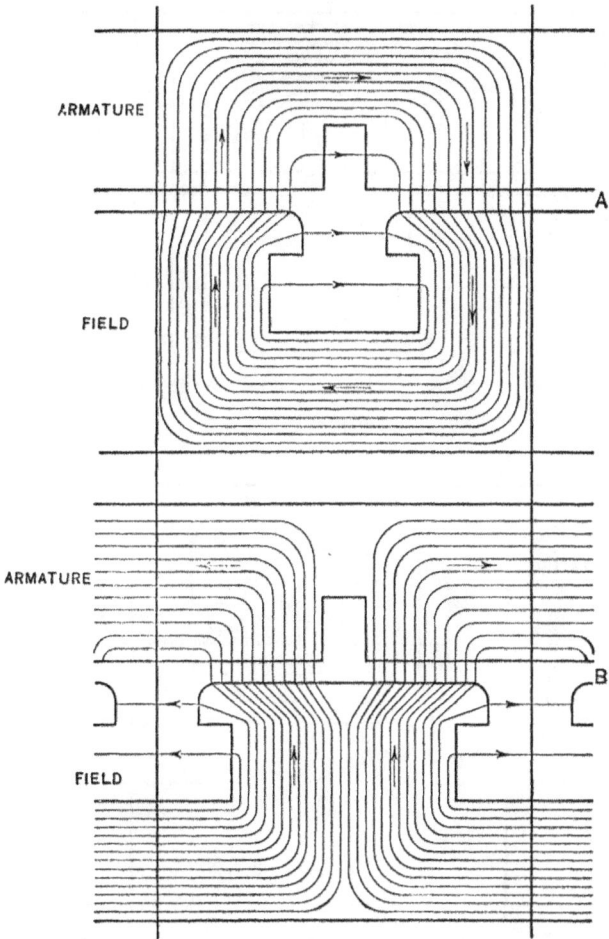

FIG. 110.

in changing flux distribution, Fig. 110A to Fig. 112A, consists in
reducing the field flux, that is, flux in the field core, increasing
the leakage flux of the field, that is, the flux which leaks from field
pole to field pole, without interlinking the armature circuit, and

still further decreasing the armature flux, that is, the flux issuing from the field and interlinking with the armature circuit.

In position 112*B*, there is no self-inductive armature flux either, but every line of force, which interlinks with the armature circuit,

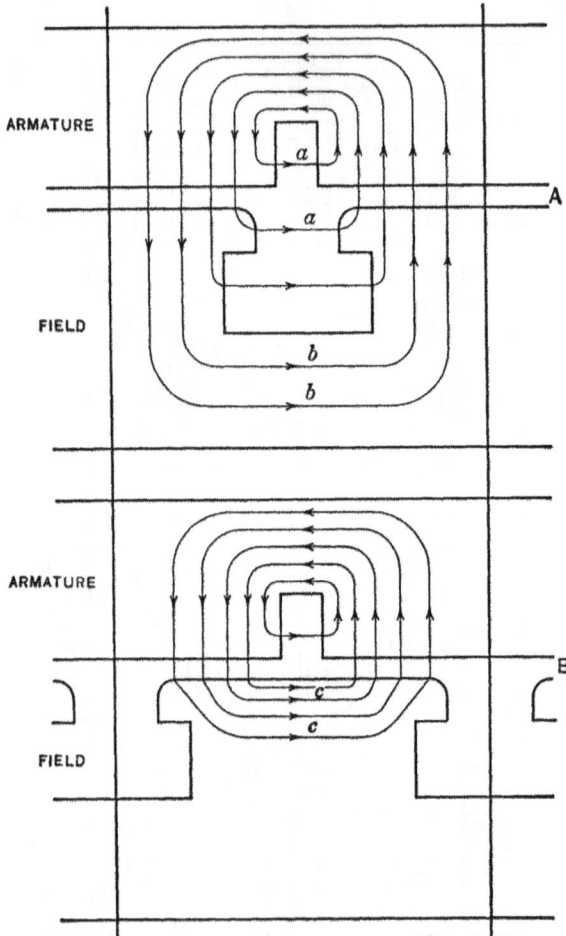

Fig. 111.

is produced by and interlinked with the field circuit. The effect of the armature current in this case is to increase the field flux and the flux entering the armature at one side of the pole, and decrease it on the other side of the pole, without changing the total field flux and the leakage flux of the field. Indirectly, a reduction of the field flux usually occurs, by magnetic saturation limiting the increase of flux at the strengthened pole corner; but this is a secondary effect.

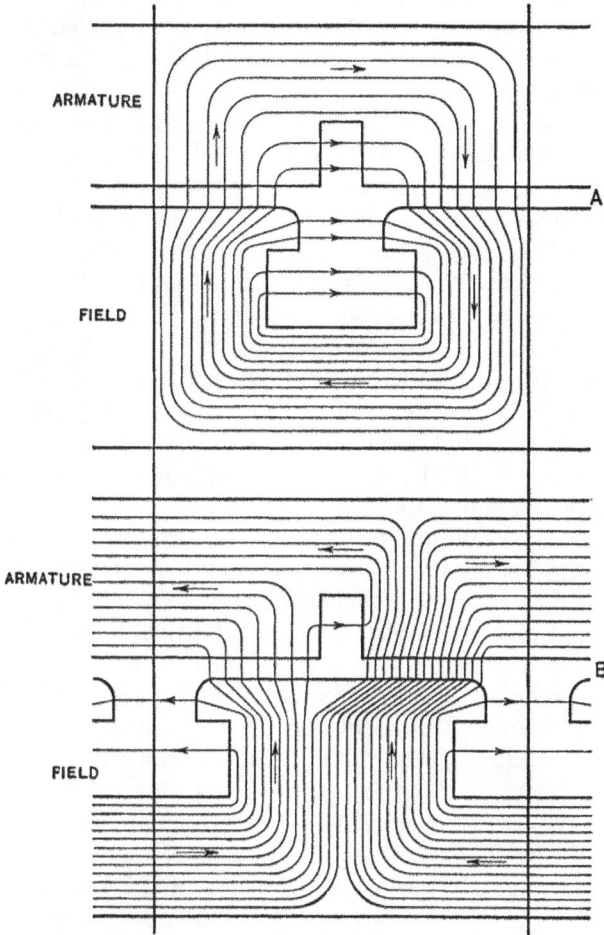

Fig. 112.

As seen, in 112*A* the armature current acts demagnetizing, in 112*B* distorting on the field flux, and in the intermediary position between *A* and *B*, a combination of demagnetization (or magnetization, in some positions) and distortion occurs.

Thus, it may be said that the armature reactance has no independent existence, is not due to a flux produced by and interlinked only with the armature circuit, but it is the electrical representation of the effect exerted on the field flux by the m.m.f. of the armature current.

Considering the magnetic disposition, an armature current, which alone would produce the flux, Fig. 111, in the presence of a field excitation which alone would give the flux, Fig. 110, has the following effect: in Fig. 112*A*, by the counter m.m.f. of the armature current the resultant m.m.f. and with it the resultant flux are reduced from that due to the m.m.f. of field excitation, to that due to field excitation minus the m.m.f. of the armature current. The difference of the magnetic potential between the field poles is increased: in Fig. 110*A* it is the sum of the m.m.fs. of the two air-gaps traversed by the flux (plus the m.m.f. consumed in the armature iron, which may be neglected as small); in Fig. 112*A* it is the sum of the m.m.fs. of the two air-gaps traversed by the flux (which is slightly smaller than in Fig. 110*A*, due to the reduced flux) plus the counter m.m.f. of the armature. The increased magnetic potential difference causes an increased magnetic leakage flux between the field poles, and thereby still further reduces the armature flux and the voltage induced by it.

In Fig. 112*B*, the m.m.f. of the armature current adds itself to the m.m.f. of field excitation on one side, and thereby increases the flux, and it subtracts on the other side and decreases the flux, and thereby causes an unsymmetrical flux distribution, that is, a field distortion.

120. Both representations of the effect of armature current are used, that by a nominal magnetic flux, Fig. 111, which gives rise to a nominal reactance, the "synchronous reactance of the armature circuit," and that by considering the direct magnetizing action of the armature current, as "armature reaction," and both have their advantages and disadvantages.

The introduction of a *synchronous reactance*, x_0, and corresponding thereto of a *nominal induced e.m.f.*, e_0, is most convenient in electrical calculations, but it must be kept in mind, that neither e_0 nor x_0 have any actual existence, correspond to actual magnetic fluxes, and for instance, when calculating efficiency and losses, the core loss of the machine does not correspond to e_0, but corresponds to the actual or resultant magnetic flux, Fig. 112. Also, in dealing with transients involving the dissipation of the magnetic energy stored in the machine, the magnetic energy of the resultant field, Fig. 112, comes into consideration, and not the—much larger—energy, which the fields corresponding to e_0 and x_0 would have. Thus the short-circuit transient of a heavily loaded machine is essentially the same as that of the same machine at no-load, with the same terminal voltage, although in the former the field excitation and the nominal induced voltage may be very much larger.

The use of the term *armature reaction* in dealing with the effect of load on the synchronous machine is usually more convenient and useful in design of the machine, but less so in the calculation dealing with the machine as part of an electric circuit.

Either has the disadvantage that its terms, synchronous reactance or armature reaction, are not homogeneous, as the different parts of the reactance field, Fig. 111, which make up the difference between Fig. 112 and Fig. 110, are very different in their action, especially in their behavior at sudden changes of circuit conditions.

121. Considering the magnetic flux of the armature current, Fig. 111A, which is represented by the synchronous reactance, x_0.

A part of this magnetic flux (lines a in Fig. 111A) interlinks with the armature circuit only, that is, is true self-inductive or leakage flux. Another part, however, (b) interlinks with the field also, and thus is mutual inductive flux of the armature circuit on the field circuit. In a polyphase machine, the resultant armature flux, that is, the resultant of the fluxes, Fig. 111, of all phases, revolves synchronously at (approximately) constant intensity, as a rotating field of armature reaction, and, therefore, is stationary with regard to the synchronously revolving field, F. Hence, the mutual inductive flux of the armature on the field, though an alternating flux, exerts no induction on the field circuit, is indeed a unidirectional or constant flux with regards to the field circuit. Therefore, under stationary conditions of load, no difference exists between the self-inductive and the mutual inductive flux of the armature circuit, and both are comprised in the synchronous reactance, x_0. If, however, the armature current changes, as by an increase of load, then with increasing armature current, the armature flux, a and b, Fig. 111, also increases. a, being interlinked with the armature current only, increases simultaneously with it, that is, the armature current can not increase without simultaneously increasing its self-inductive flux, a. The mutual inductive flux, b, however, interlinks with the field circuit, and this circuit is closed through the exciter, that is, is a closed secondary circuit with regards to the armature circuit as primary, and the change of flux, b, thus induces in the field circuit an e.m.f. and causes a current which retards the change of this flux component, b.　Or, in other words, an increase of armature current tends to increase its mutual magnetic flux, b, and thereby to decrease the field flux.　This decrease of field flux induces in the field circuit an e.m.f., which adds itself to the voltage impressed upon the field, thereby increases the field current and maintains the field flux against the demagnetizing action of the armature current, causing it to decrease only gradually.　Inversely, a decrease of armature current gives a simultaneous decrease of the self-inductive part of the flux, a in Fig. 111, but a gradual decrease of the mutual inductive part, b, and corresponding gradual increase of the resultant field flux, by inducing a transient voltage in the field, in opposition to the exciter voltage, and thereby decreasing the field current.

Every sudden increase of the armature current thus gives an equal sudden drop of terminal voltage due to the self-inductive flux, a, produced by it (and the resistance drop in the armature circuit), an equally sudden increase of the field current, and then a gradual further drop of the terminal voltage by the gradual appearance of the mutual flux, b, and corresponding gradual decrease of field current to nominal.　The reverse is the case at a sudden decrease of armature current.

The extreme case hereof is found in the momentary short-circuit currents of alternators,[1] which with some types of machines may momentarily equal many times the value of the permanent short-circuit current.　However, this phenomenon is not limited to short-circuit conditions only, but every change of current in an alternator causes a momentary overshooting, the more so, the greater and more sudden the change is.

122. That part of the synchronous reactance, x_0, which is due to the magnetic lines, a, in Fig. 111, is a true self-inductive reactance, x, and is instantaneous, but that part of x_1 representing the flux lines, b, is mutual inductive reactance with the field circuit, x', and is not instantaneous, but comes into play gradually, and whenever dealing with rapid changes of circuit conditions, the synchronous reactance, x_0, thus must be divided into a true or self-inductive reactance, x, and a mutual inductive reactance, x':

$$x_0 = x + x.'$$

[1] See "Theory and Calculation of Transient Phenomena."

The change of the flux disposition, caused by a current in the armature circuit, from that of Fig. 110 to that of Fig. 112, thus is simultaneous with the armature current and instantaneous with a sudden change of armature current only as far as it does not involve any change of the flux through the field winding, but the change of the flux through the field coils is only gradual. Thus the flux change in the armature core can be instantaneous, but that in the field is gradual.

This difference between self-inductive and mutual inductive reactance, or between instantaneous and gradual flux change, comes into consideration only in transients, and then very frequently the instantaneous or self-inductive effect is represented by a self-inductive reactance, x, the gradual or mutual inductive effect by an armature reaction.

The relation between self-inductive component, x, and mutual inductive component, x', varies from about $2 \div 1$ in the unitooth-high frequency alternators of old, to about $1 \div 20$ in some of the earlier turbo-alternators.

In those synchronous machines, which contain a squirrel-cage induction-motor winding in the field faces, for starting as motors, or as protection against hunting, or to equalize the armature reaction in single-phase machines, all the armature reactance flux, which interlinks with the squirrel-cage conductors (as the flux, c, in Fig. 111B), also is mutual inductive flux, and such machines thus have a higher ratio of mutual inductive to self-inductive armature reactance, that is, show a greater overshooting of current at sudden changing of load, and larger momentary short-circuit currents.

The mutual flux of armature reactance induces in the field circuit only under transient conditions, but under permanent circuit conditions the mutual inductance of the armature on the field has no inducing action, but is merely demagnetizing, and the distinction between self-inductive and mutual inductive reactance thus is unnecessary, and both combine in the synchronous reactance. In this respect, the synchronous machine differs from the transformer; in the latter, self-inductance and mutual inductance are always distinct in their action.

123. In permanent conditions of the circuit, the armature reactance of the synchronous machine is the synchronous reactance, $x_0 = x + x'$; at the instance of a sudden change of circuit conditions, the mutual inductive reactance, x', is still non-exist-

ing, and only the self-inductive reactance, x, comes into play.
Intermediate between the instantaneous effect and the permanent
conditions, for a time up to one or more sec., the effective reactance
changes, from x to x_0, and this may be considered as a *transient
reactance.*

During this period, mutual induction between armature circuit and field circuit occurs, and the phenomena in the synchronous machine thus are affected by the constants of the field circuit
outside of the machine.   That is, resistance and inductance of the
field circuit appear, by mutual induction, as part of the armature
circuit of the synchronous machine, just as resistance and reactance of the secondary circuit of a transformer appear, transformed by the ratio of turns, as resistance and reactance in the primary, in their effect on the primary current and its phase relation.

Thus in the synchronous machine, a high non-inductive resistance inserted into the field circuit (with an increase of the
exciter voltage to give the same field current) while without
effect on the permanent current and on the instantaneous current
in the moment of a sudden current change, reduces the duration
of the transient armature current; an inductance inserted into
the field circuit lengthens the duration of the transient and changes
its shape.

The duration of the transient reactance of the synchronous
machine is about of the same magnitude as the period of hunting
of synchronous machines—which varies from a fraction of a
second to over one sec.   The reactance, which limits the current
fluctations in hunting synchronous machines, thus is neither the
synchronous reactance, x_0, nor the true self-inductive reactance, x,
but is an intermediate transient reactance; the current change is
sufficiently slow that the mutual induction between synchronous
machine armature and field has already come into play and the
field begun to follow, but is too rapid for the complete development of the synchronous reactance.

124. In the polyphase machine on balanced load, the mutual
inductive component of the armature reactance has no inductive
effect on the field, as its resultant is unidirectional with regard
to the field flux.   In the single-phase machine, however (or
polyphase machine on unbalanced load), such inductive effect
exists, as a permanent pulsation of double frequency.   The
mutual inductive flux of the armature circuit on the field circuit
is alternating, and the field circuit, revolving synchronously

through this alternating flux, thus has an e.m.f. of double frequency induced in it, which produces a double-frequency current in the field circuit, superimposed on the direct current from the exciter. The field flux of the single-phase alternator (or polyphase alternator at unbalanced load) thus pulsates with double frequency, and, by being carried synchronously through the armature circuits, this double-frequency pulsation of flux induces a triple-frequency harmonic in the armature.

Thus, single-phase alternators, and polyphase alternators at unbalanced load, contain more or less of a third harmonic in their voltage wave, which is induced by the double-frequency pulsation of the field flux, resulting from the pulsating armature reaction, or mutual armature reactance, x'.

The statement, that three-phase alternators contain no third harmonics in their terminal voltages, since such harmonics neutralize each other, is correct only for balanced load, but at unbalanced load, three-phase alternators may have pronounced third harmonics in their terminal voltage, and on single-phase short-circuit, the not short-circuited phase of a three-phase alternator may contain a third harmonic far in excess of the fundamental.

125. Let in a Y-connected three-phase synchronous machine, the magnetic flux per field pole be Φ_0. If this flux is distributed sinusoidally around the circumference of the armature, at any time, t, represented by angle, $\phi = 2\pi ft$, the magnetic flux enclosed by an armature turn is

$$\Phi = \Phi_0 \cos \phi$$

when counting the time from the moment of maximum flux.

The voltage induced in an armature circuit of n turns then is

$$e_1 = n\frac{d\Phi}{dt} = c\Phi_0 \sin \phi$$

where

$$c = 2\pi fn$$

If, however, the flux distribution around the armature circumference is not sinusoidal, it nevertheless can, as a periodic function, be expressed by

$$\Phi = \Phi_0 [\cos \phi + a_2 \cos 2(\phi - \alpha_2) + a_3 \cos 3(\phi - \alpha_3) + a_4 \cos 4(\phi - \alpha_4) + \ldots]$$

and the voltage induced in one armature conductor, by the

16

synchronous rotation through this flux, is

$$\frac{d\Phi}{dt} = \pi f \Phi_0 [\sin \phi + 2 \, a_2 \sin 2(\phi - \alpha_2) + 3 \, a_3 \sin 3(\phi - \alpha_3) +$$
$$4 \, a_4 \sin 4(\phi - \alpha_4) + \ldots]$$

hence, the voltage induced in one full-pitch armature turn, or in two armature conductors displaced from each other on the armature surface by one pole pitch or an odd multiple thereof,

$$e = 2 \, \pi f \Phi_0 [\sin \phi + 3 \, a_3 \sin 3(\phi - \alpha_3) + 5 \, a_5 \sin 5(\phi - \alpha_5) + \ldots]$$

that is, the even harmonics cancel.

The voltage induced in one armature circuit of n effective series turns then is

$$e_1 = c \Phi_0 [\sin \phi + b_3 \sin 3(\phi - \alpha_3) + b_5 \sin 5(\phi - \alpha_5) + \ldots]$$

where

$b_3 = 3 \, a_3$, $b_5 = 5 \, a_5$, etc., if all the n turns are massed together, and are less, if the armature turns are distributed, due to the overlapping of the harmonics, and partial cancellation caused thereby. As known, by causing proper pitch of the turn, or proper pitch of the arc covered by any phase, any harmonic can be entirely eliminated.

The second and third phase of the three-phase machine then would have the voltage,

$$e_2 = c \Phi_0 [\sin (\phi - 120°) + b_3 \sin 3(\phi - \alpha_3 - 120°) +$$
$$b_5 \sin 5(\phi - \alpha_5 - 120°) + \ldots]$$
$$= c \Phi_0 [\sin (\phi - 120°) + b_3 \sin 3(\phi - \alpha_3) + b_5 \sin$$
$$(5[\phi - \alpha_5] + 120°) + \ldots]$$
$$e_3 = c \Phi_0 [\sin (\phi - 240°) + b_3 \sin 3(\phi - \alpha_3) +$$
$$b_5 \sin (5[\phi - \alpha_5] + 240°)] + \ldots]$$

As seen, the third harmonics are all three in phase with each other; the fifth harmonics are in three-phase relation, but with backward rotation; the seventh harmonics are again in three-phase relation, like the fundamentals, the ninth harmonics in phase, etc.

The terminal voltages of the machine then are

$$E_1 = e_3 - e_2 = \sqrt{3} \, c \Phi_0 [\cos \phi - b_5 \cos 5 (\phi - \alpha_5) +$$
$$b_7 \cos 7 (\phi - \alpha_7) - + \ldots]$$

and corresponding thereto $E_2 = e_1 - e_3$ and $E_3 = e_2 - e_1$, differing from E_1 merely by substituting $\phi - 120°$ and $\phi - 240°$ for ϕ.

As seen, the third harmonic eliminates in the terminal voltages of the three-phase machine, regardless of the flux distribution, provided that the flux is constant in intensity, that is, the load conditions balanced.

126. Assuming, however, that the load on the three-phase machine is unbalanced, causing a double-frequency pulsation of the magnetic flux,

$$\Phi_0 (1 + a \cos 2 \phi),$$

assuming for simplicity sinusoidal distribution of magnetic flux. The flux interlinked with a full-pitch armature turn then is

$$\Phi = \Phi_0(1 + a \cos 2 \phi) \cos (\phi - \alpha)$$
$$= \Phi_0 \left[\cos (\phi - \alpha) + \frac{a}{2} \cos (\phi + \alpha) + \frac{a}{2} \cos (3 \phi - \alpha) \right]$$

and the voltage induced in an armature circuit of n effective turns,

$$e_1 = n\frac{d\Phi}{dt} = c\Phi_0 \frac{d}{d\phi}\left[\cos (\phi - \alpha) + \frac{a}{2} \cos (\phi + \alpha) + \frac{a}{2} \cos (3 \phi - \alpha) \right]$$
$$= c\Phi_0 \left[\sin (\phi - \alpha) + \frac{a}{2} \sin (\phi + \alpha) + \frac{3 a}{2} \sin (3 \phi - \alpha) \right]$$

or, if the magnetic flux maximum coincides with the voltage maximum of the first phase, $\alpha = 0$,

$$e_1 = c\Phi_0 \left[\left(1 + \frac{a}{2}\right) \sin \phi + \frac{3 a}{2} \sin 3 \phi \right].$$

In the second phase, the flux is the same, $\Phi_0 (1 + a \cos 2 \phi)$, but the flux interlinkage 120° later, thus,

$$\Phi = \Phi_0 (1 + a \cos 2 \phi) \cos (\phi - \alpha - 120°),$$

and the voltage of the second phase thus is derived from that of the first phase, by substituting $\alpha + 120°$ for α,

$$e_2 = c\Phi_0 \left[\sin (\phi - \alpha - 120°) + \frac{a}{2} \sin (\phi + \alpha + 120°) + \right.$$
$$\left. \frac{3 a}{2} \sin (3 \phi - \alpha - 120°) \right]$$

and the third phase,

$$e_3 = c\Phi_0 \left[\sin (\phi - \alpha - 240°) + \frac{a}{2} \sin (\phi + \alpha + 240°) + \right.$$
$$\left. \frac{3 a}{2} \sin (3 \phi - \alpha - 240°) \right]$$

the terminal voltages thus are,

$$E_1 = e_3 - e_2 = \sqrt{3} \, c\Phi_0 \left[\cos(\phi - \alpha) - \frac{a}{2} \cos(\phi + \alpha) + \frac{3a}{2} \cos(3\phi - \alpha) \right]$$

and in the same manner, the other two phases,

$$E_2 = \sqrt{3} \, c\Phi_0 \left[\cos(\phi - \alpha - 120°) - \frac{a}{2} \cos(\phi + \alpha + 120°) - \frac{3a}{2} \cos(3\phi - \alpha - 120°) \right]$$

$$E_3 = \sqrt{3} \, c\Phi_0 \left[\cos(\phi - \alpha - 240°) - \frac{a}{2} \cos(\phi + \alpha + 240°) - \frac{3a}{2} \cos(3\phi - \alpha - 240°) \right].$$

For $\alpha = 0$, this gives

$$E_1 = \sqrt{3} \, c\Phi_0 \left[\left(1 - \frac{a}{2}\right) \cos\phi + \frac{3a}{2} \cos 3\phi \right]$$

$$E_2 = \sqrt{3} \, c\Phi_0 \left[\left(1 - \frac{a}{2}\right) \cos(\phi - 120°) - \frac{3a}{2} \cos(3\phi - 120°) \right]$$

$$E_3 = \sqrt{3} \, c\Phi_0 \left[\left(1 - \frac{a}{2}\right) \cos(\phi - 240°) - \frac{3a}{2} \cos(3\phi - 240°) \right].$$

As seen, all three phases have pronounced third harmonics, and the third harmonic of the loaded phase, E_1, is opposite to that of the unloaded phases.

If $a = 1$, which corresponds about to short-circuit conditions, as it makes the minimum value of Φ_0 equal zero, then the quadrature phase of the short-circuited phase, E_1, becomes

$$e_1 = \frac{3 \, c\Phi_0}{2}(\sin\phi + \sin 3\phi),$$

that is, the third harmonic becomes as large as the fundamental.

Thus, on unbalanced load, such as on single-phase short-circuit, triple harmonics appear in the terminal voltages of a three-phase generator, though at balanced loads the three-phase terminal voltage can contain no third harmonics.

SECTION III

CHAPTER XIV

CONSTANT-POTENTIAL CONSTANT-CURRENT TRANSFORMATION

127. The generation of alternating-current electric power practically always takes place at constant voltage. For some purposes, however, as for operating series arc circuits, and to a limited extent also for electric furnaces, a constant, or approximately constant alternating current is required. While constant alternating-current arcs have largely come out of use and their place taken by constant direct-current luminous arc circuits, or incandescent lamps, the constant direct current is usually derived by rectification of constant alternating-current supply circuits.

Such constant alternating currents are usually produced from constant-voltage supply circuits by means of constant or variable inductive reactances, and may be produced by the combination of inductive and condensive reactances; and the investigation of different methods of producing constant alternating current from constant alternating voltage, or inversely, constitutes a good application of the terms "impedance," admittance," etc., and offers a large number of problems or examples for the symbolic method of dealing with alternating-current phenomena.

Even outside of arc lighting, such combinations of inductance and capacity which tend toward constant-voltage constant-current transformation are of considerable importance as a possible source of danger to the system. In a constant-current circuit, the load is taken off by short-circuiting, while open-circuiting causes the voltage to rise to the maximum value permitted by the power of the generating source. Hence, where the circuit constants, with a constant-voltage supply source, are such as to approach constant-voltage constant-current transformation, as is for instance the case in very long transmission lines, open-circuiting may lead to dangerous or even destructive voltage rise.

128. With an inductive reactance inserted in series to an alter-

245

nating-current non-inductive circuit, at constant-supply voltage, the current in this circuit is approximately constant, as long as the resistance of the circuit is small compared with the series inductive reactance.

Let

$E_0 = e_0 =$ constant impressed alternating voltage;

$r =$ resistance of non-inductive receiver circuit;

$x_0 =$ inductive reactance inserted in series with this circuit.

The impedance of this circuit then is

$$Z = r + jx_0,$$

and, absolute,

$$z = \sqrt{r^2 + x_0^2},$$

and thus the current,

$$I = \frac{e_0}{Z} = \frac{e_0}{r + jx_0} \tag{1}$$

and the absolute value is

$$i = \frac{e_0}{z} = \frac{e_0}{\sqrt{r^2 + x_0^2}} \tag{2}$$

the phase angle of the supply circuit is given by

$$\tan \theta_0 = \frac{x_0}{r} \tag{3}$$

and the power factor,

$$\cos \theta_0 = \frac{r}{z} \cdot \tag{4}$$

If in this case, r is small compared with x_0, it is

$$i = \frac{e_0}{x_0} \frac{1}{\sqrt{1 + \left(\frac{r}{x_0}\right)^2}}; \tag{5}$$

or, expanded by the binomial theorem,

$$\frac{1}{\sqrt{1 + \left(\frac{r}{x_0}\right)^2}} = \left\{1 + \left(\frac{r}{x_0}\right)^2\right\}^{-\frac{1}{2}} = 1 - \frac{r^2}{2 x_0^2} + \frac{3 r^4}{8 x_0^4} - + \dots$$

hence,

$$i = \frac{e_0}{x_0}\left\{1 - \frac{r^2}{2 x_0^2} + \frac{3 r^4}{8 x_0^4} - + \dots \right\}; \tag{6}$$

that is, for small values of r, the current, i, is approximately constant, and is

$$i = \frac{e_0}{x_0}.$$

For small values of r, the power-factor

$$\cos \theta = \frac{r}{z}$$

is very low, however.

Allowing a variation of current of 10 per cent. from short-circuit or no-load, $r = 0$, to full-load, or $r = r_1$, it is, substituted in (2):

No-load current:

$$i_0 = \frac{e_0}{x_0}.$$

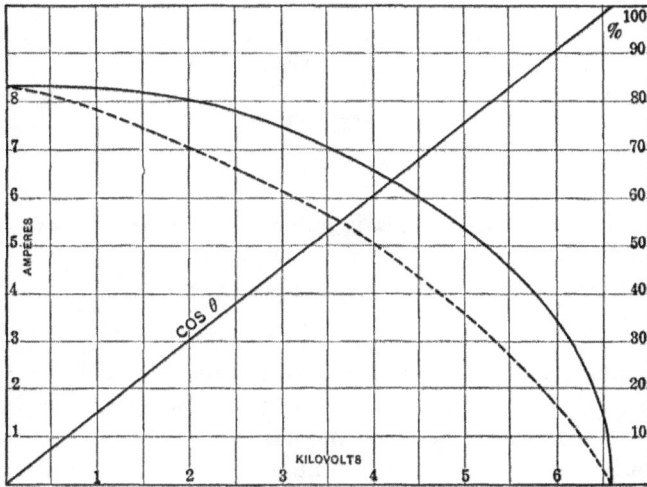

FIG. 113.

Full-load current:

$$i = \frac{e_0}{\sqrt{r_1{}^2 + x_0{}^2}} = 0.9 \, i_0.$$

Hence,

$$\frac{e_0}{\sqrt{r_1{}^2 + x_0{}^2}} = 0.9 \frac{e_0}{x_0},$$

and therefore,

$$r_1 = 0.485 \, x_0,$$

and the power-factor, from (4), is 0.437.

That is, even allowing as large a variation of current, i, as 10 per cent., the maximum power-factor only reaches 43.7 per cent., when producing constant-current regulation by series inductance reactance.

As illustrations are shown, in Fig. 113, for the constants:

$$e_0 = 6600 \text{ volts applied e.m.f.};$$
$$x_0 = 792 \text{ ohms series reactance};$$

the current:

$$i = \frac{6600}{\sqrt{r^2 + 792^2}}$$
$$= \frac{8.33}{\sqrt{1 + \left(\dfrac{r}{792}\right)^2}} \text{ amp.};$$

and the power-factor:

$$\cos \theta = \frac{r}{\sqrt{r^2 + 792^2}} = \frac{r}{792\sqrt{1 + \left(\dfrac{r}{792}\right)^2}};$$

with the voltage at the secondary terminals:

$$e = ri$$

as abscissas.

129. If the receiver circuit is inductive, that is, contains, in addition to the resistance, r, an inductive reactance, x, and if this reactance is proportional to the resistance,

$$x = kr,$$

as is commonly the case in arc circuits, due to the inductive reactance of the regulating mechanism of the arc lamp (the effective resistance, r, and the inductive reactance, x, in this case are both proportional to the number of lamps, hence proportional to each other), it is:

total impedance:

$$Z = r + j(x_0 + x) = r + j(x_0 + kr);$$

or the absolute value is

$$z = \sqrt{r^2 + (x_0 + x)^2} = \sqrt{r^2 + (x_0 + kr)^2};$$

thus, the current

$$I = \frac{e_0}{r + j(x_0 + kr)}, \qquad (7)$$

and the absolute value is

$$i = \frac{e_0}{\sqrt{r^2 + (x_0 + kr)^2}} = \frac{e_0}{x_0} \frac{1}{\sqrt{1 + \dfrac{2\,kr}{x_0} + \dfrac{r^2(1 + k^2)}{x_0{}^2}}}; \qquad (8)$$

and the power-factor:

$$\cos \theta_0 = \frac{r}{z} = \frac{r}{\sqrt{r^2 + (x_0 + kr)^2}}. \tag{9}$$

By the binomial theorem, it is

$$\frac{i}{\sqrt{1 + \dfrac{2\,kr}{x_0} + \dfrac{r^2(1 + k^2)}{x_0{}^2}}} = 1 - \frac{kr}{x_0} - \frac{r^2(2 - k^2)}{4\,x_0{}^2} + - \dots$$

Hence, the current

$$i = \frac{e_0}{x_0}\left\{1 - \frac{kr}{x_0} - \frac{r^2(2 - k)^2}{4\,x_0{}^2} + - \dots \right\} \tag{10}$$

that is, the expression of the current, i (10), contains the ratio, $\dfrac{r}{x_0}$, in the first power, with k as coefficient, and if therefore k is not very small, that is, the inductive reactance, $x = kr$, a very small fraction of the resistance, r, the current, i, is not even approximately constant, but begins to fall off immediately, even at small values of r.

Assuming, for instance,

$$k = 0.4.$$

That is, the inductive reactance, x, of the receiver circuit equals 40 per cent. of its resistance, r, and the power-factor of the receiver circuit accordingly is

$$\cos \theta = \frac{r}{r^2 + x^2}$$
$$= \frac{1}{1 + k^2}$$
$$= 93 \text{ per cent.};$$

it is, substituted in (8),

$$I = \frac{e_0}{x_0 \sqrt{\left(\dfrac{r}{792}\right)^2 + \left(1 + 0.4\,\dfrac{r}{x_0}\right)^2}},$$

As illustrations are shown, in the same Fig. 113, for the constants:

$$e_0 = 6600 \text{ volts supply e.m.f.;}$$
$$x_0 = 792 \text{ ohms series reactance;}$$

the current:

$$i = \frac{8.33}{\sqrt{\left(\dfrac{r}{792}\right)^2 + \left(1 + 0.4\dfrac{r}{792}\right)^2}} \text{ amp.}$$

This current is shown by dotted line.

In this case, in an inductive circuit, the current, i, has decreased by 10 per cent. below the no-load or short-circuit value of 8.33 amp. that is, has fallen to 7.5 amp., at the resistance $r = 187$ ohms, or at the voltage of the receiving circuit,

$$e = i \sqrt{r^2 + x^2} = ri \sqrt{1 + k^2} = 1.077 \, ri = 1500 \text{ volts};$$

while, in the case of a non-inductive load, the current has fallen off to 7.5 amp., or by 10 per cent. at the resistance $r = 395$ ohms, or at the voltage of the receiving circuit: $e = 2950$ volts.

130. As seen, a moderate constant-current regulation can be produced in a non-inductive circuit, by a constant series inductive reactance, at a considerable sacrifice, however, of the power-factor, while in an inductive receiver circuit, the constant-current regulation is not even approximate.

To produce constant alternating current, from a constant-potential supply, by a series inductive reactance, over a wide range of load and without too great a sacrifice of power-factor, therefore requires a variation of the series inductive reactance with the load. That is, with increasing load, or increasing resistance of the receiver circuit, the series inductive reactance has to be decreased, so as to maintain the total impedance of the circuit, and thereby the current, constant.

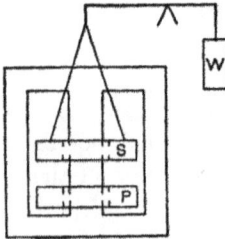

Fig. 114.

In constant-current apparatus, as transformers from constant potential to constant current, or regulators, this variation of series inductive reactance with the load is usually accomplished automatically by the mechanical motion caused by the mechanical force exerted by the magnetic field of the current, upon the conductor in which the current exists.

For instance, in the constant-current transformer, as shown diagrammatically in Fig. 114, the secondary coils, S, are arranged so that they can move away from the primary coils, P, or inversely. Primary and secondary currents are proportional to each other, as in any transformer, and the magnetic field between primary and secondary coils, or the magnetic stray field, in which the secondary coils float, is proportional to either current. The magnetic repulsion between primary coils and secondary coils is proportional to the current (or rather its ampere-turns), and to the magnetic stray field, hence is proportional to the square of the current, but independent of the voltage. The secondary coils, S, are counter-balanced by a weight, W, which is adjusted so that this weight, W, plus the repulsive thrust between secondary coils, S, and primary coils, P (which, as seen above, is proportional to the square of the current), just balances the weight of the secondary coils. Any increase of secondary current, as, for instance, caused by short-circuiting a part of the secondary load, then increases the repulsion between primary and secondary coils, and the secondary coils move away from the primary; hence more of the magnetic flux produced by the primary coils passes between primary and secondary, as stray field, or self-inductive flux, less passes through the secondary coils, and therefore the secondary generated voltage decreases with the separation of the coils, and also thereby the secondary current, until it has resumed the same value, and the secondary coil is again at rest, its weight balancing counterweight plus repulsion.

Inversely, an increase of load, that is, of secondary impedance, decreases the secondary current, so causes the secondary coils to move nearer the primary, and to receive more of the primary flux; that is, generate higher voltage.

In this manner, by the mechanical repulsion caused by the current, the magnetic stray flux, or, in other words, the series inductive reactance of the constant-current transformer, varies automatically between a maximum, with the primary and secondary coils at their maximum distance apart, and a minimum with the coils touching each other. Obviously, this automatic action is independent of frequency, impressed voltage, and character of load.

If the two coils P and S in Fig. 114 are wound with the same number of turns and connected in series with each other and with the circuit, Fig. 114 is a constant-current regulator, or a regulating reactance, that is, a reactance which varies with the load so as to maintain constant current. If P is primary and S secondary circuit, Fig. 114 is a constant-current transformer.

Assuming then, in the constant-current transformer or regulator or other apparatus, a device to vary the series inductive reactance so as to maintain the current constant. Let

$E_0 = e_0 = $ constant $=$ impressed e.m.f.,

$Z = r + jx$,

$= r (1 + jk)$ the impedance of the load, and let

$x_0 = $ inductive series reactance, as the self-inductive internal reactance of the constant-current transformer.

The current in the circuit then is

$$I = \frac{e_0}{r + j(x_0 + x)},$$

or, the absolute value,

$$i = \frac{e_0}{\sqrt{r^2 + (x_0 + x)^2}};$$

and, to maintain the current, i, constant $(i = i_0)$, then requires

$$i_0 = \frac{e_0}{\sqrt{r^2 + (x_0 + x)^2}};$$

or, transposed,

$$x_0 = \sqrt{\left(\frac{e_0}{i_0}\right)^2 - r^2} - x \tag{11}$$

or, for

$$x = kr,$$

$$x_0 = \sqrt{\left(\frac{e_0}{i_0}\right)^2 - r^2} - kr \tag{12}$$

that is, to produce perfectly constant current by means of a variable series inductive reactance, this series reactance must be varied with the load on the circuit, according to equation (11) or (12).

For non-inductive load, or $x = 0$, it is

$$x_0 = \sqrt{\left(\frac{e_0}{i_0}\right)^2 - r^2} \tag{13}$$

the maximum load, which can be carried, is given by

$$x_0 = 0$$

and is

$$z = \sqrt{r^2 + x^2} = r\sqrt{1 + k^2} = \frac{e_0}{i_0} \tag{14}$$

As seen from equation (13), the decrease of inductive reactance, x_0, required to maintain constant current with non-inductive load, is small for small values of resistance, r, when the r^2 under the root is negligible. With inductive load, equation (11), the inductive reactance, x_0, has still further to be decreased by the inductive reactance of the load, x.

Substituting:

$$x_{00} = \frac{e_0}{i_0}$$

as the value of the series inductive reactance at no-load or short-circuit, equations (11), (12), (13) assume the form:

General inductive load:

$$x_0 = \sqrt{x_{00}^2 - r^2} - x, \tag{14}$$

Inductive load of $\dfrac{x}{r} = k$:

$$x_0 = \sqrt{x_{00}^2 - r^2} - kr \tag{15}$$

Non-inductive load:

$$x_0 = \sqrt{x_{00}^2 - r^2}. \tag{16}$$

131. As seen, a constant series inductive reactance gives an approximately constant-current regulation with non-inductive load, but if the load is inductive this regulation is spoiled. Inversely it can be shown, that condensive reactance, that is, a source of leading current in the load, improves the constant-current regulation.

With a non-inductive load, series condensive reactance exerts the same effect on the current regulation as series inductive reactance; the equations discussed in the preceding paragraphs remain the same, except that the sign of x_0 is reversed and the current always leading.

With series condensive reactance, condensive reactance in the load spoils, inductive reactance in the load improves the constant-current regulation.

That is, in general, a constant series reactance gives approximately constant-current regulation in a non-inductive circuit, and with a reactive load this regulation is impaired if the reactance of the load is of the same sign as the series reactance, and the regulation is improved if the reactance of the load is of opposite sign as the series reactance.

Since a constant-current load is usually somewhat inductive, it follows that a constant series condensive reactance gives a better constant-current regulation, in the average case of a somewhat inductive arc circuit, than the constant series inductive reactance.

Let

$E_0 = e_0 =$ constant $=$ impressed, or supply voltage.

$Z = r + jx =$ impedance of the load, or the receiver circuit, and

$$x = kr,$$

that is,

$$Z = r(1 + jk)$$

or, absolute,

$$z = r\sqrt{1 + k^2}.$$

Let now a constant condensive reactance be inserted in series with this circuit, of the reactance, $-x_c$, then the total impedance of the circuit is

$$Z' = r - j(x_c - kr). \tag{17}$$

The current is

$$I = \frac{e_0}{r - j(x_c - kr)}, \tag{18}$$

or, the absolute value is

$$i = \frac{e_0}{\sqrt{r^2 + (x_c - kr)^2}} \tag{19}$$

the phase angle is

$$\tan \theta_0 = -\frac{x_c - kr}{r} \tag{20}$$

and the power-factor is

$$\cos \theta_0 = \frac{r}{\sqrt{r^2 + (x_c - kr)^2}} \tag{21}$$

for

$$k = 0,$$

or non-inductive load, equations (19) and (21) assume the form:

$$i = \frac{e_0}{\sqrt{r^2 + x_c^2}} \text{ and } \cos \theta = \frac{r}{\sqrt{r^2 + x_c^2}},$$

that is, the same as with series inductive reactance.

From equation (19) it follows, that with increasing current, i, from no-load:

$$r = 0, \text{ hence: } i_0 = \frac{e_0}{x_c} \tag{22}$$

the current, i_0, first increases, reaches a maximum, and then decreases again. When decreasing, it once more reaches the value, i_0, for the resistance, r_1, of the load, which is given by

$$i_0 = \frac{e_0}{\sqrt{r_1^2 + (x_c - kr_1)^2}} = \frac{e_0}{x_c};$$

hence, expanded,

$$r_1 = \frac{2 kx_c}{1 + k^2} \tag{23}$$

and the maximum value through which i passes between $r = 0$ and $r = r_1$, is given by

$$\frac{di}{dr} = 0,$$

or

$$\frac{d}{dr}\{r^2 + (x_c - kr)^2\} = 0 = 2\,r - 2\,k(x_c - kr);$$

hence,

$$r_2 = \frac{kx_c}{1 + k^2} = \frac{r_1}{2}. \tag{24}$$

This maximum value is given by substituting (24) in (19), as

$$i_2 = \frac{e_0}{x_c}\sqrt{1 + k^2},$$

for

$$= i_0\sqrt{1 + k^2} \tag{25}$$
$$k = 0.4,$$

this value is

$$i^2 = 1.077\,i_0,$$

that is, the current rises from no-load to a maximum 7.7 per cent. above the no-load value, and then decreases again.

As an example, let

$$e_0 = 6600 \text{ volts impressed e.m.f.}$$

and

$$x_c = 880 \text{ ohm condensive reactance,}$$

x_c being chosen so as to give

$$i_0 = \frac{e_0}{x_c} = 7.5 \text{ amp.};$$

for

$$k = 0.4,$$

then,

$$i = \frac{6600}{\sqrt{r^2 + (880 - 0.4\,r)^2}},$$

$$\cos\theta_0 = \frac{r}{\sqrt{r^2 + (880 - 0.4\,r)^2}},$$

$$e = zi = 1.077\,ri.$$

These values of current and power-factor are plotted, with the receiver voltage as abscissæ, in Fig. 115.

132. The conclusions from the preceding are that a constant series reactance, whether condensive or inductive, when inserted in a constant-potential circuit, tends toward a constant-current regulation, at least within a certain range of load. That is, at varying resistance, r, and therefore varying load, the current is approximately constant at light load, and drops off only gradually with increasing load.

This constant-current regulation, and the power-factor of the circuit, are best if the reactance of the receiver circuit is of opposite sign to the series reactance, and poorest if of the same sign. That is, series condensive reactance in an inductive circuit, and series inductive reactance in a circuit carrying leading current,

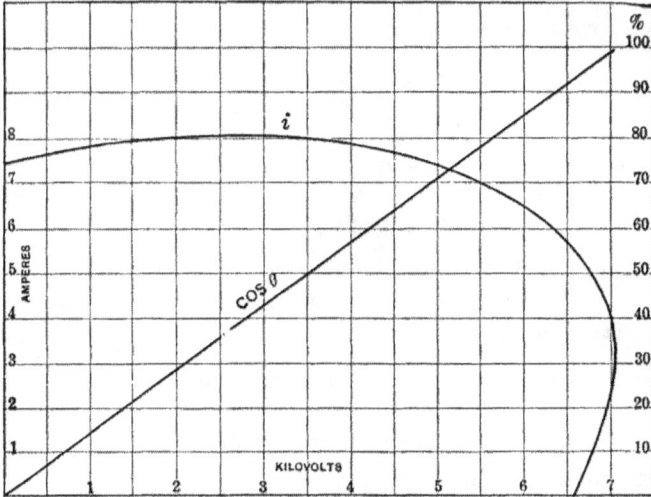

FIG. 115.

give the best regulation; series inductive reactance with an inductive, and series condensive reactance with leading current in the circuit, give the poorest regulation.

Since the receiver circuit is usually inductive, to get best regulation, either a series condensive reactance has to be used, as in Fig. 115, or, if a series inductive reactance is used, the current in the receiver circuit is made leading, as, for instance, by shunting the receiver circuit by a condensive reactance.

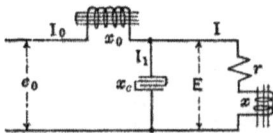

FIG 116.

Assuming, then, as sketched diagrammatically in Fig. 116, in a circuit of constant impressed e.m.f., $E_0 = e_0 = $ constant, a constant inductive reactance, x_0, inserted in series; and the receiver circuit, of impedance,

$$Z = r + jx = r(1 + jk)$$

where

$$k = \text{tangent of the angle of lag} = \frac{x}{r};$$

let the receiver circuit be shunted by a constant condensive reactance, x_c; let then:

E = potential difference of receiver circuit or the condenser terminals,

I = current in the receiver circuit, or the "secondary current,"

I_1 = current in the condenser,

I_0 = total supply current, or "primary current."

Then
$$I_0 = I + I_1 \tag{26}$$

and the e.m.f. at receiver circuit is
$$E = ZI \tag{27}$$

at the condenser,
$$E = -jx_cI_1 \tag{28}$$

hence,
$$I_1 = j\frac{Z}{x_c}I \tag{29}$$

and, in the main circuit, the impressed e.m.f. is
$$E_0 = e_0 = E + jxI_0 \tag{30}$$

Hence, substituting (26), (27) and (29) in (30),

$$e_0 = ZI + jx_0\left(I + j\frac{Z}{x_c}I\right)$$
$$= \left\{Z + jx_0 - \frac{x_0}{x_c}Z\right\}I$$

or

$$e_0 = \left\{Z\frac{x_c - x_0}{x_c} + jx_0\right\}I \tag{31}$$

and

$$I = \frac{e_0}{Z\dfrac{x_c - x_0}{x_c} + jx_0} \tag{32}$$

If $x_c = x_0$, that is, if the shunted condensive reactance equals the series inductive reactance, equations (32) assume the form,

$$I = +\frac{e_0}{jx_0} = -j\frac{e_0}{x_0} \tag{33}$$

and the absolute value is

$$i = \frac{e_0}{x_0} \tag{34}$$

that is, the current, i, is constant, independent of the load and the power-factor.

17

That is, if in a constant-potential circuit, of impressed e.m.f., e_0, an inductive reactance, x_0, and a condensive reactance, x_c, are connected in series with each other, and if

$$x_c = x_0, \tag{35}$$

that is, the two reactances are in resonance condition with each other, any circuit shunting the capacity reactance is a constant-current circuit, and regardless of the impedance of this circuit, $Z = r + jx$, the current in the circuit is

$$i = \frac{e_0}{x_0}.$$

133. Such a combination of two equal reactances of opposite sign can be considered as a transforming device from constant potential to constant current.

Substituting, therefore, (35) in the preceding equation gives: (33) substituted in (29):

Current in shunted capacity

$$I_1 = \frac{Z}{x_0{}^2} e_0 \tag{36}$$

or, absolute,

$$i_1 = \frac{z e_0}{x_0{}^2} \tag{37}$$

and, substituting (33) and (36) in (26):
primary supply current is

$$I_0 = \frac{Z - j x_0}{x_0{}^2} e_0 \tag{38}$$

or the absolute value is

$$i_0 = \frac{e_0}{x_0{}^2} \sqrt{r^2 + (x_0 - x)^2} \tag{39}$$

and the power-factor of the supply current is

$$\tan \theta_0 = -\frac{x_0 - x}{r}, \quad \cos \theta_0 = \frac{r}{\sqrt{r^2 + (x_0 - x)^2}} \tag{40}$$

In this case, the higher the inductive reactance, x, of the receiving circuit the lower is the supply current, i_0, at the same resistance, r, and the higher is the power-factor, and if $x = x_0$

$$I_0 = \frac{e^2 r}{x_0{}^2} \text{ and } \cos \theta = 1 \tag{41}$$

that is, the primary, or supply circuit is non-inductive, and the primary current is in phase with the supply e.m.f., and the power-factor is unity, while the secondary or receiver current (33) is 90° in phase behind the primary impressed e.m.f., e_0.

Inserting, therefore, an inductive reactance, $x_1 = x_0 - x$, in series in the receiver circuit of impedance, $Z = r + jx$, raises the power-factor of the supply current, i_0, to unity, and makes this current, i_0, a minimum. Or, if the inductive reactance, x_0, is inserted in the receiver circuit, thus giving a total impedance, $Z + jx_0 = r + j(x + x_0)$ by equation (38), substituting $Z + jx_0$ instead of Z, gives the primary supply current as

$$I_0 = \frac{Ze_0}{x_0^2} \tag{42}$$

or the absolute value as

$$i_0 = \frac{ze_0}{x_0^2} \tag{43}$$

and the tangent of the primary phase angle

$$\tan \theta_0 = \frac{x}{r} = \tan \theta,$$

that is, the primary power-factor equals that of the secondary.

Hence, as shown diagrammatically in Fig. 117, a combination of two equal inductive reactances in series with each other and with the receiver circuit, and shunted midway between the inductive reactances by a condensive reactance equal to the inductive reactance,

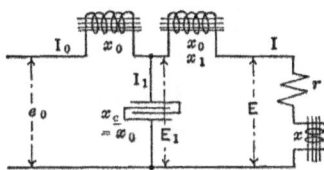

FIG. 117.

transforms constant potential into constant current, and inversely, without any change of power-factor, that is, the primary supply current has the same power-factor as the secondary current.

With an inductive secondary circuit, the primary power-factor can in this case be made unity, by reducing the inductive reactance of the secondary side, by the amount of secondary reactance.

134. Shunted condensive reactance, x_c, and series inductive reactance, x_0, therefore transforms from constant potential, e_0, to constant current, i, and inversely, if their reactances are equal, $x_c = x_0$, and in this case, the main current is leading, with non-inductive load, and the lead of the main current decreases, with increasing inductive reactance, that is, increasing lag, of the

receiving circuit. The constant secondary current, i, lags 90°
behind the constant primary e.m.f., e_0.

Inversely, by reversing the signs of x_0 and x_c in the preceding
equations, that is, exchanging inductive and condensive reactances, it follows that shunted inductive reactance, x_0, and
series condensive reactance, x_c, if of equal reactance, $x_c = x_0$,
transform constant potential, e_0, into constant current, i, and
inversely. In this case, the main current lags the more the
higher the inductive reactance of the receiving circuit, and
the constant secondary current, i, is 90° ahead of the constant
primary e.m.f., e_0.

In general, it follows that, if equal inductive and condensive
reactances, $x_0 = x_c$, that is, in resonance conditions, are connected in series across a constant-potential circuit of impressed

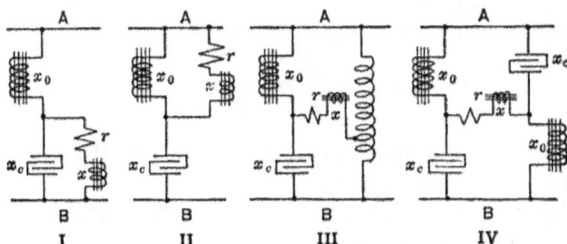

Fig. 118.

e.m.f., e_0, any circuit connected to the common point between
the reactances is a constant-current circuit, and carries the
current, $i = \dfrac{e_0}{x_0}$.

Instead of connecting this secondary or constant-current
circuit with its other terminal to line, A, so shunting the condensive reactance with it, and causing the main current to lead
(I in Fig. 118), or to line, B, so shunting the inductive reactance
with it, and causing the main current to lag (II in Fig. 118), it
can be connected to any point intermediate between A and B,
by a autotransformer as in III, Fig. 118. If connected to the middle point between A and B, the main current is neither lagging nor
leading, that is, is non-inductive, with non-inductive, load, and
with inductive load, has the same power-factor as the load.

The two arrangements, I and II, can also be combined, by
connecting the constant-current circuit across, as in IV, Fig. 118,
and in this case the two inductive reactances and two conden-

sive reactances diagrammatically form a square, with the constant potential, e_0, as one, the constant current, i, as the other diagonal, as shown in Fig. 119.   This arrangement has been called the *monocyclic square.*

The insertion of an e.m.f. into the constant-current circuit, in such arrangements, obviously, does not exert any effect on the constancy of the secondary current, i, but merely changes the primary current, i_0, by the amount of power supplied or consumed by the e.m.f. inserted in the secondary circuit.

While theoretically the secondary current is absolutely constant, at constant primary e.m.f., practically it can not be perfectly constant, due to the power consumed in the reactances, but falls off slightly with increase of load, the more, the greater the loss of power in the reactances, that is, the lower the efficiency of the transforming device.

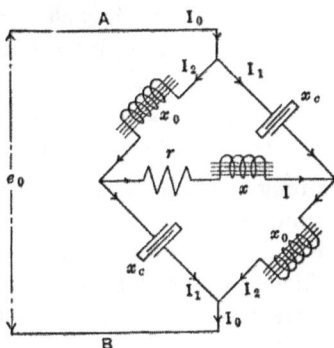

FIG. 119.

Two typical arrangements of such constant-current transforming devices are the *T-connection* or the *resonating-circuit,* diagram Fig. 117, and the *monocyclic square,* diagram Fig. 119. From these, a very large number of different combinations of inductive and condensive reactances, with addition of autotransformers, and of impressed e.m.fs., can be devised to transform from constant potential to constant current and inversely, and by the use of quadrature e.m.fs. taken from a second phase of the polyphase system, the secondary output, for the same amount of reactances, increased.

These combinations afford very convenient and instructive examples for accustoming oneself to the use of the symbolic method in the solution of alternating-current problems.

Only two typical cases, the T-connection and the monocyclic square will be more fully discussed.

A. T-Connection or Resonating Circuit

135. *General.*—A combination, in a constant-potential circuit, of an inductive and a condensive reactance in series with each

other in resonance condition, that is, with the condensive reactance equal to the inductive reactance, gives constant current in
a circuit shunting the capacity. This circuit thus can be called
the "secondary circuit" of the constant potential constant-
current transforming device, while the constant-potential supply
circuit may be called the "primary circuit."

If the total inductive reactance in the constant-current circuit is equal to the condensive reactance, the primary supply
current is in phase with the impressed e.m.f.

Let, as shown diagrammatically in Fig. 117,

x_0 = value of the inductive and the condensive reactances which
are in series with each other.

x_1 = the additional inductive reactance inserted in the constant-
current circuit.

$Z = r + jx$, or $z = \sqrt{r^2 + x^2}$ = the absolute value of the impedance of the constant-current load.

Assuming now in the constant-current circuit the inductive
reactance and the resistance as proportional to each other, as
for instance is approximately the case in a series arc circuit,
in which, by varying the number of lamps and therewith the
load, reactance and resistance change proportionally. Let, then,

$k = \dfrac{x}{r}$ = ratio of inductive reactance to resistance of the load,

or tangent of the angle of lag of the constant-current circuit.
It is then

$$Z = r(1 + jk)$$

and
$$z = r\sqrt{1 + k^2} \tag{1}$$

let, then,

$E_0 = e_0$ = constant = primary impressed e.m.f., or supply voltage,

E_1 = potential difference at condenser terminals,

E = secondary e.m.f., or voltage at constant-current
circuit,

I_0 = primary supply current,

I_1 = condenser current,

I = secondary current,

then, in the secondary or receiver circuit,

$$E = ZI \tag{2}$$

at the condenser terminals

$$E_1 = E + jx_1 I$$
$$= (Z + jx_1) I \tag{3}$$

and, also,

$$E_1 = - jx_0 I_1 \tag{4}$$

hence,

$$I_1 = j \frac{Z + jx_1}{x_0} I \tag{5}$$

and the primary current is

$$I_0 = I + I_1 = \left\{ j\frac{Z + jx_1}{x_0} + 1 \right\} I$$

hence, expanded,

$$I_0 = j\frac{Z - j(x_0 - x_1)}{x_0} I \tag{6}$$

and the primary supply voltage is

$$e_0 = E_1 + jx_0 I_0;$$

hence, substituting (3) and (6),

$$e_0 = [(Z + jx_1) - \{Z - j(x_0 - x_1)\}]I,$$

or, expanded,

$$e_0 = + jx_0 I \tag{7}$$

or, the secondary current is

$$I = - \frac{je_0}{x_0} \tag{8}$$

and, substituting (8) in (6) and (5): the primary current is

$$I_0 = \frac{Z - j(x_0 - x_1)}{x_0{}^2} e_0 \tag{9}$$

the condenser current is

$$I_1 = \frac{Z + jx_1}{x_0{}^2} e_0 \tag{10}$$

or, the absolute value is

$$i = \frac{e_0}{x_0} \tag{11}$$

$$i_0 = \frac{\sqrt{r^2 + (x_0 - x_1 - x)^2}}{x_0{}^2} e_0 \tag{12}$$

$$i_1 = \frac{\sqrt{r^2 + (x + x_1)^2}}{x_0{}^2} e_0 \tag{13}$$

$$\tan \theta = \frac{x}{r} = k \text{ gives the secondary phase angle} \tag{14}$$

and

$$\tan \theta_0 = - \frac{x_0 - x_1 - x}{r} \text{ gives the primary phase angle} \tag{15}$$

This phase angle $\theta_1 = 0$, that is, the primary supply current is non-inductive, if

$$x_0 - x_1 - x = 0,$$

that is,

$$x_1 = x_0 - x. \tag{16}$$

The primary supply can in this way be made non-inductive for any desired value of secondary load, by choosing the reactance, x_1, according to equation (16).

If $x = 0$, that is, a non-inductive secondary circuit (series incandescent lamps for instance), $x_1 = x_0$, that is, with a non-inductive secondary circuit, the primary supply current is always non-inductive, if the secondary reactance, x_1, is made equal to the primary reactance, x_0.

In this case $x_1 = x_0$, with an inductive secondary circuit $\tan \theta_0 = \dfrac{x}{r} = \tan \theta$; that is, the primary supply current has the same phase angle as the secondary load, if all three reactances (two inductive and one condensive reactance) are made equal.

In general, x_1 would probably be chosen so as to make I_0 non-inductive at full-load, or at some average load.

136. *Example.*—A 100-lamp arc circuit of 7.5 amp. is to be operated from a 6600-volt constant-potential supply $e_0 = 6600$ volts, and $i = 7.5$ amp.

Assuming 75 volts per lamp, including line resistance, gives a maximum secondary voltage, for 100 lamps, of $e' = 7500$ volts.

Assuming the power-factor of the arc circuit as 93 per cent. lagging, gives

$$\cos \theta = 0.93, \text{ or } \tan \theta = 0.4;$$

hence,

$$k = \frac{x}{r} = 0.4, \text{ and } Z = r(1 + 0.4j),$$

or

$$z = 1.077 \, r \text{ at full-load,}$$

if

$$e' = 7500 \text{ volts,}$$

$$z' = \frac{e'}{i} = 1000 \text{ ohms,}$$

hence

$$r' = 0.93 \, z' = 930 \text{ ohms,}$$
$$x' = 0.4 \, r' = 372 \text{ ohms,}$$

and

$$i = \frac{e_0}{x_0}, \text{ or } x_0 = \frac{e_0}{i} = \frac{6600}{7.5} = 880 \text{ ohms.}$$

To make the primary current i_0 non-inductive at full-load, or for $x' = 372$ ohms, this requires
$$x_1 = x_0 - x' = 508 \text{ ohms.}$$
This gives the equations
$$i = 7.5 \text{ amp.,}$$
$$e = 7.5 z = 8.08 r \text{ volts.}$$
$$i_0 = \sqrt{r^2 + (372 - 0.4 r)^2} \times \frac{6600}{880^2}$$
$$= 7.5 \sqrt{\left(\frac{r}{880}\right)^2 + \left(0.423 - \frac{r}{2200}\right)^2}$$
$$\tan \theta_0 = -\frac{372 - 0.4 r}{r}$$
$$= 0.4 - \frac{372}{r},$$

hence, leading current below full-load, non-inductive at full-load and lagging current at overload.

137. *Apparatus Economy.*—Denoting by z', r', x' the respective full-load values, the volt-ampere output at full-load is
$$Q_0 = i^2 z' = \frac{e_0^2 z'}{x_0^2} = \frac{e_0^2 r' \sqrt{1 + k^2}}{x_0^2} \tag{17}$$
volt-ampere input,
$$Q_1 = i_0 e_0 = \frac{e_0^2 r'}{x_0^2} \tag{18}$$

That is, the volt-ampere input is less than the volt-ampere output, since the input is non-inductive, while the output is not.

The power output is
$$P = i^2 r' = \frac{e_0^2 r'}{x_0^2} \tag{19}$$
which is equal to the volt-ampere input, since the losses of power in the reactances were neglected in the preceding equations.

The volt-amperes at the condenser are
$$Q' = i_1^2 x_0;$$
hence, substituting (13),
$$Q' = \frac{r'^2 + (x' + x_1)^2}{x_0^3} e_0^2 = \frac{r'^2 + (kr' + x_1)^2}{x_0^3} \tag{20}$$

The volt-ampere consumption of the first, or primary inductive reactance, x_0, is
$$Q'' = i_0^2 x_0;$$

hence, substituting (12),

$$Q'' = \frac{r'^2 + (x_0 - x' - x_1)^2}{x_0^3} e_0^2 = \frac{r'^2 + (x_0 - kr' - x_1)^2}{x_0^3} e'_0^2 \quad (21)$$

the volt-ampere consumption of the second, or secondary inductive reactance, x_1, is

$$Q''' = i^2 x_1,$$

or

$$Q''' = \frac{x_1}{x_0^2} e_0^2 \quad (22)$$

The total volt-ampere rating of the reactances required for the transformation from constant potential to constant current then is

$$Q = Q' + Q'' + Q'''$$

$$= \frac{2\, r'^2(1 + k^2) + 2\, kr'(2\, x_1 - x_0) + (x_0^2 - x_0 x_1 + 2\, x_1^2)}{x_0^3} e_0^2 \quad (23)$$

and the apparatus economy, or the ratio of volt-amperes output to the volt-ampere rating of the apparatus is

$$f = \frac{Q_0}{Q} = \frac{r' x_0 \sqrt{1 + k^2}}{2\, r'^2(1 + k^2) + 2\, kr'(2\, x_1 - x_0) + (x_0^2 - x_0 x_1 + 2\, x_1^2)} \quad (24)$$

this apparatus economy depends upon the load, r', the power-factor or phase angle of the load, k, and the secondary additional inductive reactance, x_1.

To determine the effect of the secondary inductive reactance, x_1: The apparatus economy is a maximum for that value of secondary inductive reactance, x_1, for which $\dfrac{df}{dx_1} = 0$.

Instead of directly differentiating f, it is preferable to simplify the function f first, by dropping all those factors, terms, etc., which inspection shows do not change the position of the maximum or the minimum value of the function. Thus the numerator can be dropped, the denominator made numerator, and its first term dropped, leaving

$$f' = 2\, kr'\, (2\, x_1 - x_0) + (x_0^2 - x_0 x_1 + 2\, x_1^2)$$

as the simplest function, which has an extreme value for the same value of x_1, as f. Then

$$\frac{df'}{dx_1} = 4\, kr' - x_0 + 4\, x_1 = 0,$$

and

$$x_1 = \frac{x_0 - 4\, kr'}{4} \quad (25)$$

substituting (25) in (24), gives

$$f_1 = \frac{8\,r'x_0\sqrt{1+k^2}}{16\,r'^2 - 8\,kr'x_0 + 7\,x_0^2}. \tag{26}$$

To determine the effect of the load r':
f_1 becomes a maximum for that load, r', which makes

$$\frac{df_1}{dr'} = 0,$$

or, simplified,

$$f'_1 = \frac{16\,r'^2 - 8\,kr'x_0 + 7\,x_0^2}{r'},$$

hence

$$\frac{df'_1}{dr'} = r'(32\,r' - 8\,kx_0) - (16\,r'^2 - 8\,kr'x_0 + 7\,x_0^2) = 0,$$

hence

$$r' = \frac{x_0\sqrt{7}}{4} \tag{27}$$

and, substituting (27) in (26),

$$f_2 = \frac{\sqrt{1+k^2}}{\sqrt{7}-k} \tag{28}$$

hence, for $k = 0$:

$$f_2 = \frac{1}{\sqrt{7}} = 0.378,$$

$$r' = \frac{x_0\sqrt{7}}{4} = 0.662\,x_0,$$

$$x_1 = \frac{x_0}{4} = 0.25\,x_0,$$

for $k = 0.4$:

$$f_2 = \frac{\sqrt{1.16}}{\sqrt{7}-0.4} = 0.478$$

$$r' = \frac{x_0\sqrt{7}}{4} = 0.662\,x_0$$

$$z' = \sqrt{1.16}\,\frac{x_0\sqrt{7}}{4} = 0.712\,x_0$$

$$x_1 = \frac{x_0}{4}(1 - 0.4\sqrt{7}) = -0.016\,x_0$$

$$= \text{approximately zero.}$$

At non-inductive load

$$k = 0$$

and with non-inductive primary supply, that is,

$$x_1 = x_0,$$

by substituting these values in (24), the apparatus economy is

$$f = \frac{r'x_0}{2(r'^2 + x_0^2)} \tag{29}$$

which is a maximum for

$$r' = x_0 \tag{30}$$

$$f_0 = \frac{1}{4} = 0.25 \tag{31}$$

which is rather low:

That is, non-inductive load and supply circuit do not give very high apparatus economy, but inductive reactance of the load, and phase displacement in the supply circuit, gives far higher apparatus economy, that is, more output with the same volt-amperes in reactance.

By inserting in (23), with the quantities, Q', Q'', and Q''', coefficients n_1, n_2, n_3, which are proportional respectively to the cost of the reactances per kilovolt-ampere, the expression

$$\frac{n_1Q' + n_2Q'' + n_3Q'''}{P} \tag{32}$$

then represents the commercial economy, that is, the maximum of this expression, derived by analogous considerations as before, gives the arrangement for minimum cost at given output.

138. *Power Losses in Reactances.*—

In the preceding equations, the losses of power in the reactances have been neglected. However small these may be, in accurate investigations, they require consideration as to their effect on the regulation of the transforming device, and on the efficiency.

Let

a = power-factor of inductive reactance, that is, loss of power, as fraction of total volt-amperes.

b = power-factor of condensive reactance, that is, loss of power, as fraction of total volt-amperes.

Here a and b are very small quantities, in general b, the loss in the condensive reactance, being far smaller than the loss in the inductive reactance.

Approximately, the inductive reactances are $(a + j)x_0$ and $(a + j)x_1$ respectively, and the condensive reactance is $(b - j)x_0$.

Assuming the same denotations as in the preceding paragraphs, receiver circuit

$$E = ZI \tag{33}$$

at condenser terminals

$$E_1 = E + (a + j)x_1 I$$
$$= \{Z + (a + j)x_1\}I \tag{34}$$

and also

$$E_1 = (b - j)x_0 I_1 \tag{35}$$

hence

$$I_1 = \frac{Z + (a + j)x_1}{(b - j)x_0} I \tag{36}$$

and

$$I_0 = I + I_1$$
$$= \frac{Z + (b - j)x_0 + (a + j)x_1}{(b - j)x_0} I$$
$$= \frac{Z - j(x_0 - x_1) + (bx_0 + ax_1)}{(b - j)x_0} I \tag{37}$$

and the impressed e.m.f.

$$e_0 = E_1 + (a + j)x_0 I_0;$$

hence, substituting (35) and (37),

$$e_0 = \frac{x_0 + \{Z(a+b) - jx_0(a-b) + jx_1(a+b)\} + \{x_0ab + x_1a(a+b)\}}{b - j} I \tag{38}$$

Since a and b are very small quantities, their products and squares can be neglected, then

$$e_0 = \frac{x_0 + \{Z(a + b) - jx_0(a - b) + jx_1(a + b)\}}{b - j} I \tag{39}$$

or

$$I = \frac{(b - j)e_0}{x_0 + \{Z(a + b) - jx_0(a - b) + jx_1(a + b)\}} \tag{40}$$

this can be written

$$I = -\frac{je_0}{x_0} \frac{1 + jb}{1 + \left\{\frac{Z}{x_0}(a + b) - j(a - b) + j\frac{x_1}{x_0}(a + b)\right\}};$$

hence

$$I = -\frac{je_0}{x_0}\left\{1 + ja - j\frac{x_1}{x_0}(a + b) - \frac{Z}{x_0}(a + b)\right\} \tag{41}$$

that is, due to the loss of power in the reactances, the secondary current is less than it would be otherwise, and decreases with increasing load still further.

Equation (41) can also be written

$$I = -\frac{je_0}{x_0}\left\{\left[1 - \frac{r}{x_0}(a+b)\right] - j\left[\frac{x+x_1}{x_0}(a+b) - a\right]\right\}(42)$$

here the imaginary component is very small in the parenthesis, that is, the secondary current remains practically in quadrature with the primary voltage.

The absolute value is, neglecting terms of secondary order,

$$i = \frac{e_0}{x_0}\left\{1 - \frac{r}{x_0}(a+b)\right\}. \tag{43}$$

The primary current is, by equation (37) and (40),

$$I_0 = \frac{Z - j(x_0 - x_1) + (bx_0 + ax_1)}{x_0 + Z(a+b) - jx_0(a-b) + jx_1(a+b)}\frac{e_0}{x_0} \tag{44}$$

$$= \frac{e_0}{x_0}\frac{\frac{Z}{x_0} - j\left(1 - \frac{x_1}{x_0}\right)\left(b + a\frac{x_1}{x_0}\right)}{1 + \frac{Z - jx_1}{x_0}(a+b) - j(a-b)}.$$

139. *Example.—*

Considering the same example as before: a constant-potential circuit of $e_0 = 6600$ volts supplying a 100-lamp series arc circuit, with $i' = 7.5$ amp., and $e' = 7500$ volts at full-load of 93 per cent. power-factor, that is, $k = 0.4$, and $Z = (1 - 0.4\,j)r$. Assuming now, however, the loss in the inductive reactance as 3 per cent., and in the capacity as 1 per cent., that is, $a = 0.03\ b = 0.01$, the full-load value of the secondary load impedance is: $z' = 1000$ ohms, $r' = 930$ ohms and $x' = 372$ ohms.

To give non-inductive primary supply at full-load, the following equation must be fulfilled:

$$x_1 = x_0 - x' = x_0 - 372.$$

From equation (43), the secondary current, at full-load, is

$$i' = \frac{e_0}{x_0}\left\{1 - \frac{r}{x_0}(a+b)\right\}$$

or

$$7.5 = \frac{6600}{x_0}\left\{1 - \frac{930 \times 0.04}{x_0}\right\};$$

hence

$$x_0 = 840 \text{ ohms, and } x_1 = 468 \text{ ohms.}$$

Substituting in (42), (43), (44),

$$I = -7.86j\left\{\left(1 - 0.04\frac{r}{840}\right) + j\left(0.052 - 0.016\frac{r}{840}\right)\right\}$$

$$i = 7.86\left(1 - 0.04\frac{r}{840}\right)$$

$$e = iz = 1.077\,ri$$

$$= 8.46\,r\left(1 - 0.04\frac{r}{840}\right)$$

$$I_0 = 7.86\,\frac{\left(\frac{r}{840} + 0.027\right) - j\left(0.443 - \frac{0.4\,r}{840}\right)}{\left(1 + \frac{0.04\,r}{840}\right) + j\left(\frac{0.016\,r}{840} + 0.002\right)}$$

and herefrom the power-factor, efficiency, etc.

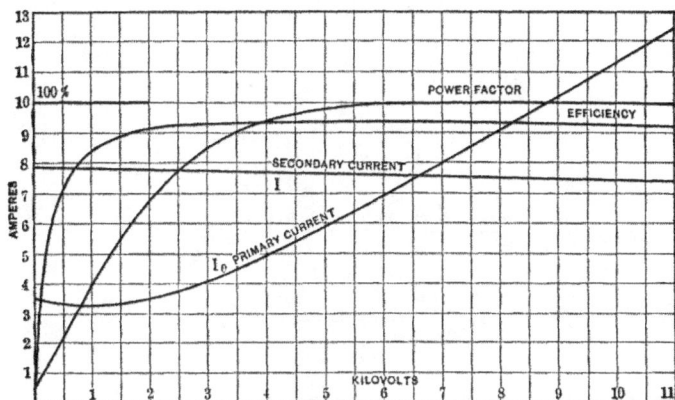

FIG. 120.

In Fig. 120, there are plotted, with the secondary, e.m.f., e, as abscissæ, the values: secondary current, i; primary current, i_0; primary power-factor, $\cos\theta$, and efficiency.

140. In alternating-current circuits small variations of frequency are unavoidable, as for instance, caused by changes of load, etc., and the inductive reactance is directly proportional, the condensive reactance inversely proportional to the frequency. Wherever inductive and condensive reactances are used in series with each other and of equal or approximately equal reactance, so more or less neutralizing each other, even small changes of frequency may cause very large variations in the result, and in such cases it is therefore necessary to investigate the effect of a change of frequency on the result: for instance, in a resonating circuit of very small power loss, a small change of frequency at constant impressed e.m.f. may change the current over an enormous range.

Since in the preceding, constant-current regulation is produced by inductive and condensive reactances in series with each other, the effect of a variation of frequency requires investigation.

Let, then, the frequency be increased by a small fraction, s.

The inductive reactance thereby changes to $x_0(1 + s)$ and $x(1 + s)$, and $Z = r + j(1 + s)x$ respectively, and the condensive reactance to $\dfrac{x_0}{1 + s}$.

Leaving all the other denotations the same, and neglecting the loss of power in the reactances,

$$E = ZI$$
$$E_1 = \{Z + j(1 + s)x_1\}I$$
$$= -\frac{jx_0I_1}{1 + s},$$

hence,

$$I_1 = j\frac{(1 + s)\,\{Z + j(1 + s)x_1\}}{x_0}\,I$$

and

$$I_0 = I + I_1 = j\frac{(1 + s)Z - j\{x_0 - (1 + s)^2x_1\}}{x_0}$$

thus

$$e_0 \approx E_1 + j(1 + s)x_0I_0$$
$$= [Z + j(1 + s)x_1 - (1 + s)^2Z + j(1 + s)\,\{x_0 - (1 + s)^2x_1\}]I$$

hence, expanding and dropping terms of higher order,

$$e_0 = +jI\,\{x_0 + s(x_0 - 2\,x_1 + 4\,Zj) - s^2(3\,x_1 + 2\,j)\},$$

or

$$I = -\frac{je_0}{x_0}\left\{1 - s\left(1 - 2\frac{x_1}{x_0} + 4\frac{jZ}{x_0}\right)\right\}. \tag{45}$$

Hence, the current is not greatly affected by a change of frequency. That is, the constant-current regulation of the above-discussed device does not depend, or require, a constancy of frequency beyond that available in ordinary alternating-current circuits.

B. Monocyclic Square

141. *General.—*

A combination of four equal reactances, two condensive and two inductive, arranged in a square as shown diagrammatically in Fig. 119, page 261, transforms a constant voltage, impressed upon one diagonal, into a constant current across the other diagonal, and inversely.

Let, then,

$E_0 = e_0$ = constant = primary impressed e.m.f., or supply voltage,

E = secondary terminal voltage,

E_1 = voltage across the condensive reactance,

E_2 = voltage across the inductive reactance,

and

I_0 = primary supply current,

I = secondary current,

I_1 = current in condensive reactance,

I_2 = current in inductive reactance,

these currents and e.m.fs. being assumed in the direction as indicated by the arrows in Fig. 119.

Let

x_0 = condensive and inductive reactances;

hence,

$$Z_1 = -jx_0 = \text{condensive reactance} \tag{1}$$

$$Z_2 = +jx_0 = \text{inductive reactance} \tag{2}$$

Then, at the dividing points,

$$I_0 = I_1 + I_2 \tag{3}$$

and

$$I = I_2 - I_1 \tag{4}$$

hence,

$$I_1 = \frac{I_0 - I}{2} \tag{5}$$

and

$$I_2 = \frac{I_0 + I}{2} \tag{6}$$

In the e.m.f. triangles,

$$e_0 = Z_1 I_1 + Z_2 I_2 \tag{7}$$

and
$$E = Z_1 I_1 - Z_2 I_2 \tag{8}$$
and
$$E = ZI \tag{9}$$
substituting (1) and (2) in (7) and (8) gives
$$e_0 = - jx_0 (I_1 - I_2) \tag{10}$$
and
$$ZI = - jx_0 (I_1 + I_2) \tag{11}$$
and, substituting herein the current,
$$e_0 = + jx_0 I \tag{12}$$
and
$$ZI = - jx_0 I_0 \tag{13}$$
hence, the secondary current is
$$I = - \frac{je_0}{x_0} \tag{14}$$
the primary current,
$$I_0 = \frac{e_0 Z}{x_0^2} \tag{15}$$
the condenser current,
$$I_1 = \frac{Z + jx_0}{2 x_0^2} \tag{16}$$
and the current in the inductive reactance,
$$I_2 = \frac{Z - jx_0}{2 x_0^2}. \tag{17}$$

The secondary terminal voltage is
$$E = - je_0 \frac{Z}{x_0} \tag{18}$$
the condenser voltage,
$$E_1 = - jx_0 I_1 = - \frac{j(Z + jx_0)}{2 x_0} e_0 \tag{19}$$
and the inductive reactance voltage,
$$E_2 = + jx_0 I_2 = + \frac{j(Z - jx_0)}{2 x_0} e_0. \tag{20}$$

The tangent of the primary phase angle is
$$\tan \theta_0 = \frac{x}{r} = \tan \theta \tag{21}$$
hence, the absolute value of the secondary current is
$$i = \frac{e_0}{x_0} \tag{22}$$

of the primary current,

$$i_0 = \frac{e_0 z}{x_0{}^2} \qquad (23)$$

of the condenser current,

$$i_1 = \frac{\sqrt{r^2 + (x_0 + x)^2}}{2\,x_0{}^2}\, e_0 \qquad (24)$$

and of the inductive reactance current

$$i_2 = \frac{\sqrt{r^2 + (x_0 - x)^2}}{2\,x_0{}^2}\, e_0. \qquad (25)$$

The secondary terminal voltage is

$$e = \frac{z}{x_0}\, e_0 \qquad (26)$$

the condenser voltage,

$$e_1 = \frac{\sqrt{r^2 + (x_0 + x)^2}}{2\,x_0}\, e_0 \qquad (27)$$

and the inductive reactance voltage,

$$e_2 = \frac{\sqrt{r^2 + (x_0 - x)^2}}{2\,x_0}\, e_0. \qquad (28)$$

142. From these equations follow the apparent powers, or volt-amperes of the different circuits as:

Output,

$$Q_0 = ei = \frac{e_0{}^2 z}{x_0{}^2}, \qquad (29)$$

Input,

$$Q_i = e_0 i_0 = \frac{e_0{}^2 z}{x_0{}^2}, \qquad (30)$$

Hence the input is the same as the output. This is obvious, since the losses of power in the reactances are neglected, and it was found (21), that the phase angle or the power-factor of the primary circuit equals that of the secondary circuit.

Apparent power of the condensive reactance,

$$Q_1 = e_1 i_1 = \frac{r^2 + (x_0 + x)^2}{4\,x_0{}^3}\, e_0{}^2. \qquad (31)$$

Inductance,

$$Q_2 = e_2 i_2 = \frac{r^2 + (x_0 - x)^2}{4 x_0{}^3}\, e_0{}^2; \qquad (32)$$

and, therefore, total volt-ampere capacity of the reactances is

$$Q = 2\,(Q_1 + Q_2)$$

$$= \frac{r^2 + x^2 + x_0{}^2}{x_0{}^3} e_0{}^2;$$

hence

$$Q = \frac{z^2 + x_0{}^2}{x_0{}^3} e_0{}^2 \tag{33}$$

and,
apparatus economy,

$$f = \frac{Q_0}{Q} = \frac{zx_0}{z^2 + x_0{}^2} \tag{34}$$

hence a maximum for $\qquad z = x_0 \tag{35}$

and this maximum is equal to $f_0 = \frac{1}{2}$, or 50 per cent. $\tag{36}$

That is, the maximum apparatus economy of the monocyclic square, as discussed here, is 50 per cent., or in other words, for every kilovolt-ampere output, 2 kv.-amp. in reactances have to be provided.

This apparatus economy is higher than that of the T-connection, in which under the same conditions, that is for $x_1 = x_0$, the apparatus economy was only 25 per cent.

The commercial, or cost economy would be given by

$$g = \frac{Q_0}{2\,(n_1 Q_1 + n_2 Q_2)} = \text{maximum} \tag{37}$$

where
$n_1 =$ price per kilovolt-ampere of condensive reactance, $n_2 =$ price per kilovolt-ampere of inductive reactance.

143. *Example.—*

Considering the same problem as under A. From a constant impressed e.m.f. $e_0 = 6600$ volts, a 100-lamp arc circuit, of 93 per cent. power-factor, is to be operated, requiring

$$i = 7.5 \text{ amp.}$$
$$Z = r + jx$$
$$= r\,(1 + jk)$$

where

$$k = \frac{x}{r} = 0.4;$$

hence

$$Z = r\,(1 + 0.4\,j),$$

and at full-load

$$e' = 7500 \text{ volts.}$$

Then, from (22),

$$x_0 = \frac{e_0}{i} = 880 \text{ ohms,} \qquad z' = \frac{e'}{i} = 1000 \text{ ohms;}$$

hence
$$r' = 930 \text{ ohms}, \qquad x' = kr' = 372 \text{ ohms},$$
and, therefore,
$$i = 7.5 \text{ amp.,}$$
$$i_0 = 7.5 \frac{z}{880} \text{ amp.,}$$
$$e = 7.5 z,$$
and at full-load, or $r = 930$, when denoting full-load values by prime,
$$i' \quad = 7.5 \text{ amp.,}$$
$$i'_0 \quad = 7.93 \text{ amp.,}$$
$$i'_1 \quad = 6.65 \text{ amp.,}$$
$$i'_2 \quad = 4.52 \text{ amp.,}$$
$$e' \quad = 7500 \text{ volts,}$$
$$e'_1 \quad = 5850 \text{ volts,}$$
$$e'_2 \quad = 3980 \text{ volts,}$$
$$\left. \begin{array}{l} P'_{ai} = \\ P'_{a_0} = \end{array} \right\} 56.25 \text{ kv.-amp.}$$
$$P'_{a_2} = 38.9 \text{ kv.-amp.}$$
$$P'_a = 18.0 \text{ kv.-amp.}$$
$$P'_a{}^2 = 113.8 \text{ kv.-amp.}$$
$$f' = 0.4943$$

or 49.43 per cent. that is, practically the maximum.

144. *Power Loss in Reactances.*—

In the preceding, as first approximation, the loss of power in the reactances has been neglected, and so the constancy of current, i, was perfect, and the output equal to the input. Considering, however, the loss of power in the reactances, it is found that the current, i, varies slightly, decreasing with increasing load, and the input exceeds the output.

Let, then,
$$Z_1 = (b - j) x_0 = \text{condensive reactance,}$$
$$Z_2 = (a + j) x_0 = \text{inductive reactance,}$$

otherwise retaining the same denotations as in the preceding paragraphs,

Then, substituting in (7) and (8),

$$\frac{e_0}{x_0} = (b - j)I_1 + (a + j)I_2 \qquad (38)$$

$$\frac{Zi}{x_0} = (b - j)I_1 - (a + j)I_2 \qquad (39)$$

Assuming

$$a = c_1 + c_2 \atop b = c_1 - c_2 \Big\} \tag{40}$$

$$c_1 = \frac{a + b}{2}, \; c_2 = \frac{a - b}{2}. \tag{41}$$

Substituting in (38) and (39)

$$\frac{e_0}{x_0} = -j(I_1 - I_2) + c_1(I_1 + I_2) - c_2(I_1 - I_2)$$

$$\frac{ZI}{x_0} = -j(I_1 + I_2) + c_1(I_1 - I_2) - c_2(I_1 + I_2),$$

substituting herein from equations (3) and (4) gives

$$\frac{e_0}{x_0} = (c_2 + j)I + c_1 I_0 \tag{42}$$

and

$$\frac{ZI}{x_0} = -c_1 I - (c_2 + j)I_0 \tag{43}$$

and from these two equations with the two variables, I and I_0, it follows from (43) that

$$I_0 = -\frac{Z - c_1 x_0}{(j + c_2)x_0} I \tag{44}$$

Substituting (44) in (42), transposing, and dropping terms of secondary order, that is, products and squares of c_1 and c_2, gives

$$I = -\frac{je_0}{x_0}\left\{1 + jc_2 - c_1\frac{Z}{x_0}\right\} \tag{45}$$

substituting (45) in (44), and transposing,

$$I_0 = \frac{e_0}{x_0^2}\left\{Z + c_1\frac{x_0^2 - Z^2}{x_0} + 2jc_2 Z\right\} \tag{46}$$

then, substituting (45) and (46) in (5) and (6),

$$I_1 = \frac{e_0}{x_0^2}\left\{\frac{Z_1 + jx_0}{2} + \frac{c_1 - c_2}{2}x_0 + jZ\frac{2c_2 - c_1}{2} - \frac{c_1 Z^2}{2\,x_0}\right\} \tag{47}$$

$$I_2 = \frac{e_0}{x_0^2}\left\{\frac{Z_1 - jx_0}{2} + \frac{c_1 + c_2}{2}x_0 + jZ\frac{2c_2 + c_1}{2} - \frac{c_1 Z^2}{2\,x_0}\right\} \tag{48}$$

and the absolute value is

$$i = \frac{e_0}{x_0}\sqrt{\left(1 - c_1\frac{r}{x_0}\right)^2 + \left(c_2 - c_1\frac{x}{x_0}\right)^2}$$

or, approximately,

$$i = \frac{e_0}{x_0}\left(1 - c_1 \frac{x_0}{r}\right), \text{ etc.} \tag{49}$$

145. *Example.—*

Considering the same example as before, of a 7.5-amp. 100-lamp arc circuit operated from a 6600-volt constant-potential supply, and assuming again as in paragraph 139:

3 per cent. power-factor of inductive reactance, or $a = 0.03$.

1 per cent. power-factor of condensive reactance, or $b = 0.01$. It is then,

$$c_1 = 0.02, c_2 = 0.01,$$

and at full-load,

$$i' = \frac{e_0}{x_0}\left(1 - c_1 \frac{r'}{x_0}\right)$$

or,

$$7.5 = \frac{6600}{x_0}\left(1 - 0.02 \frac{930}{x_0}\right);$$

hence,

$$x_0 = 861, \text{ and } i = 7.66\left(1 - 0.02\frac{r}{861}\right),$$

and we have, approximately,

$$I_0 = 7.66\left\{\frac{r}{861} + 0.02 + \frac{0.42\,jr}{861}\right\},$$

$$I_1 = 3.83\left\{\frac{r}{861} + j\left(1 + \frac{0.4\,r}{861}\right)\right\},$$

$$I_2 = 3.83\left\{\frac{r}{861} - j\left(1 - \frac{0.4\,r}{861}\right)\right\},$$

$$e = zi = 1.077\,ri.$$

In Fig. 121 are plotted, with the secondary terminal voltage, e, as abscissæ, the values of secondary current, i; primary current, i_0; condenser current, i_1; inductive reactance current, i_2, and efficiency.

As seen, with the monocyclic square, the current regulation is closer, and the efficiency higher than with the T connection. This is due to the lesser amount of reactance required with the monocyclic square.

The investigation of the effect of a variation of frequency on the current regulation by the monocyclic square, now can be carried out in the analogous manner as in A with the T connection.

C. General Discussion of Constant-potential Constant-current Transformation

146. In the preceding methods of transformation between constant potential and constant current by reactances, that is, by combinations of inductive and condensive reactances, the constant alternating current is in quadrature with the constant e.m.f. Even in constant-current control by series inductive reactances, the constancy of current is most perfect for light loads, where the reactance voltage is large and thus the constant-current voltage almost in quadrature, and the constant-current control is impaired in direct proportion to the shift of phase of the constant current from quadrature relation.

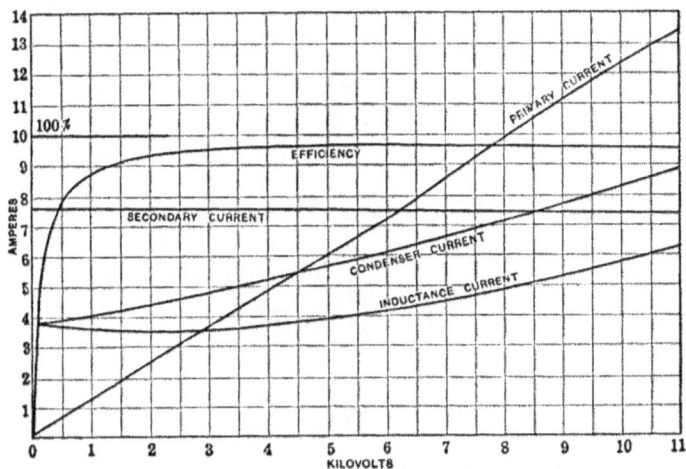

FIG. 121.

The cause hereof is the storage of energy required to change the character of the flow of energy.   That is, the energy supplied at constant potential in the primary circuit, is stored in the reactances, and returned at constant current, in the secondary circuit.

The storage of the total transformed energy in the reactances allows a determination of the theoretical minimum of reactive power, that is, of inductive and condensive reactances required for constant-potential to constant-current transformation, since the energy supplied in the constant-current circuit must be stored for a quarter period after being received from the constant-potential circuit.

Let

$$p = P(1 + \cos 2\ \theta)$$

= Power supplied to the constant-current circuit; thus, neglecting losses,

$$p_0 = P(1 - \cos 2\ \theta)$$

= Power consumed from the constant-potential circuit, and

$$p_0 - p = 2\ P \cos 2\ \theta$$

= Power in the reactances.

That is, to produce the constant-current power, P, from a single-phase constant-potential circuit, the apparent power, $2\ P$, must be used in reactances; or, in other words, per kilowatt constant-current power produced from a single-phase constant-potential circuit, reactances rated at 2 kv.-amp. as a minimum are required, arranged so as to be shifted 45° against the constant-potential and the constant-current circuit.

The reactances used for the constant-potential constant-current transformation may be divided between inductive and condensive reactances in any desired proportion.

The additional wattless component of constant-potential power is obviously the difference between the wattless volt-amperes of the inductive and that of the condensive reactances. That is, if the wattless volt-amperes of reactance is one-half of inductive and one-half of condensive, the resultant wattless volt-amperes of the main circuit is zero, and the constant-potential circuit is non-inductive, at non-inductive load, or consumes current proportional to the load.

If A is the condensive and B the inductive volt-amperes, the resultant wattless volt-amperes is $B-A$; that is, a lagging wattless volt-amperes of $B-A$ (or a leading volt-ampere of $A-B$) exist in the main circuit, in addition to the wattless volt-amperes of the secondary circuit, which reappear in the primary circuit.

147. These theoretical considerations permit the criticism of the different methods of constant-potential to constant-current transformation in regard to what may be called their *apparatus economy*, that is, the kilovolt-ampere rating of the reactance used, compared with the theoretical minimum rating required.

1. *Series inductive reactance*, that is, a reactive coil of constant inductive reactance in series with the circuit. This arrangement obviously gives only imperfect constant-current control. Per-

mitting a variation of 5 per cent. in the value of the current (that is, full-load current in 5 per cent. less than no-load current) and assuming 4 per cent. loss in the reactive coil, a reactance rated at 2.45 kv.-amp. is required per kilowatt constant-current load. This apparatus operates at 87.9 per cent. economy and 30 per cent. power-factor.

Assuming 10 per cent. variation in the value of the current, reactance rated at 2.22 kv.-amp. is required per kilowatt constant-current load.   This arrangement operates at an economy of 91.8 per cent., and a power-factor of 49.5 per cent.

In the first case, the apparatus economy, that is, the ratio of the theoretical minimum kilovolt-ampere rating of the reactance to the actual rating of the reactance is 88 per cent., and in the last case 92 per cent., thus the objection to this method is not the high rating of the reactance and the economy, but the poor constant-current control, and especially the very low power-factor.

2. *Inductive and condensive reactances in resonance condition*, the condensive reactance being shunted by the constant-current circuit.   In this case, condensive reactance rated at 1 kv.-amp. and inductive reactance rated at 2 kv.-amp. are required per kilowatt constant-current load, and the main circuit gives a constant wattless lagging apparent power of 1 kv.-amp.   Assuming again 4 per cent. loss in the inductive and 2 per cent. loss in the condensive reactances, gives a full-load efficiency of 91 per cent. and a power-factor (lagging) of 74 per cent.   The apparatus economy by this method is 66.7 per cent.

3. *Inductive and condensive reactances in resonance condition*, the *inductive reactance shunted* by the constant-current circuit. In this case, as a minimum, per kilowatt constant-current load, condensive reactance rated at 2 kv.-amp. and inductive reactance rated at 1 kv.-amp. is required, and the main circuit gives a constant wattless leading apparent power of 1 kv.-amp.   The efficiency of transformation is at full-load 92.5 per cent., the power-factor (leading) 73 per cent., the apparatus economy 66.7 per cent.

4. *T-connection*, that is, two equal inductive reactances in series to the constant-current circuit and shunted midway by an equal condensive reactance.   In this case per kilowatt constant-current load, condensive reactance rated at 2 kv.-amp. and inductive reactance rated at 2 kv.-amp. are required.

The main circuit is non-inductive at all non-inductive loads, that is, the power-factor is 100 per cent.

The full-load efficiency is 89.3 per cent. apparatus economy 50 per cent.

5. *The monocyclic square.* In this case a condensive reactance rated at 1 kv.-amp. and inductive reactance rated at 1 kv.-amp. are required per kilowatt constant-current load. The main circuit is non-inductive at all non-inductive loads, that is, the power-factor is 100 per cent. The full-load efficiency is 94.3 per cent., the apparatus economy 100 per cent.

6. The *monocyclic square* in combination with a *constant-potential polyphase system* of impressed e.m.f. In this case, per kilowatt constant-current load, condensive reactance rated at 0.5 kv.-amp. and inductive reactance rated at 0.5 kv.-amp. are required. The main circuits are non-inductive at all loads, that is, the power-factor is 100 per cent. The full-load efficiency is over 97 per cent. the apparatus economy 200 per cent.

148. In the preceding, the constant-potential to constant-current transformation with a single-phase system of constant impressed e.m.f., has been discussed; and shown that as a minimum in this case, to produce 1 kw. constant-current output, reactances rated at 2 kv.-amp. are required for energy storage. The constant current is in quadrature with the main or impressed e.m.f., but can be either leading or lagging. Thus the total range available is from 1 kw. leading, to zero, to 1 kw. lagging. Hence if a constant-quadrature e.m.f. is available by the use of a polyphase system, the range of constant current can be doubled, that is, reactance rated at 2 kv.-amp. can be made to control the potential for 2 kw. constant-current output in the way shown in Fig. 122 for a three-phase, and Fig. 123 for a quarter-phase system of impressed e.m.f.

In this case, one transformer feeds a monocyclic square, the other transformer inserts an equal constant e.m.f. in quadrature with the former, which from no-load to half-load is subtractive, from half-load to full-load is additive, that is, at full-load, both phases are equally loaded; at half-load only one phase is loaded and at no-load one phase transforms energy into the other phase.

The monocyclic e.m.f. square in this case, when passing from full-load to no-load, gradually collapses to a straight line at half-load, then overturns and opens again to a square in the opposite direction at no-load. That is, at full-load the transformation is from constant potential to constant current, and at no-load the transformation is from constant current to constant potential.

Obviously with this arrangement the efficiency is greatly increased by the reduction of the losses to one-half, and the constant-current control improved.

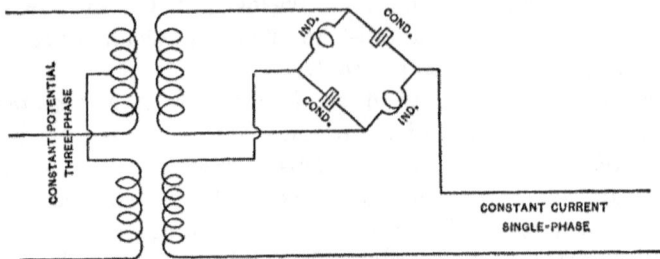

FIG 122.

At the same time, the sensitiveness of the arrangement for distortion of the wave shape, as will be discussed later, is greatly reduced, due to the insertion of a constant-potential e.m.f. into the constant-current circuit.

Obviously the arrangments in Figs. 122 and 123 are not the only ones, but many arrangements of inserting a constant-quadrature e.m.f. into the monocyclic square or triangle are suitable.

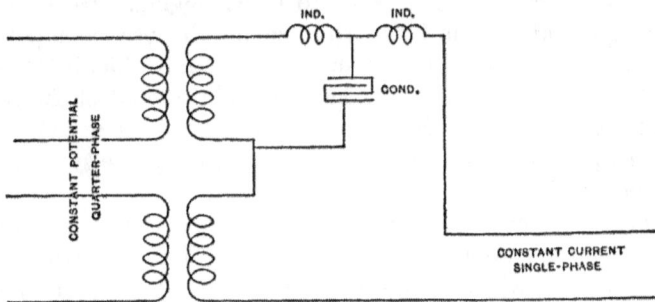

FIG. 123.

Different arrangements can also be used of the constant-current control, for instance, the inductive and condensive reactances in resonance condition with their common connection connected to the center of an autotransformer or transformer, with the insertion of the constant-potential quadrature e.m.f. in the latter circuit as shown in Fig. 124, or the T-connection, shown applied to a quarter-phase system in Fig. 125.

Constant-potential apparatus and constant-current single-phase circuits can also be operated from the same transformer secondaries in a similar manner, as indicated in Fig. 124 for a three-phase secondary system.

In Figs. 122 to 125 the arrangement has been shown as applied to step-down transformers, but in the estimate of the efficiency

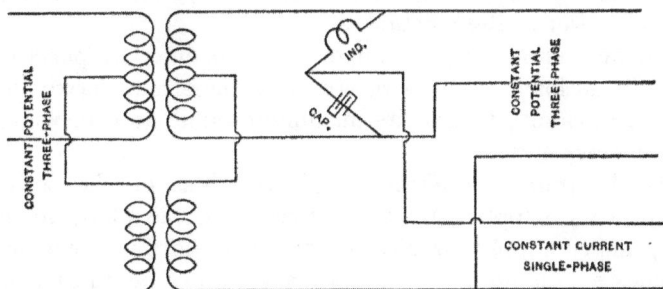

FIG. 124.

the losses in these transformers have not been included, since these transformers are obviously not essential but merely for the convenience of separating electrically the constant-current circuit from the high-potential line. It is evident, for instance, in Fig. 124, that the constant-current and constant-potential cir-

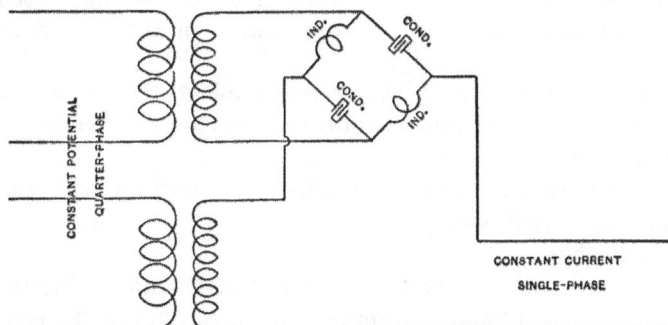

FIG. 125.

cuits instead of being operated from the three-phase secondaries of the step-down transformers can be operated directly from the three-phase primaries by replacing the central connection of the one transformer by the central connection of the auto-transformer.

D. Problems

149. In the following problems referring to constant-potential to constant-current transformation by reactances, it is recommended:

(a) To derive the equation of all the currents and e.m.fs., in complex quantities as well as in absolute terms, while neglecting the loss of power in the reactances.

(b) To determine the volt-amperes in the different parts of the circuit, as load, reactances, etc., and therefrom derive the apparatus economy, to find its maximum value, and on which condition it depends.

(c) To determine the effect of inductive load on the power of the primary supply circuit, to investigate the phase angle of the primary supply circuit, and the conditions under which it becomes a minimum, or the primary supply becomes non-inductive.

(d) To redetermine the equations of the problem, while considering the power lost in the reactances, and apply these equations to a numerical example, plotting all the interesting values.

(e) To investigate the effect of a change of frequency on the equations, more particularly on the constant-current regulation.

(f) To investigate the effect of distortion of wave shape, that is, the existence of higher harmonics in the impressed e.m.f., and their suppression or reappearance in the secondary circuit.

(g) To study the reversibility of the problem, that is, apply (a) to (f) to the reversed problem of transformation from constant current to constant potential.

Some of the transforming devices between constant potential and constant current are:

A. Single-phase.

(a) The resonating circuit, or condensive and inductive reactances, of equal values, in series with each other in the constant-potential circuit, and the one reactance shunted by the constant-current circuit.

(b) T-connection, as partially discussed in (A).

(c) The monocyclic square, as partially discussed in (B).

(d) The monocyclic triangle: a condensive reactance and an inductive reactance of equal values, in series with each other across the constant-potential circuit, the constant-current circuit connecting between the reactance neutral, or the common connection between the two (opposite) reactances, and the neutral of a compensator or autotransformer connected across the constant-potential circuit. Instead of the compensator neutral, the constant-current circuit can be carried back to the neutral of the transformer connected to the constant-potential circuit.

B. *Polyphase.*

(a) In the two-phase system the two phases of e.m.fs., e_0 and je_0, are connected in series with each other, giving the outside terminals, A and B, and the neutral or common connection, C. A condensive reactance and an inductive reactance of equal values, in series with each other and with their neutral or common connection, D, are connected either between A and B, and the constant-current circuit between C and D, or the reactances are connected between A and C, and the constant-current circuit between B and D. In either case, several arrangements are possible, of which only a few have a good apparatus economy.

(b) In a three-phase system, a condensive reactance, an inductive reactance equal in value to that of the condensive reactance and the constant-current circuit, are connected in star connection between the three-phase, constant-potential terminals. Here also two arrangements are possible, of which one only gives good apparatus economy.

(c) In a constant-potential three-phase system, each of the three terminals, A, B, C, connects with a condensive and an inductive reactance, and all these reactances are of equal value, and joined together in pairs to three terminals, a, b, c, so that each of these terminals, a, b, c, connects an inductive with a condensive reactance. a, b, c, then, are constant-current three-phase terminals, that is, the three currents at a, b, c, are constant and independent of the load or the distribution of load, and displaced from each by one-third of a period. This arrangement is especially suitable for rectification of the constant alternating-current, to produce constant direct current.

150. Some further problems are:

1. In a single-phase, constant-current transforming device, as the monocyclic square, the constant current, i, is in quadrature with the constant impressed e.m.f., e_0. By inserting a constant-potential e.m.f., E_3, into the constant-current circuit, the appa-

ratus economy can be greatly increased, in the maximum can be doubled; that is, the e.m.f., E_3 gives constant-power output, and from no-load to half-load, the transformation is from constant current to constant potential, that is, a part of the power supply, E_3, is transformed into the circuit, of e.m.f., e_0, that is, the circuit, e_0, receives power. Above half-load the circuit of e_0 transforms power from constant potential to constant current, into the circuit of e.m.f. E_3.

Since i is in time quadrature with e_0, with non-inductive secondary load, that is, the secondary terminal voltage, E, in phase with the secondary current, i, E_3 should also be in phase with i, that is, $E_3 = je_3$. With inductive secondary load, of phase angle, θ, E_3 should be in phase with E, that is, leading i by angle θ, or should be: $E_3 = je_3 (1 + kj)$.

It is interesting, therefore, to investigate how the equation of the constant-potential constant-current devices are changed by the introduction of such an e.m.f., E_3, at non-inductive as well as at inductive load, if $E_3 = je_3$, or $E_3 = j(e_3 - je'_3)$, in either case, and also to determine how such an e.m.f., E_3, of the proper phase relation, can be derived directly or by transformation from a two-phase or three-phase system.

2. If in the constant-potential constant-current transforming device one of the reactances is gradually changed, increased or decreased from its proper value, then in either case the regulation of the system is impaired. That is, the ratio of full-load current to no-load current falls off, but at the same time, the no-load current also changes.

With increase of load, the frequency of the system decreases, due to the decreasing speed of the prime mover, if the output of the system is an appreciable part of the rated output. If, therefore, the reactances are adjusted for equality of the frequency of full-load, at the higher frequency of no-load, the inductive reactance is increased, and thereby the no-load current decreased below the value which it would have at constant reactance, and in this manner the increase of current from full-load to no-load is reduced.

Such a drop of speed and therefore of frequency, s, can therefore be found, that the current at full-load, with perfect equality between the reactances, equals the current at no-load, where the reactances are not quite equal. That is, the variation of frequency compensates for the incomplete regulation of the current, caused by the energy loss in the reactances. Furthermore, with a given variation of frequency, s, from no-load to full-load, the reactances can be chosen so as to be slightly unequal at full-load, and more unequal at no-load; the change of current caused hereby compensates for the incomplete current regulation, that is, with a given frequency variation, s (within certain limits), the current regulation can be made perfect from no-load to full-load, by the proper degree of inequality of the reactances.

It is interesting to investigate this, and apply to an example, a, to determine the proper s, for perfect equality of reactance at full-load; b, with a given value of $s = 0.04$, to determine the inequality of reactance required. Assuming $a = 0.03$; $b = 0.01$.

3. If one point of the constant-current circuit, either a terminal or an intermediate point, connects to a point of the constant-potential circuit, either a terminal or some intermediate point (as inside of a transformer winding), the constant current is not changed hereby, that is, the regulation of the system is not impaired, and no current exists in the cross between the two circuits. The distribution of potential between the reactances, however, may be considerably changed, some reactances receiving a higher, others a lower voltage.

It follows herefrom, that a ground on a constant-current system does not act as a ground on the constant-potential system, but electrically the two systems, although connected with each other, are essentially independent, just as if separated from each other by a transformer. So, for instance, in the monocyclic square, one side may be short-circuited without change of current in the secondary, but with an increase of current in the other three sides. It is interesting to investigate how far this independence of the circuits extends.

In general, as an example, the following constants may be chosen: In the constant-potential circuit: $e_0 = 6600$ volts and $i'_0 = 10$ amp. at full-load.

In the constant-current circuit: $i = 7.5$ amp., $e' = 7500$ volts at full-load.

Or, especially in polyphase systems, e', respectively, i'_0 corresponding to the maximum economy point,

and $\qquad\qquad a = 0.03; b = 0.01.$

E. Distortion of Voltage Wave

151. It is of interest to investigate what effect the distortion of the voltage wave, that is, the existence of higher harmonics in the wave of supply voltage, has on the regulation of the constant-potential constant-current transformation systems discussed in the preceding.

Where constant current is produced by inductive reactance only, higher harmonics in the voltage wave naturally are suppressed the more, the larger the inductive reactance and the higher the order of the harmonic.

An increase of the intensity of the harmonics in the current wave, over that in the voltage wave, and with it an impairment of the constant-current regulation, could thus be expected only with devices using capacity reactance.

As example may be investigated the effect of the distortion of the impressed voltage wave on the T connection, and on the monocyclic square.

The symbolic method of treating general alternating waves may be used, as discussed in Chapter XXVII, of "Theory and Calculation of Alternating-current Phenomena," fifth edition, page 379. That is, the voltage wave is represented by

$$E = \sum_{1}^{\infty} {}^{2n-1} (e'_n - j_n e'_n)$$

and the impedance by

$$Z = r + j_n \left(n x_m + x_0 + \frac{x_c}{n} \right)$$

where

$$n = \text{order of harmonic.}$$

A. T Connection or Resonating Circuit

152. Assuming the same denotation as before, we have, for the nth harmonic:

primary inductive reactance,

$$Z_0 = + j n x_0;$$

secondary inductive reactance,

$$Z_1 = + j n x_1;$$

condensive reactance,

$$Z_2 = - \frac{j x_0}{n};$$

when neglecting the energy losses in the reactances, load

$$Z = r(1+jnk)$$

therefore, also for the nth harmonic.

$$E = r(1+jnk) I_1$$

$$E_1 = E + Z_1 I$$

$$= [r (1+jnk) + jnx_1] I,$$

and also

$$E_1 = - j \frac{x_0}{n} I_1;$$

hence,

$$I_1 = j \frac{n[r (1 + jnk) + jnx_1]}{x_0} I,$$

and

$$I_0 = I + I_1$$

$$= \frac{jx_0 - jn^2x_1 - nr(1 + jnk)}{jx_0} I;$$

hence,

$$E_1 = E_1 + Z_0 I_0$$
$$= \{[r (1 + jnk) + jnx_1] + n [jx_0 - jn^2x_1 - nr (1 + jnk)]\} I$$
$$= \{ - (n^2 - 1)r (1 + jnk) - jnx_1 (n^2 - 1) + jnx_0\} I;$$

hence,

$$I = \frac{- jE_0}{nx_0 - nx_1(n^2 - 1) + j (n^2 - 1) r (1 + jnk)}$$

$$= \frac{- jE_0}{nx_0 - (n^2 - 1)[n (x_1 + kr) + jr]};$$

hence, approximately, for higher values of n,

$$I = + \frac{jE_0}{n^2(x_1 + kr)};$$

that is, for larger values of n, $I = 0$, or the higher harmonics in the current wave disappear.

Herefrom, by substituting in the preceding equations, the supply current, I_0, the condenser current, I_1, their respective e.m.fs., etc., are derived.

It is then, in general expression:

If

$$E_0 = \sum_1^\infty (e_n - j_n e_n{}^1) = \text{impressed e.m.f.},$$

$$I = \sum_1^{\infty} \frac{j_n(e_n + j_n e_n{}')}{n x_0 - (n^2 - 1)\left[n(x_1 + kr) + jr\right]}$$

$$= -\frac{j(e_0 - je_0{}')}{x_0} - \sum_3^{\infty} \frac{j_n(e_n - je_n{}')}{n^3(x_1 + kr)},$$

the equation of the secondary current.

For instance, let

$$E_0 = 6600 \; \{1_1 - 0.20_3 - 0.15_5 + 0.06_7 - 0.25 \, j_3\}$$
$$= \text{constant-impressed c.m.f.}$$

or, absolute,

$$e_0 = 6600 \; \sqrt{1 + 0.20^2 + 0.15^2 + 0.06^2 + 0.25^2}$$
$$= 6600 \times 1.062$$
$$= 7010 \text{ volts,}$$

and choosing the same values as before, in paragraph 143,

$$x_0 = 880 \text{ ohms,}$$
$$x_1 = 508 \text{ ohms,}$$
$$r' = 930 \text{ ohms,}$$
$$k = 0.4;$$

it is, substituting,

$$I = -7.5 \, j + \frac{60.0_3 - 48.8 \, j_3 - 8.0 \, j_5 + 1.2 \, j_7}{508 + 0.4 \, r},$$

or, absolute,

$$i = \sqrt{7.5^2 + \frac{60.0^2 + 48.8^2 + 8.0^2 + 1.2^2}{(508 + 0.4 \, r)^2}}$$

$$= \sqrt{7.5^2 + \frac{604{,}600}{(508 + 0.4 \, r)^2}};$$

hence, at no-load,

$$i = 7.5 \times 1.00021$$

and, at full-load,

$$r = 930,$$
$$i = 7.5 \times 1.00003.$$

That is, the current wave is as perfect a sine wave as possible, regardless of the distortion of the impressed e.m.f., which, for instance, in the above example, contains a third harmonic of 32 per cent. Or in other words, in the T connection or the resonating circuit, all harmonics of e.m.f. are wiped out in the current wave, and this method indeed offers the best and most convenient means of producing perfect sine waves of current from any shape of e.m.f. waves.

153. **B. Monocyclic Square**

Assuming the same denotation as before, we have for the nth harmonic:

inductive reactance,
$$Z_2 = + jnx_0;$$

condensive reactance,
$$Z_1 = - j\,\frac{x_0}{n};$$

load,
$$Z = r(1 + jnk);$$

currents,
$$I_1 = \frac{I_0 - I}{2},$$

and
$$I_2 = \frac{I_0 + I}{2};$$

e.m.fs.,
$$E_0 = Z_1 I_1 + Z_2 I_2,$$
$$ZI = Z_1 I_1 - Z_2 I_2;$$

hence, substituting, we have
$$E_0 = - jx_0\left(\frac{I_1}{n} - xI_2\right),$$
$$r(1 + jnk)I = - jx_0\left(\frac{I_1}{n} + nI_2\right);$$

thus,
$$E_0 = - \frac{jx_0}{2}\left\{ I_0\left(\frac{1}{n} - n\right) - I\left(\frac{1}{n} + n\right) \right\},$$
$$r(1 + jnk)I = - \frac{jx_0}{2}\left\{ I_0\left(\frac{1}{n} + n\right) - I\left(\frac{1}{n} - n\right) \right\};$$

then, combining, we obtain
$$E_0\left(n + \frac{1}{n}\right) + r(1 + jnk)I\left(n - \frac{1}{n}\right) = + \frac{jx_0}{2}I\left\{ \left(n + \frac{1}{n}\right)^2 - \left(n - \frac{1}{n}\right)^2 \right\}$$
$$= + 2jx_0I,$$

and
$$I = \frac{- jE_0\left(x + \frac{1}{n}\right)}{2\,x_0 + jr(1 + jnk)\left(n - \frac{1}{n}\right)}$$
$$= \frac{- jE_0\,(n^2 + 1)}{2\,nx_0 + jr(n^2 - 1)(1 + jnk)},$$

and herefrom I_1, I_1, I_2, etc.

Approximately, for higher values of n, and for high loads, r,

$$I = \frac{jE_0}{nkr}.$$

That is, the higher harmonics of current decrease proportionally to their order, at heavy loads—that is, large values of r. For light loads, however, or small values of r, and in the extreme case, at no-load, or $r = 0$, it is

$$I = -\frac{jE_0\,(n^2 + 1)}{2\,nx_0},$$

and, approximately,

$$I = -\frac{jE_0 n}{2\,x_0}.$$

That is, the current is increased proportional to the order of the harmonics, or in other words, at no-load, in the monocyclic square, the higher harmonics of impressed e.m.fs. produce increased values of the higher harmonics of current, that is, the wave-shape distortion is increased the more, the higher the harmonics.

In general expression:

If

$$E_0 = \sum_1^\infty (e_n - j_n e'_n) = \text{impressed e.m.f.},$$

$$I = \sum_1^\infty \frac{j_n(n^2 + 1)(e_n - j_n e'_n)}{2\,nx_0 + j_n r(n^2 - 1)(1 + j_n nk)},$$

and herefrom I_0, I_1, I_2, etc.

For instance, let

$$E_0 = 6600\ \{1_1 - 0.20_3 - 0.25\,j_3 - 0.15_5 + 0.06_7\}$$
$$= \text{constant-impressed e.m.f.},$$

or, absolute,

$$e_0 = 7010 \text{ volts},$$

and, choosing the same values as before,

$$x_0 = 880 \text{ ohms},$$
$$r' = 930 \text{ ohms},$$
$$k = 0.4;$$

it is, substituted,

$$I = 7.5 - \frac{(2.5 - 2\,j_3)\,6600}{5280 - (9.6 - 8\,j)r} \quad \frac{25{,}740}{8800 - (48 - 24\,j_5)r}$$
$$+ \frac{19{,}800}{12{,}320 - (134.4 - 48\,j)r};$$

herefrom follows,

at no-load, $r = 0$,
$$I = 7.5 - (3.12 - 2.5 j_3) - 2.92 + 1.61_7.$$

That is, at no-load, the secondary current contains excessive higher harmonics, for instance, a third harmonic,

$$\sqrt{3.12^2 + 2.5^2} = 4.0, \text{ or } 53.3 \text{ per cent. of the fundamental.}$$

Absolute, the no-load current is

$$i = \sqrt{7.5^2 + 3.12^2 + 2.5^2 + 2.92^2 + 1.61^2} = 9.13 \quad \text{amp.}$$

At full-load, or $r = 930$, it is

$$I = 7.5 + (2.18 + 1.07 j_3) + (0.51 + 0.32 j_5) - (0.14 + 0.06 j_7);$$

that is, at full-load, the harmonics, while still intensified, are less than at no-load, and decrease with their order, n, more rapidly. The absolute value is

$$i = \sqrt{7.5^2 + 2.18^2 + 1.07^2 + 0.51^2 + 0.32^2 + 0.14^2 + 0.06^2}$$
$$= 7.91 \text{ amp.}$$

Instead of 7.5 amp., the value which the current would have at all loads if no higher harmonics were present, the higher harmonics of impressed e.m.f. raise the current to 9.13 amp., or by 21.7 per cent. at no-load, and to 7.91 amp., or by 5.5 per cent. at full-load, while the impressed e.m.f. is increased by 6.2 per cent. by its higher harmonics.

It follows also that the constant-current regulation of the system is seriously impaired, and between no-load and full-load the current decreases from 9.13 to 7.91 amp., or by 15.4 per cent., which as a rule is too much for an arc circuit.

154. It follows herefrom:

While the T connection of transformation from constant potential to constant current suppresses the higher harmonics of impressed e.m.f. and makes the constant current a perfect sine wave, the monocyclic square intensifies the higher harmonics so that the higher harmonics of impressed e.m.f. appear at greatly increased intensity in the constant-current wave. The increase of the higher harmonics is different for the different harmonics and for different loads, and the distortion of wave shape produced hereby is far greater at no-load, and the constant-current regulation of the system is thereby greatly impaired, and at load the distortion is less, and very high harmonics are fairly well suppressed, and the operation of an arc circuit so feasible.

Assuming, then, that in the monocyclic square of constant-potential constant-current transformation, with a distorted wave of impressed e.m.f., we insert in series to the monocyclic square into the main circuit, I_0, two reactances of opposite sign, which are equal to each other for the fundamental frequency,

that is, a condensive reactance, $Z_3 = -j\dfrac{x_3}{n}$, and an inductive reactance, $Z_4 = +jnx_3$. Then for the fundamental, these two reactances together offer no resultant impedance, but neutralize each other, and the only drop of voltage produced by them is that due to the small loss of power in them. At the nth harmonic, however, the resultant reactance is

$$Z_3 + Z_4 = +jx_3\left(n + \frac{1}{n}\right),$$

or, approximately,

$$= +jx_3n,$$

and two such impedances so obstruct the higher harmonics, the more, the higher their order while passing the fundamental sine wave.

Such a pair of equal reactances of opposite sign so can be called a "*wave screen.*"

Further problems for investigation by the student then are:

1. The investigation of the effect of the distortion of the wave of impressed e.m.f. on the constant current, with other transforming devices, and also the reverse problem, the investigation of the effect of the distortion of the constant-current wave, as caused by an arc, on the system of transformation.

2. What must be the value, x_1, of the reactance of a wave screen, to reduce the wave-shape distortion of the secondary current in the monocyclic square to the same percentage as the distortion of the impressed e.m.f. wave, or to any desired percentage, or to reduce the variation of the constant current with the load, as due to the wave-shape distortion, below a given percentage?

3. Determination of efficiency and regulation in the monocyclic square with interposed wave screen, x_1, assuming again 3 per cent. loss in the inductances, 1 per cent. loss in the capacities and choosing x_4 so as to fill given conditions, regarding wave-shape distortion, or regulation, or efficiency, etc.

CHAPTER XV

CONSTANT-VOLTAGE SERIES OPERATION

155. Where a considerable number of devices, distributed over a large area, and each consuming a small amount of power, are to be operated in the same circuit, low-voltage supply—110 or 220 volts—usually is not feasible, due to the distances, and high-voltage distribution—2300 volts—with individual step-down transformers at the consuming devices, usually is uneconomical, due to the small power consumption of each device.

In such a case, series connection of the devices is the most economical arrangement, and therefore commonly used.

Such for instance is the case in lighting the streets of a city, etc.

Most of the street lighting has been done by arc lamps operated on constant-current circuits, and as the universal electric power supply today is at constant voltage, transformation from constant voltage to constant current thus is of importance, and has been discussed in Chapter XIV.

The constant-current system thus is used in this case:

(*a*) Because by series connection of the consuming devices, as the arc lamps in street lighting, it permits the use of a sufficiently high voltage to make the distribution economical.

(*b*) The dropping volt-ampere characteristic of the arc makes it unstable on constant voltage, as further discussed in Chapters II and X, and a constant-current circuit thus is used to secure stability of operation of series arc circuits.

The condition (*b*), the use of constant current, thus applies only where the consuming devices are arcs, and ceases to be pertinent when the consuming devices are incandescent lamps or other constant-voltage devices.

The modern incandescent lamp, however, is primarily a constant-voltage device, that is, at constant-voltage supply, the life of the lamp is greater than at constant-current supply, assuming the same percentage fluctuation from constancy. The reason is: a variation of voltage at the lamp terminals, by p per cent., gives a variation of current of about $0.6p$ per cent., and thus a variation

297

of power of about $1.6p$ per cent., while a variation of current in the lamp, by p per cent., gives a variation of voltage of about $\dfrac{p}{0.6}$ per cent., and thus a variation of power of about $(1 + \dfrac{1}{0.6})p = 2.67\,p$ per cent.

Thus, with the increasing use of incandescent lamps for street illumination, series operation in a constant-voltage circuit becomes of increasing importance.

If e = rated voltage, i = rated current of lamp or other consuming device, and e_0 = supply voltage, $n = \dfrac{e_0}{e}$ lamps can be operated in series on the constant-voltage supply e_0. If now one lamp goes out by the filament breaking, all the lamps of the series circuit would go out, if e_0 is small; if e_0 is large, an arc will hold in the lamp or the fixture, and more or less destroy the circuit.

Thus in series connection, especially at higher supply voltage, e_0, some shunt protective device is necessary to maintain circuit in case of one of the consuming devices open-circuiting

On constant-current supply, a short-circuiting device, such as a film cutout, takes care of this. With series connection on constant-voltage supply, it is not permissible, however, to short-circuit a disabled consuming device, as this would increase the voltage on the other devices. Thus the shunt protective device in the constant-voltage series system must be such, that in case of one lamp burning out, the shunt consumes such a voltage as to maintain the voltage on the other devices the same as before. A film cutout, with another lamp in series, would accomplish this: if a lamp burns out, its shunting film cutout punctures and puts the second lamp in circuit. However, in general such arrangement is too complicated for use.

As practically all such circuits would be alternating-current circuits, and thus alternating currents only need to be considered, the question arises, whether a reactance shunting each lamp would not give the desired effect. Suppose each lamp, of resistance, r, is shunted by a reactance, x, which is sufficiently large not to withdraw too much current from the lamp: assuming the current shunted by x is 20 per cent. of the current in the lamp, or $x = 5\,r$. With 6.6 amp. in r, x thus would take 1.32 amp., and the total, or line current would be: $i = \sqrt{6.6^2 + 1.32^2} = 6.73$ amp., thus only 2 per cent. more than the lamp current. If now a lamp burns out, the total current flows through x, instead of 20 per cent. only, and the voltage consumed by x is increased fivefold—assuming x as constant—this voltage, however, is in quadrature with the current, thus combines vectorially with the voltages of the other consuming devices, which are practically in phase with the current, and the question then arises, whether, and under what conditions such a reactance shunt would maintain constant voltage on the other consuming devices, or, what amounts to the same, constant current in the series circuit.

Such a reactance shunting the consuming device could at the same time be used as autotransformer (compensator), to change the current, so that consuming devices of different current requirements, as lamps of various sizes, could be operated in series on the same circuit, from constant-voltage supply.

156. Let n lamps of voltage, e_1, and current, i_1, thus conductance

$$g = \frac{i_1}{e_1} \tag{1}$$

be connected in series into a circuit of supply voltage,

$$e_0 = ne_1 \tag{2}$$

and each lamp be shunted by a reactance of susceptance, b.

In each consuming device, comprising lamp and reactance, the admittance thus is, vectorially,

$$Y_1 = g - jb \tag{3}$$

if, then,

I = current in the series circuit, the voltage consumed by the device comprising lamp and reactance, thus is

$$E_1 = \frac{I}{Y_1} = \frac{I}{g - jb} \tag{4}$$

in a consuming device, however, in which the lamp is burned out, and only the reactance remains, the admittance is

$$Y_2 = -jb \tag{5}$$

hence, the voltage, with the entire current, I, passing through the admittance, Y_2,

$$E_2 = \frac{I}{Y_2} = j\frac{I}{b}. \tag{6}$$

If, then, of the n series lamps, the fraction, p, is burned out, leaving $n(1 - p)$ operative lamps, it is:

voltage consumed by operative devices:

$$n(1 - p)E_1 = \frac{n(1 - p)\dot{I}}{g - jb}$$

voltage consumed by devices with burned-out lamps:

$$npE_2 = \frac{jnp\dot{I}}{b}$$

thus, total circuit voltage:

$$e_0 = n(1 - p) E_1 + npE_2 \qquad (7)$$

$$= n\dot{I}\left\{\frac{1 - p}{g - jb} + \frac{jp}{b}\right\}$$

or,

$$e_0 = \frac{n\dot{I}(b + jpg)}{b(g - jb)} \qquad (8)$$

or, absolute,

$$e_0 = \frac{ni}{y}\sqrt{1 + \left(\frac{pg}{b}\right)^2} \qquad (9)$$

where

$$y = \sqrt{g^2 + b^2} = \text{admittance of operative device, absolute,} \qquad (10)$$

hence,

$$i = \frac{e_0 y}{n\sqrt{1 + \left(\frac{pg}{b}\right)^2}} \qquad (11)$$

is the current in the circuit, and the current in the lamps thus is

$$i_1 = \frac{g}{y} i \qquad (12)$$

hence,

$$i_1 = \frac{e_0 g}{n\sqrt{1 + \left(\frac{pg}{b}\right)^2}} \qquad (13)$$

for

$p = 0$, or all devices operative ("full-load," as we may say), it is

$$i_1 = \frac{e_0 g}{n}$$

for

$p = 1$, or all lamps out ("no-load"), it is

$$i_1 = \frac{e_0 g}{n\sqrt{1 + \left(\frac{g}{b}\right)^2}}$$

$$= \frac{e_0 g b}{n y}$$

(14)

thus smaller than at full-load.

As seen from equation (13), the current steadily decreases, from $p = 0$ or full-load, to $p = 1$ or no-load, and no value of shunted reactance, b, exists, which maintains constant current.   With decreasing load, the current, i_1, decreases the slower, the higher b is, that is, the more current is shunted by the reactive susceptance, b, and the poorer therefore the power-factor is.

Thus shunted constant reactance can not give constant-voltage regulation.

However, with $b = 0.2\ g$, at no-load the shunted reactance would get five times as much current as at load, and thus have five times as high a voltage at its terminals.

The latter, however, is not feasible, except by making the reactance abnormally large and therefore uneconomical.

In general, long before five times normal voltage is reached, magnetic saturation will have occurred, and the reactance thereby decreased, that is, the susceptance, b, increased, as more fully discussed in Chapter VIII.

This actual condition would correspond to a value, b_1, of the shunted susceptance when shunted by the lamp, and a different, higher value, b_2, of the shunted susceptance when the lamp is burned out.

The question then arises, whether such values of b_1 and b_2 can be found, as to give voltage regulation.   The increase of b_2 over b_1 naturally depends on the degree of magnetic saturation in the reactance, that is, on the value of magnetic density chosen, and thus can be made anything, depending on the design.

157. Let then, as heretofore,

$E_0 = e_0 = $ constant-supply voltage.

I = current in series circuit.

n = number of consuming devices (lamps) in series.   (15)

p = fraction of burned-out lamps.

g = conductance of lamp.

and let

b_1 = shunted susceptance with the lamp in circuit, that is, exciting susceptance of reactor or auto-transformer, and

y = $\sqrt{g^2 + b_1{}^2}$ = admittance of complete consuming device.

b_2 = shunted susceptance with the lamp burned out

$$(16)$$

and let

$c = \dfrac{b_1}{g}$ = exciting current as fraction of load current: $c < 1$.

$a = \dfrac{g}{b_2}$ = saturation factor of reactor or autotransformer: $a > 1$.

$$(17)$$

it is, then:

voltage of lamp and reactor:

$$E_1 = \frac{\dot{I}}{g - jb_1} \qquad (18)$$

voltage of reactor with lamp burned out:

$$E_2 = \frac{\dot{I}}{-jb_2} = j\frac{\dot{I}}{b_2} \qquad (19)$$

thus,

with pn lamps burned out, and $(1-p)n$ lamps burning, it is total voltage,

$$e_0 = n(1-p)E_1 + np\,E_2 \qquad (20)$$

$$= n\dot{I}\left\{\frac{1-p}{g - jb_1} + j\frac{p}{b_2}\right\}$$

substituting (17),

$$e_0 = \frac{n\dot{I}}{g}\left\{\frac{1-p}{1 - jc} + jap\right\} \qquad (21)$$

or,

$$e_0 = \frac{n\dot{I}}{g}\frac{1 - p(1 - ac) + jap}{1 - jc};$$

hence, absolute,

$$e_0 = \frac{ni}{y}\sqrt{[1 - p(1 - ac)]^2 + a^2p^2} \qquad (22)$$

since,

$$y = g\sqrt{1 + c^2}$$

thus, the *current* in the series circuit,

$$i = \frac{e_0 y}{n\sqrt{[1 - p(1 - ac)]^2 + a^2p^2}} \qquad (24)$$

158. For, $p = 0$, or full-load, it is

$$i_0 = \frac{e_0 y}{n} \tag{25}$$

thus,

$$i = \frac{i_0}{\sqrt{[1 - p(1 - ac)]^2 + a^2 p^2}} \tag{26}$$

The same value of i_0 as at full-load is reached again for the value $p = p_0$, where the square root in (24) becomes one, that is,

$$[1 - p_0(1 - ac)]^2 + a^2 p_0^2 = 1,$$

hence,

$$p_0 = \frac{2(1 - ac)}{a^2 + (1 - ac)^2} \tag{27}$$

for, $p = 1$, or no-load, it is

$$i = \frac{i_0}{a\sqrt{1 + c^2}} \tag{28}$$

The current is a maximum, $i = i_m$, for the value of $p = p_m$, given by

$$\frac{di}{dp} = 0,$$

or, from (26),

$$\frac{d}{dp}\{[1 - p(1 - ac)]^2 + a^2 p^2\} = 0,$$

this gives

$$p_m = \frac{1 - ac}{a^2 + (1 - ac)^2} \tag{29}$$

$$= \frac{p_0}{2},$$

as was to be expected.

Substituting (29) into (26), gives as the value of maximum current

$$i_m = i_0 \sqrt{1 + \left(\frac{1 - ac}{a}\right)^2} \tag{30}$$

and the regulation q, that is, the excess of maximum current over full-load current, as fraction of the latter, thus is

$$q = \frac{i_m - i_0}{i_0}$$

$$= \sqrt{1 + \left(\frac{1 - ac}{a}\right)^2} - 1 \tag{31}$$

If q is small (31) resolved by the binomial, gives

$$q = \frac{1}{2}\left(\frac{1 - ac}{a}\right)^2 \tag{32}$$

As seen, with the shunted susceptance increased by saturation at open circuit, the current and thus lamp voltage are approximately constant over a range of p. That is, with decreasing load, from full-load $p = 0$, the current i, and proportional thereto the lamp voltage increases from i_0 to a maximum value i_m, at $p = \frac{p_0}{2}$, then decreases again, to i_0 at $p = p_0$, and decreases further, to i_1 at no-load, $p = 1$.

Thus, there exists a regulating range from $p = 0$ to p a little above p_0, where the current is approximately constant.

Instance:

Saturation: $a =$	1.5	1.5	1.5	2.0	2.0	2.0	2.5	2.5	2.5
Excitation: $c =$	0.1	0.2	0.3	0.1	0.2	0.3	0.1	0.2	0.3
Regulation: $q =$	0.147	0.103	0.067	0.077	0.044	0.020	0.044	0.020	0.005
Range: $p_0 =$	0.573	0.510	0.432	0.345	0.275	0.192	0.220	0.154	0.079

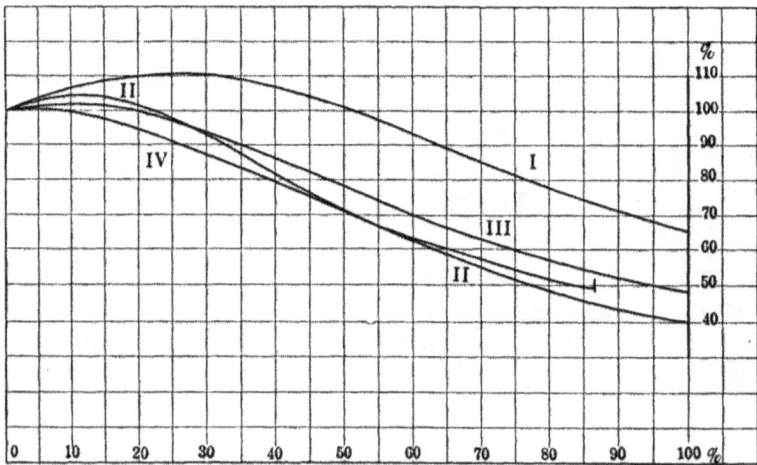

Fig. 126.

As illustrations are shown, in Fig. 126, the regulation curves, $\frac{i}{i_0}$, from equation (26), for:

$a = 1.5$ $c = 0.2$ Curve I

     $= 2.5$ $= 0.1$ II

     $= 2.0$ $= 0.3$ III

159. By the preceding equations, it is possible now to calculate the values of exciting susceptance b_1, and saturation b_2, required by the shunting reactors to give desired values of regulation within a given range.

From (32) follows:

$$c = \frac{1}{a} - \sqrt{2\,q} \tag{33}$$

Substituting (33) into (27) gives:

$$\left.\begin{aligned}
a &= \frac{2\sqrt{2\,q}}{p_0(1 + 2\,q)} \\[2mm]
c &= \frac{p_0(1 + 2\,q) - 4\,q}{2\sqrt{q}}
\end{aligned}\right\} \tag{34}$$

From chosen values of q and p_0, a and c thus can be calculated, from a and c and the conductance g of the consuming device, b_1, b_2, i, etc., follow.

Instance:

$n = 100$ lamps of $i_1 = 6$ amp. and $e_1 = 50$ volts, are to be operated in series on constant-voltage supply, with negligible line resistance and reactance. The regulation shall be within 4 per cent. in a range of 30 per cent. That is, $q = 0.04$ and $p_0 = 0.30$. It thus is:

$$i_1 = 6$$
$$e_1 = 50$$
$$g = \frac{i_1}{e_1} = 0.12$$
$$n = 100$$
$$q = 0.04$$
$$p_0 = 0.30$$

From (34) follows:

$$a = 1.75$$
$$c = 0.287$$

Hence by (17):

$$b_1 = 0.0345$$
$$b_2 = 0.0685$$

and by (16):

$$y = 0.1248$$

by (2):

$$e_0 = 5000 \text{ volts}$$

20

and by (25):

$$i_0 = 6.24 \text{ amp.}$$

thus, by (26)·

$$i = \frac{6.24}{\sqrt{1 - p + 3.31 \, p^2}}. \tag{35}$$

Fig. 127 shows, as curve I, the values of $q = \dfrac{i}{i_0} - 1$, in per cent., that is, the regulation, with p as abscissæ.

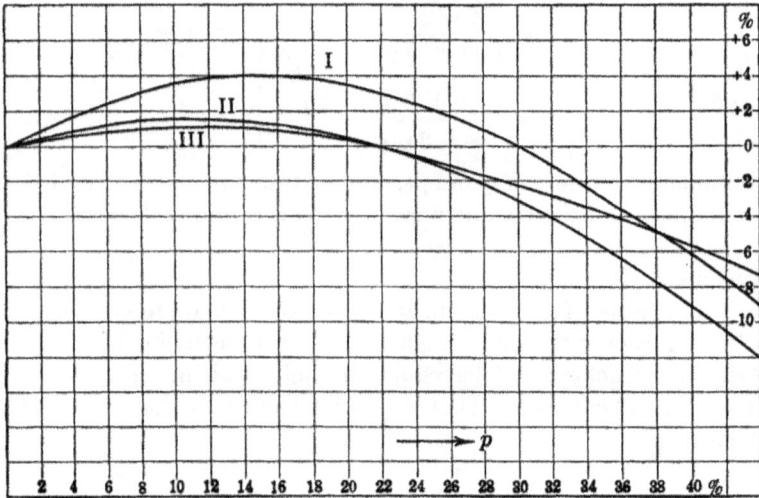

FIG. 127.

160. In general, the resistance and reactance of the circuit or line is not negligible, as assumed in the preceding, and the reactors, especially if used at the same time as autotransformers, contain a leakage reactance, which acts as a series reactance in the circuit, and the lamp circuit of conductance g also may contain a small series reactance.

Let then:

$$r_0 = \text{line resistance;}$$
$$x_0 = \text{line reactance;}$$
$$x = \text{series or leakage reactance per autotransformer}$$
$$\text{or consumption device.}$$

The most convenient way is to represent r_0, x_0 and x by their equivalent in lamps or reactors. The admittance of each consumption device, comprising lamp and reactor or autotransformer, is

$$Y_1 = g - jb_1 = g(1 - jc),$$

thus the impedance,

$$Z_1 = \frac{1}{Y_1} = \frac{1}{g(1 - jc)} = \frac{1 + jc}{g(1 + c^2)},$$

and by (23),

$$Z_1 = \frac{1 + jc}{y\sqrt{1 + c^2}}.$$

If, then, we add to the resistance r_0 a part cr_0 of the reactance, we get an impedance,

$$Z = r_0(1 + jc),$$

which has the same phase angle as Z_1, and thus can be expressed as a multiple of Z_1,

$$Z = n_1 Z_1,$$

where

$$n_1 = \frac{Z}{Z_1} = r_0 y\sqrt{1 + c^2} \qquad (36)$$

thus is the "lamp equivalent" of the line resistance r_0 plus the part cr_0 of the reactance.

This leaves the reactance,

$$x_1 = x_0 + n(1 - p)x - cr_0,$$

and as the reactance of a reactor without lamp is

$$x_2 = \frac{1}{b_2},$$

the reactance x_1 can be expressed as multiple of x_2,

$$x_1 = n_2 x_2$$

$$= \frac{n_2}{b_2}$$

where

$$n_2 = x_1 b_2 = b_2 \left[x_0 + n(1 - p)x - cr_0 \right] \qquad (37)$$

thus is the "lamp equivalent" of the line reactance x_0 and leakage reactances x_1 in burned-out lamps.

Thus the addition of the line impedance $r_0 + jx_0$, and the leakage reactances x, is represented by n_1 lamps with reactors, and n_2 burned-out lamps, or a total of $n_1 + n_2$ lamps.

Thus the circuit can carry $n_1 - (n_1 + n_2)$ lamps, and its regulation curve starts at the point $p = \dfrac{n_2}{n}$ and ends at $p = 1 - \dfrac{n_1 + n_2}{n}$ of the complete regulation curve.

However, in this case, the full-load current, for $p = \dfrac{n_2}{n}$, would already be slightly higher than in a circuit without line impedance, and all the current values would thus have to be proportionally reduced.

Instance:

In the case

$a = 2.0$

$c = 0.3$

given as curve III of Fig. 126 let:

$n = 100$

$g = 0.12$

$r_0 = 50$

$x_0 = 0.5$

hence,

$$b_1 = cg = 0.036$$

$$y = \sqrt{g^2 + b_1{}^2} = 0.125$$

$$b_2 = \frac{g}{a} = 0.06,$$

thus,

$$n_1 = r_0 y \sqrt{1 + c^2} = 6.54$$
$$n_2 = b_2 [x_0 + n (1-p) x - cr_0] = 7.0$$
$$n_1 + n_2 = 13.54.$$

Thus, the regulation curve starts at $p = \dfrac{n_2}{n} = 0.07$ of curve III, Fig. 126, and ends at $p = 1 - \dfrac{n_1}{n} = 0.935$ of this curve.

For $p = 0.07$ it is, by equation (26),

$$\frac{i}{i_0} = 1.017,$$

thus, all values of curve III, Fig. 126, are reduced by dividing with 1.017, and then plotted from $p = 0.07$ on, and then give the regulation curve inclusive line resistance shown as curve IV.

As seen, the regulation range is reduced, but the regulation greatly improved by the line impedance. This is done essentially by the line reactance and leakage reactance, but not by the resistance.

161. Instead of approximating the effect of line impedance and leakage reactance by equivalent lamps and reactors, it can be directly calculated, as follows:

Let

$$r_0 = \text{line resistance}$$
$$x_0 = \text{line reactance}$$
$$x = \text{leakage or series reactance per autotransformer}$$ (38)

the other symbols being the same as (15), (16) and (17).

It is then:

voltage consumed by line resistance r_0:

$$r_0 I$$

voltage consumed by line reactance x_0:

$$j x_0 I$$

voltage consumed by leakage reactances x of the $n(1 - p)$ lamp devices:

$$j x n(1 - p) I,$$

thus, total circuit voltage:

$$e_0 = I\left\{ \frac{(1 - p)n}{g - jb_1} + j\frac{pn}{b_2} + r_0 + jx_0 + j(1 - p)nx \right\} \quad (39)$$

substituting the abbreviation,

$$h_1 = \frac{r_0 g}{n}$$
$$h_2 = \frac{x_0 g}{n}$$ (40)
$$h_3 = xg$$

and substituting (17) and (40) into (39), gives

$$e_0 = \frac{in}{g}\left\{ \frac{1 - p}{1 - jc} + jpa + h_1 + jh_2 + j(1 - p)h_3 \right\}$$

$$= \frac{in}{g}\left\{ \left[\frac{1 - p}{1 + c^2} + h_1 \right] + j\left[\frac{(1-p)c}{1+c^2} + pa + h_2 + (1 - p)h_3 \right] \right\} \quad (41)$$

hence, absolute,

$$e_0 = \frac{in}{g}\sqrt{\left[\frac{1 - p}{1 + c^2} + h_1 \right]^2 + \left[\frac{(1 - p)c}{1 + c^2} + pa + h_2 + (1 - p)h_3 \right]^2} \quad (42)$$

hence, the current,

$$i = \frac{ge_0}{n\sqrt{\left[\frac{1 - p}{1 + c^2} + h_1 \right]^2 + \left[\frac{(1-p)c}{1+c^2} + pa + h_2 + (1 - p)h_3 \right]^2}} \quad (43)$$

for $p = 0$ (42) and (43) gives the *full-load current and voltage,*

$$e_0 = \frac{i_0 n}{g} \sqrt{\left[\frac{1}{1+c^2} + h_1\right]^2 + \left[\frac{c}{1+c^2} + h_2 + h_3\right]^2} \qquad (44)$$

where (12)

$$i_0 = i_1 \frac{y}{g} \qquad (45)$$

is the full-load line current, for i_1 = full-load lamp current.

162. Let, in the instance paragraph 159 and Fig. 126;

$$r_0 = 50$$
$$x_0 = 75$$
$$x = 0.5$$

the other constants remaining the same as in paragraph 159, that is:

$$i_1 = 6$$
$$n = 100$$
$$g = 0.12$$
$$b_1 = 0.0345$$
$$b_2 = 0.0685$$
$$y = 0.1248$$
$$a = 1.75$$
$$c = 0.287$$

It is then (40),

$$h_1 = 0.06$$
$$h_2 = 0.09$$
$$h_3 = 0.06$$

hence, by (45),

$$i_0 = 1.04 \times 6 = 6.24 \text{ amp.}$$

by (44),

$$e_0 = 5200 \sqrt{(0.923 + 0.06)^2 + (0.264 + 0.09 + 0.06)^2}$$
$$= 5200 \sqrt{0.983^2 + 414^2}$$
$$= 5200 \sqrt{1.137}$$
$$= 5200 \times 1.066$$
$$e_0 = 5540 \text{ volts}$$

and by (43),

$$i = \frac{6.65}{\sqrt{(0.983 - 0.923 p)^2 + (0.414 + 1.426 p)^2}}$$

$$= \frac{6.65}{\sqrt{1.137 - 0.634\,p + 2.885\,p^2}}$$

$$i = \frac{6.24}{\sqrt{1 - 0.558\,p + 2.54\,p^2}} \tag{46}$$

Fig. 127 shows, as curve II, the values of $\frac{i}{6.24} - 1$ from equation (46), that is, the regulation, as modified by line impedance and leakage reactance, with p as abscissæ.

The regulating range, p_0, of equation (46) is given by

$$1 - 0.558\,p_0 + 2.54\,p_0{}^2 = 1,$$

hence,

$$p_0 = 0.22.$$

Thus the regulation range is reduced by the line impedance and leakage reactance, from 30 per cent. to 22 per cent.

The maximum value of current, i_m, occurs at

$$p_m = \frac{p_0}{2} = 0.11$$

and is given by substitution into (46), as,

$$\frac{i_m}{6.24} = 1.015,$$

or,

$$q = 0.015.$$

That is, the regulation is improved, by the line and leakage reactance, from $q = 4$ per cent. to $q = 1.5$ per cent. as seen in Fig. 127.

163. In paragraph 161 and the preceding, the shunted reactances, b_1 and b_2, have been assumed as constant and independent of p. However, with the change of p, the wave-shape distortion between current and voltage changes, as with increasing p, more and more saturated reactors are thrown into the circuit and distort the current wave.

As b_1 is shunted by g, and carries a small part of the current only, and g is non-inductive, the change of wave shape in b_1 will be less, and as b_1 carries only a part of the current, the effect of the change of wave shape in b_1 thus is practically negligible, so that b_1 can be assumed as constant and independent of p.

b_2, however, carries the total current, at fairly high saturation, and thus exerts a great distorting effect.

At and near full-load, with all or nearly all conductances, g, in circuit, the entire circuit is practically non-inductive, that is, the current has the same wave shape as the voltage. Assuming a sine wave of impressed voltage, e_0, the current, i, at and near full-load thus is practically a sine wave, and the shunting reactance, b_2, thus has the value corresponding to a sine wave of current traversing it, that is, the value denoted as "constant-current reactance," x_c, in Chapter VIII.

At no-load, with all or nearly all conductances, g, open-circuited, the entire circuit consists of a series of n reactive susceptances, b_2. If, then, the impressed voltage, e_0, is a sine wave, each susceptance, b_2, receives $1/n$ of the impressed voltage, thus also a sine wave. That is, at and near no-load, the shunted reactance, b_2, has the value corresponding to an impressed sine wave of voltage, that is, the value denoted as "constant-potential reactance," x_p, in Chapter VIII.

x_c, however, is materially larger than x_p, and the shunting reactance thus decreases, that is, the shunting susceptance, b_2, increases from full-load to no-load, or with increasing p.

Due to the changing wave-shape distortion, b_2 thus is not constant, but increases with increasing p, thus can be denoted by

$$b_2 = b_0(1 + sp) \tag{47}$$

this gives

$$a = \frac{a_0}{1 + sp}. \tag{48}$$

Substituting (48) into (43) gives, as the equation of current, allowing for the change of wave-shape distortion,

$$i = \frac{ge_0}{n\sqrt{\left[\dfrac{1 - p}{1 + c^2} + h_1\right]^2 + \left[\dfrac{(1 - p)c}{1 + c^2} + \dfrac{pa_0}{1 + sp} + h_2 + (1 - p)h_3\right]^2}} \tag{49}$$

Assume, in the instance paragraphs 159 and 161, and Fig. 127, that the shunted susceptance, b_2, increases from full-load to no-load by 40 per cent. That is,

$$s = 0.4;$$

it is, then,

$$a = \frac{a_0}{1 + 0.4\,p}.$$

Assuming now, that at the end of the regulating range,

$$p = p_0 = 0.22,$$

a has the same value as before,

$$a = 1.75,$$

this gives

$$1.75 = \frac{a_0}{1 + 0.4 \times 0.22}$$

$$a = 1.90$$

and

$$a = \frac{1.9}{1 + 0.4\ p} \tag{50}$$

Substituting now the numerical values in equation (49), gives

$$\left. \begin{aligned} i &= \frac{6.65}{\sqrt{(0.983 - 0.923\ p)^2 + (0.414 + (a - 0.324)p)^2}} \\ &= \frac{6.24}{\sqrt{[0.928 - 0.866\ p]^2 + [0.388 + (0.938\ a - 0.304)\ p]^2}} \end{aligned} \right\} \tag{51}$$

Fig. 127 shows, as curve III, the values of $\frac{i}{6.24}$ from equation (51), that is, the regulation as modified by the changing wave shape caused by the saturated reactance.

The maximum value of current, i_m, occurs at $p_m = \frac{p_0}{2} = 0.11$, and is given by substitution into (50) and (51), as,

$$a = 1.82$$

$$\frac{i_m}{6.24} = 1.011$$

that is,

$$q = 0.011$$

thus, the regulation is still further improved, by changing wave shape, to 1.1 per cent.

CHAPTER XVI

LOAD BALANCE OF POLYPHASE SYSTEMS

163. The total flow of power of a balanced symmetrical poly-phase system is constant. That is, the sum of the instantaneous values of power of all the phases is constant throughout the cycle. In the single-phase system, however, or in a polyphase system with unbalanced load, that is, a system in which the different phases are unequally loaded, the total flow of power is pulsating, with double frequency. To balance an unbalanced polyphase system thus requires a storage of energy, hence can not be done by any method of connection or transformation. Thus mechanical momentum acts as energy-storing device in the use as phase balancer, of the induction or the synchronous machine. Electrically, energy is stored by inductance and by capacity. The question then arises, whether by the use of a reactor, or a condenser, connected to a suitable phase of the system, an unequally loaded polyphase system can be balanced, so as to give constant power during the cycle.

In interlinked polyphase circuits, such as the three-phase system, with unbalanced load carried over lines of appreciable impedance, the voltages of the three phases become unequal. This makes voltage regulation more complicated than in a balanced system. A great unbalancing of the load, such as produced by operating a heavy single-phase load, as a single-phase railway or electric furnace, greatly reduces the power capacity of lines, transformers and generators. Unbalanced load on the generators causes a pulsating armature reaction: at single-phase load, the armature reaction pulsates between more than twice the average value, and a small reversed value, between $F(\cos \alpha + 1)$ and $F(\cos \alpha - 1)$, where $\cos \alpha$ is the power-factor of the single-phase load. Especially in alternators of very high armature reaction, as modern steam-turbine alternators, a pulsation of the armature reaction is very objectionable. It causes a pulsation of the field flux, leading to excessive eddy-current losses and consequent reduction of the output. The use of a squirrel-cage winding in the field pole faces of the single-phase alternator reduces the pulsation of the field flux, but also increases the momentary short-circuit stresses.

Thus, it is of interest to study the question of balancing unbalanced polyphase circuits by stationary energy-storing devices, as reactor or condenser.

164. Let a voltage,

$$e = E \cos \phi \tag{1}$$

be impressed upon a non-inductive load, giving the current

$$i = I \cos \phi \tag{2}$$

The power then is

$$p = ei = EI \cos^2 \phi$$
$$= \frac{EI}{2} (1 + \cos 2\phi)$$
$$= Q + Q \cos 2\phi \tag{3}$$

where

$$Q = \frac{EI}{2} \tag{4}$$

that is,
in a non-inductive single-phase circuit, the power consists of a constant component,

$$Q = \frac{EI}{2},$$

and an alternating component,

$$Q = \frac{EI}{2} \cos 2\phi,$$

of twice the frequency of the supply voltage, and a maximum value equal to that of the constant component. The instantaneous power thus pulsates between zero and $2Q$, by equation (3).

If the circuit is inductive, of lag angle α, the current is

$$i = I \cos (\phi - \alpha) \tag{5}$$

and the instantaneous power thus,

$$p = EI \cos \phi \cos (\phi - \alpha)$$
$$= \frac{EI}{2} \Big[\cos \alpha + \cos (2\phi - \alpha) \Big]$$
$$= P + Q \cos (2\phi - \alpha),$$

thus consists of a constant component,

$$P = \frac{EI}{2} \cos \alpha = Q \cos \alpha \tag{7}$$

and an alternating component,

$$Q \cos (2 \phi - \alpha);$$

it thus pulsates between a small negative and a large positive value, $P - Q$ and $P + Q$.

If the circuit is completely inductive, that is, the current lags 90° or $\frac{\pi}{2}$ behind the voltage, the current is

$$i = I \cos\left(\phi - \frac{\pi}{2}\right) \tag{8}$$

and the instantaneous power thus,

$$p = EI \cos \phi \cos\left(\phi - \frac{\pi}{2}\right)$$
$$= \frac{EI}{2} \sin 2 \phi$$
$$= Q \cos\left(2 \phi - \frac{\pi}{2}\right) \, {}^1 \tag{9}$$

Thus, the power comprises only an alternating component, but no continuous component; in other words, no power is consumed, but the power surges or alternates between $+Q$ and $-Q$, that is, power is stored and then again returned to the circuit.

If the circuit is closed by a capacity, C, the current leads the impressed voltage by $\frac{\pi}{2}$, thus is

$$i = I \cos\left(\phi + \frac{\pi}{2}\right) \tag{10}$$

and the instantaneous power thus,

$$p = EI \cos \phi \cos\left(\phi + \frac{\pi}{2}\right)$$
$$= Q \cos\left(2 \phi + \frac{\pi}{2}\right) \tag{11}$$

thus, comprises only an alternating component, surging between $-Q$ and $+Q$, with double frequency.

The power consumed by a condenser, equation (11), is opposite in sign and thus in direction, from that consumed by a reactor (9),

$$Q \cos\left(2 \phi + \frac{\pi}{2}\right) = - Q \cos\left(2 \phi - \frac{\pi}{2}\right).$$

165. If a number of voltages,

$$e_i = E_i \cos (\phi - \gamma_i) \tag{12}$$

---

[1] "Engineering Mathematics," Chapter III, paragraphs 66 to 75.

of a polyphase system, produce currents,

$$i_i = I_i \cos (\phi - \gamma_i - \alpha_i) \tag{13}$$

the instantaneous power of each voltage e_i is

$$
\begin{aligned}
p_i &= e_i i_i \\
&= Q_i \{ \cos \alpha_i + \cos (2\phi - 2\gamma_i - \alpha_i) \}
\end{aligned} \tag{14}
$$

and the total instantaneous power of the system thus is

$$
\begin{aligned}
p &= \Sigma p_i \\
&= \Sigma Q_i \cos \alpha_i + \Sigma Q_i \cos (2\phi - 2\gamma_i - \alpha_i) \\
&= P + Q \cos (2\phi - \alpha)
\end{aligned} \tag{15}
$$

where

$$P = \Sigma Q_i \cos \alpha_i \tag{16}$$

is the total effective power of the system, and

$$Q = \Sigma Q_i \cos (2\phi - 2\gamma_i - \alpha_i) \tag{17}$$

is the total resultant alternating component of power, or the resultant power pulsation of the system.

Thus, the power of the polyphase system pulsates, with double frequency, between $P - Q$ and $P + Q$.

In this case, P may be greater than Q, and frequently is, and the power thus pulsates between two positive values, while in the single-phase circuit (6) it pulsated between positive and negative value.

It thus can be seen, that in any system, polyphase or single-phase, with any kind of load, the total instantaneous power of the system can be expressed,

$$p = P + Q \cos (2\phi - \alpha) \tag{18}$$

where P is the constant component of power, and Q the amplitude of the double-frequency alternating component of power, and Q may be larger or smaller than P.

It must be noted, that Q is *not* the total reactive power of the system—which would have to be considered, for instance, in power-factor compensation etc.—but Q is the *vector* resultant of the reactive powers of the individual circuits, while the total reactive power of the system is the algebraic sum of the individual reactive powers (see "Theory and Calculation of Alternating-current Phenomena," Chapter XVI).

Thus, for instance, in a system of balanced load, even if the load is reactive, $Q = 0$. Thus, Q is the unbalanced reactive power of the system, and does not include the reactive power, which is balanced between the phases and thereby gives zero as vector resultant.

166. The expression of the power of a polyphase system of general unbalanced load is by (15)

$$p = P + Q \cos (2 \phi - \alpha) \tag{19}$$

this also is the expression of power of the single-phase load of lag angle α, of the impressed voltage and current,

$$\left. \begin{array}{l} e = E \cos \phi \\ i = I \cos (\phi - \alpha) \end{array} \right\} \tag{20}$$

where, from (20),

$$\left. \begin{array}{l} P = Q \sin \alpha \\ Q = \dfrac{EI}{2} \end{array} \right\} \tag{21}$$

while in the general case (19) P and Q may have any values.

Suppose now we select from the polyphase system a voltage,

$$e' = E' \cos (\phi - \beta) \tag{22}$$

and load it with an inductive load of zero power-factor,

$$i' = I' \cos \left(\phi - \beta - \frac{\pi}{2} \right) \tag{23}$$

that is, we connect a reactor of $x = \dfrac{E'}{I'}$ into the phase e'.

The power of (22) (23) then is

$$p' = Q' \cos \left(2 \phi - 2 \beta - \frac{\pi}{2} \right) \tag{24}$$

where

$$Q' = \frac{E'I'}{2} \tag{25}$$

and the total power of the system, comprising (19) and (25), thus is

$$\begin{aligned} p_0 &= p + p' \\ &= P + Q \cos (2 \phi - \alpha) + Q' \cos \left(2 \phi - 2 \beta - \frac{\pi}{2} \right) \end{aligned}$$

and this would become constant, and the double-frequency term eliminated, that is, the system would be balanced, if Q' and β are chosen so that

$$Q \cos (2 \phi - \alpha) + Q' \cos \left(2 \phi - 2 \beta - \frac{\pi}{2} \right) = 0 \tag{26}$$

hence,

$$Q' = Q \tag{27}$$

$$2\phi - 2\beta - \frac{\pi}{2} = 2\phi - \alpha - \pi$$

or,

$$\beta = \frac{\alpha}{2} + \frac{\pi}{4} \tag{28}$$

$$\frac{E'I'}{2} = Q$$

$$I' = \frac{2Q}{E'} \tag{29}$$

$$x = \frac{E'}{I'} = \frac{E'^2}{2Q} \tag{30}$$

thus,

$$e' = E' \cos\left[\phi - \left(\frac{\alpha}{2} + \frac{\pi}{4}\right)\right] \tag{31}$$

is the voltage, which, impressed upon a reactor of reactance,

$$x = \frac{E'^2}{2Q} \tag{30}$$

balances the power,

$$p = P + Q \cos(2\phi - \alpha) \tag{24}$$

of an unbalanced polyphase system. That is,

$$e' = E' \cos\left[\phi - \left(\frac{\alpha}{2} + \frac{\pi}{4}\right)\right] \tag{31}$$

impressed upon the reactance, x, gives the current,

$$i' = \frac{2Q}{E'} \cos\left[\phi - \left(\frac{\alpha}{2} + \frac{3\pi}{4}\right)\right] \tag{32}$$

and thus the power,

$$p' = Q \cos\left[\phi - \left(\frac{\alpha}{2} + \frac{\pi}{4}\right)\right] \cos\left[\phi - \left(\frac{\alpha}{2} + \frac{3\pi}{4}\right)\right]$$
$$= -Q \cos(2\phi - \alpha) \tag{33}$$

and this reactive power, p', added to the unbalanced polyphase power, p, gives the balanced power,

$$p = p + p'$$
$$= P.$$

167. Comparing (31) with (20) or (24), it follows:
The unbalanced load of a single-phase voltage,

$$e = E \cos \phi,$$

of lag angle, α, or in general, the unbalanced load of a polyphase system with the resultant instantaneous power of lag angle, α,

$$p = P + Q \cos (2 \phi - \alpha)$$

can be balanced by a wattless reactive load, p', having the same volt-amperes, Q', as the alternating component, Q, of the unbalanced load, and having a phase of voltage lagging by

$$\frac{\alpha}{2} + \frac{\pi}{4}$$

or by 45° plus half the lag angle, α, of the unbalanced load or unbalanced single-phase current.

Just as the unbalanced polyphase load, p, (24) may be single-phase load on one phase, or the vector resultant of the loads on different phases, so the wattless reactive compensating volt-amperes (33) may be due to a single reactor connected into the compensating voltage, e', (31), or may be the vector resultant of several voltages, e'_1, loaded by reactances, x_1, so that their vector resultant is p' (33).

If a capacity is used for energy storage in balancing unbalanced load (24), the compensating voltage (22),

$$e' = E' \cos^9(\phi - \beta),$$

impressed upon the capacity gives the reactive leading current,

$$i' = I' \cos\left(\phi - \beta + \frac{\pi}{2}\right) \tag{34}$$

hence the compensating reactive power,

$$p' = E'I' \cos\left(2 \phi - 2 \beta + \frac{\beta}{2}\right) \tag{35}$$

and therefrom, by the same reasoning as before,

$$\beta = \frac{\alpha}{2} + \frac{3 \pi}{4} \tag{36}$$

$$\left. \begin{array}{l} e' = E'\left[\cos \phi - \left(\frac{\alpha}{2} + \frac{3 \pi}{4}\right)\right] \\[2mm] i' = \frac{2 Q}{E'} \cos\left[\phi - \left(\frac{\alpha}{2} + \frac{\pi}{4}\right)\right] \end{array} \right\} \tag{37}$$

That is, when using a capacity for balancing the load, the compensating voltage, e', has the phase,

$$\frac{\alpha}{2} + \frac{3 \pi}{4},$$

or, what is the same as regards to the power expression,

$$\frac{\alpha}{2} - \frac{\pi}{4},$$

thus lags by half the phase angle, α, minus 45° (or plus 135°).

168. As instance may be considered a quarter-phase system with one phase loaded.

Let

$$e_1 = E \, \cos \, \phi$$
$$e_2 = E \, \cos\left(\phi - \frac{\pi}{2}\right) \qquad (38)$$

be the two phase-voltages of the quarter-phase system.

Let the first phase, e_1, be loaded by a current lagging by phase angle, α,

$$i_1 = I \, \cos \, (\phi - \alpha) \qquad (39)$$

while the second phase, e_2, is not loaded.

The power then is

$$p = e_1 i_1$$
$$= \frac{EI}{2}\{\cos \, \alpha + \cos(2 \, \phi - \alpha)\} \qquad (40)$$

and is compensated or balanced by a reactance connected to a compensating phase,

$$e' = E' \cos \, (\phi - \beta) \qquad (41)$$

and consuming the reactive current,

$$i' = I' \cos\left(\phi - \beta \mp \frac{\pi}{2}\right) \qquad (42)$$

where the $-\frac{\pi}{2}$ represents inductive reactance, the $+\frac{\pi}{2}$ capacity reactance.

The compensating reactive power then is

$$p' = e'i'$$
$$= \frac{E'I'}{2} \, \cos\left(2 \, \phi - 2 \, \beta \mp \frac{\pi}{2}\right) \qquad (43)$$

and this becomes equal to

$$- \frac{EI}{2} \, \cos \, (2 \, \phi - \alpha),$$

for

$$E'I' = EI$$
$$\beta = \frac{\alpha}{2} + \frac{\pi}{4} \qquad (44)$$

21

and the compensating circuit thus is

$$e' = E' \cos\left(\phi - \frac{\alpha}{2} \mp \frac{\pi}{4}\right) \Big\}$$

$$i' = I' \cos\left(\phi - \frac{\alpha}{2} \mp \frac{3}{4}\pi\right) \Big\}$$

(45)

it is, then,

$$p' = e'i'$$
$$= E'I' \cos\left(2\phi - \alpha \mp \pi\right)$$
$$= -EI \cos\left(2\phi - \alpha\right)$$

hence,

$$p_0 = p + p'$$
$$= \frac{EI}{2} \cos \alpha,$$

for

$$\alpha = 0, \text{ or non-inductive load, it is}$$

$$e' = E' \cos\left(\phi \mp \frac{\pi}{4}\right)$$
$$= \frac{E'}{\sqrt{2}}\left\{\cos\phi \pm \cos\left(\phi - \frac{\pi}{2}\right)\right\},$$

hence, if we choose,

$$E' = E\sqrt{2},$$

hence,

$$I' = \frac{I}{\sqrt{2}},$$

it is

$$e' = e_1 \pm e_2$$

(46)

that is, connecting the two phases in series, gives the compensating voltage for non-inductive load. Or:

"Non-inductive single-phase load, on one phase of a quarter-phase system, can be balanced by connecting a reactance across both phases in series, of such value as to consume a current equal to the single-phase load current divided by $\sqrt{2}$, that is, having the same volt-ampere as the single-phase load."

169. In the general case of inductive load of power-factor, α, the compensating voltage (45) can be written,

$$e' = E'\left\{\cos\left(\frac{\alpha}{2} \pm \frac{\pi}{4}\right)\cos\phi + \sin\left(\frac{\alpha}{2} \pm \frac{\pi}{4}\right)\sin\phi\right\}$$
$$= E'\left\{\cos\left(\frac{\alpha}{2} \pm \frac{\pi}{4}\right)\cos\phi \pm \cos\left(\frac{\phi}{2} \mp \frac{\pi}{4}\right)\cos\left(\phi - \frac{\pi}{2}\right)\right\},$$

or, choosing,

$$E' = E,$$

thus,

$$I' = I,$$

it is, by (38),

$$e' = a_1 e_1 \pm a_2 e_2$$

where

$$a_1 = \cos\left(\frac{\alpha}{2} \pm \frac{\pi}{4}\right)$$

$$a_2 = \cos\left(\frac{\alpha}{2} \mp \frac{\pi}{4}\right)$$

(47)

and the upper sign applies to the reactor, the lower to the condenser as compensating circuit.

The current then is

$$i' = I \cos\left(\phi - \frac{\alpha}{2} \mp \frac{3\,\pi}{4}\right).$$

(48)

The compensating voltage e' thus can be produced by connecting a transformer of ratio a_1 into the first phase, e_1, a transformer of ratio, a_2, into the second phase, e_2, and connecting their secondaries in series across a reactor or condenser of suitable reactance.

The current, i', in the compensating circuit consumes a current, $a_1 i'$, in the first phase, e_1, and a current, $a_2 i'$, in the second phase, e_2. As the latter phase has no load, the total current in the second phase is

$$i_2 = a_2 i' = I \cos\left(\frac{\alpha}{2} \mp \frac{\pi}{4}\right) \cos\left(\phi - \frac{\alpha}{2} \mp \frac{3\,\pi}{4}\right)$$

the total current in the first phase is
$i_1{}^0 = i_1 + a_1 i'$

$$= I\left\{\cos\left(\phi - \alpha\right) + \cos\left(\frac{\alpha}{2} \pm \frac{\pi}{4}\right) \cos\left(\phi - \frac{\alpha}{2} \mp \frac{3\,\pi}{4}\right)\right\}$$

$$= I\left\{\cos(\phi - \alpha) + 0.5 \cos\left(\phi \mp \frac{\pi}{2}\right) + 0.5 \cos\left(\phi - \alpha \mp \pi\right)\right\}$$

$$= 0.5\,I\left\{\cos\left(\phi - \alpha\right) + \cos\left(\phi \mp \frac{\pi}{2}\right)\right\}$$

$$= I\left\{\cos\left(\frac{\alpha}{2} \mp \frac{\pi}{4}\right) \cos\left(\phi - \frac{\alpha}{2} \mp \frac{\pi}{4}\right)\right\},$$

hence has the same value as i_2, but differs from it by $\frac{\pi}{2}$ or 90° in phase, thus has to its voltage, e_1, the same phase relation as i_2

has to its voltage, e_2. That is, the system is balanced in load, in phase and in armature reaction.

In the unbalanced single-phase load, the power-factor is

$$a_1 = \cos \alpha$$

in the balanced load, the power-factor is

$$a_1 = \cos\left(\frac{\alpha}{2} \pm \frac{\pi}{4}\right)$$

thus, is materially reduced for a reactor as compensator, $+\frac{\pi}{4}$;

it is in general increased for a condenser as compensator, $-\frac{\pi}{4}$.

170. Instead of varying the phase angle of the compensating voltage, e', with varying phase angle, α, of the single-phase load, compensation can be produced by compensating voltages of constant-phase angle, utilizing two such voltages and varying the proportions of their reactive currents, with changes of α.

Thus, if

$$i_1 = I \cos (\phi - \alpha),$$

is the load on phase,

$$e_1 = E \cos \phi,$$

and the second phase

$$e_2 = E \cos\left(\phi - \frac{\pi}{2}\right),$$

is not loaded, thus giving the unbalanced power,

$$p = \frac{EI}{2}\{ \cos \alpha + \cos (2 \phi - \alpha)\} \tag{49}$$

as compensating voltage may be used,
the voltage of both phases connected in series,

$$e = e_1 + e_2$$
$$= E\sqrt{2} \cos\left(\phi - \frac{\pi}{4}\right) \tag{50}$$

and the voltage of the second phase,

$$e_2 = E \cos\left(\phi - \frac{\pi}{2}\right). \tag{51}$$

Let, then,

$$i' = I' \cos\left(\phi - \frac{3\pi}{4}\right),$$

be the reactive current of the compensating phase, e, and

$$i'_2 = I'_2 \cos (\phi - \pi),$$

$$= - I'_2 \cos \phi$$

the reactive current of the compensating phase, e_2.

The powers of the two compensating circuits then are

$$p' = ei'$$

$$= \frac{EI'\sqrt{2}}{2} \cos (2\phi - \pi)$$

$$= - \frac{EI'\sqrt{2}}{2} \cos 2\phi \tag{52}$$

and

$$p'_2 = e_2 i'_2$$

$$= - \frac{EI'_2}{2} \cos \left(2\phi - \frac{\pi}{2}\right) \tag{53}$$

and the condition of compensation thus is

$$\frac{EI}{2} \cos (2\phi - \alpha) = \frac{EI'\sqrt{2}}{2} \cos 2\phi + \frac{EI'_2}{2} \cos \left(2\phi - \frac{\pi}{2}\right) \tag{54}$$

or, resolved,

$$(I \cos \alpha - I'\sqrt{2}) \cos 2\phi + (I \sin \alpha - I'_2) \sin 2\phi = 0,$$

and as this must be an identity, the individual coefficients must vanish, that is,

$$I' = \frac{I \cos \alpha}{\sqrt{2}}$$

$$I'_2 = I \sin \alpha = I \cos \left(\alpha - \frac{\pi}{2}\right) \tag{55}$$

thus, the compensating voltages and currents, which balance the single-phase load,

$$e_1 = E \cos \phi$$
$$i_1 = I \cos (\phi - \alpha) \tag{56}$$

are

$$e = e_1 + e_2$$

$$= E\sqrt{2} \cos \left(\phi - \frac{\pi}{4}\right)$$

$$i' = \frac{I \cos a}{\sqrt{2}} \cos \left(\phi - \frac{3\pi}{4}\right) \tag{57}$$

and

$$e_2 = E \cos \left(\phi - \frac{\pi}{2}\right)$$

$$i'_2 = - I \cos \left(\alpha - \frac{\pi}{2}\right) \cos \phi \tag{58}$$

$$= I \sin \alpha \cos \phi$$

As seen, this means loading the second phase with a reactor giving the same volt-amperes,

$$e_2 i'_2 = \frac{EI}{2} \sin \alpha,$$

as the unbalanced single-phase load (56), and thereby balancing the reactive component of load, and then balancing the energy component of the load by the compensating voltage $e_1 + e_2$, as given by (46).

If the single-phase load is connected across both phases of the quarter-phase machine in series,

$$\left.\begin{aligned}
e &= e_1 + e_2 \\
&= E\sqrt{2}\ \cos\left(\phi \mp \frac{\pi}{4}\right) \\
i &= \frac{I}{\sqrt{2}} \cos\left(\phi \mp \frac{\pi}{4} - \alpha\right)
\end{aligned}\right\} \tag{59}$$

in the same manner the conditions of compensation can be derived, and give the compensating circuit,

$$\left.\begin{aligned}
e' &= E' \cos\left(\phi - \frac{\alpha}{2}\right) \\
i' &= I' \cos\left(\phi - \frac{\alpha}{2} - \frac{\pi}{2}\right)
\end{aligned}\right\} \tag{60}$$

where

$$E'I' = EI.$$

For non-inductive load,

$$\alpha = 0,$$

this gives

$$e' = e_1,$$

that is, one of the two phases is compensating phase for the resultant.

171. As further instance may be considered the balancing of single-phase load on one phase of a three-phase system. Let

$$\left.\begin{aligned}
e_1 &= E \cos \varphi \\
e_2 &= E \cos\left(\phi - \frac{2\pi}{3}\right) \\
e_3 &= E \cos\left(\phi - \frac{4\pi}{3}\right)
\end{aligned}\right\} \tag{61}$$

be the three voltages between the three lines and the neutral.

The voltage from line 1 to line 2, then, is

$$e_1{}^2 = e_1 - e_2$$

$$= E\sqrt{3}\,\cos\left(\phi + \frac{\pi}{6}\right) \tag{62}$$

and if $\alpha =$ lag of current behind the voltage, the current produced by voltage, e, is

$$i = I \cos\left(\phi + \frac{\pi}{6} - \alpha\right),$$

thus the power,

$$p = \frac{EI\sqrt{3}}{2}\left\{\cos\alpha + \cos\left(2\phi + \frac{\pi}{3} - \alpha\right)\right\} \tag{63}$$

and this is balanced by the compensating voltage and current, as discussed before,

$$\left. \begin{array}{l} e' = E\sqrt{3}\,\cos\left(\phi - \left(\dfrac{\alpha}{2} + \dfrac{\pi}{12}\right)\right) \\[2mm] i' = I \cos\left(\phi - \left(\dfrac{\alpha}{2} + \dfrac{7\,\pi}{12}\right)\right) \end{array} \right\} \tag{64}$$

it is

$$p' = e'i'$$

$$= \frac{EI\sqrt{3}}{2}\cos\left(2\phi - \alpha - \frac{2\,\pi}{3}\right)$$

$$= -\frac{EI\sqrt{3}}{2}\cos\left(2\phi - \alpha + \frac{\pi}{3}\right) \tag{65}$$

thus,

$$p_0 = p + p'$$

$$= \frac{EI\sqrt{3}}{2}\cos\alpha,$$

thus balanced.

The balancing voltage (64),

$$e' = E\sqrt{3}\,\cos\left(\phi - \frac{\alpha}{2} - \frac{\pi}{12}\right),$$

lags behind the load voltage, e (62), by

$$\frac{\alpha}{2} + \frac{\pi}{4},$$

or by half the lag angle of the load, plus 45°.

If

$$\alpha = \frac{\pi}{6},$$

or 30° lag, it is

$$e' = E\sqrt{3}\,\cos\left(\phi - \frac{\pi}{6}\right) \tag{66}$$

thus the compensating voltage, e', is displaced in phase from the load voltage, e_{12} (62), by 60°, if the lag angle of the load is 30°, and in this case, the second phase of the three-phase system thus can be used as compensating voltage,

$$
\begin{aligned}
e_{13} &= e_1 - e_3 \\
&= E\sqrt{3}\,\cos\left(\phi - \frac{\pi}{6}\right) \\
&= e'.
\end{aligned}
$$

In the general case, for any lag angle, α, the compensating voltage (64) can be produced by the combination of the two-phase voltages, e_1 and e_3, as

$$e' = a_1 e_1 - a_2 e_3$$

similar as was discussed in the quarter-phase system.

The second phase, e_{13}, as compensating voltage, loaded by a reactor, balances the load of phase angle, $\alpha = \frac{\pi}{6}$, or 30°. For other angles of lag, either another phase angle of the balancing voltage is necessary, or, if using the same balancing voltage, the balance is incomplete.

Let thus:
the load

$$e_{12} = E\sqrt{3}\,\cos\left(\phi + \frac{\pi}{6}\right),$$

$$i = I\cos\left(\phi + \frac{\pi}{6} - \alpha\right),$$

be balanced by reactive load on the second phase,

$$e_{13} = E\sqrt{3}\,\cos\left(\phi - \frac{\pi}{6}\right),$$

$$i' = I\cos\left(\phi - \frac{2\pi}{3}\right),$$

it is:
power of the load,

$$
\begin{aligned}
p &= e_{12}i \\
&= \frac{EI\sqrt{3}}{2}\left\{\cos\alpha + \cos\left(2\phi + \frac{\pi}{3} - \alpha\right)\right\};
\end{aligned}
$$

balancing power,

$$
\begin{aligned}
p' &= e_{13}i' \\
&= -\frac{EI\sqrt{3}}{2}\cos\left(2\phi + \frac{\pi}{6}\right)
\end{aligned}
$$

thus, total power,

$$p_0 = p + p'$$

$$= \frac{EI\sqrt{3}}{2} \left\{ \cos \alpha + \cos \left(2\phi + \frac{\pi}{3} - \alpha \right) - \cos \left(2\phi + \frac{\pi}{6} \right) \right\}$$

$$= \frac{EI\sqrt{3}}{2} \left\{ \cos \alpha + \sin \left(\frac{\pi}{6} - \frac{\alpha}{2} \right) \cos \left(2\phi - \frac{\alpha}{2} + \frac{\pi}{4} \right) \right\};$$

and

$$q = \frac{\sin \left(\frac{\pi}{6} - \alpha \right)}{\cos \alpha}.$$

is the ratio of the remaining alternating component of power, to the constant power, and may be called the coefficient of unbalancing.

CHAPTER XVII

CIRCUITS WITH DISTRIBUTED LEAKAGE

172. If an uninsulated electric circuit is immersed in a high-resistance conducting medium, such as water, the current does not remain entirely in the "circuit," but more or less leaks through the surrounding medium. The current, then, is not the same throughout the entire circuit, but varies from point to point: the currents at two points of the circuit differ from each other by the current which leaks from the circuit between these two points.

Such circuits with distributed leakage are the rail return circuit of electric railways; the lead armors of cables laid directly in the ground; water and gas pipes, etc. With lead-armored cables in ducts, with railway return circuits where the rails are supported above the ground by sleepers, as in interurban roads, the leakage may be localized at frequently recurring points; the breaks in the ducts, the sleepers supporting the rails, etc., but even then an assumption of distributed leakage probably best represents the conditions. The same applies to low-voltage distributing systems, telephone and telegraph lines, etc.

The current in the conductor with distributed leakage may be the result of a voltage impressed upon a circuit of which the leaky conductor is a part, as is the case with the rail return of electric railways, or occurs when a cable conductor grounds on the cable armor, and the current thereby returns over the armor; or it may be induced in the leaky conductor, as in the lead armor of a single-conductor cable traversed by an alternating current; or it may enter the conductor as leakage current, as is the case in cable armors, gas and water pipes, etc., in those cases where they pick up stray railway return currents, etc.

When dealing with direct-current circuits, the inductance and the capacity of the conductor do not come into consideration except in the transients of current change, and in stationary conditions such a circuit thus is one of distributed series resistance and shunted conductance.

Inductance also is absent with the current induced in the cable armor by an alternating current traversing the cable conductor, and with all low- and medium-voltage conductors, with the commercial frequencies of alternating currents, the capacity effects are so small as to be negligible.

In high-voltage conductors, such as transmission lines, etc., in general, capacity and inductance require consideration as well as resistance and shunted conductance. This general case is fully discussed in "Theory and Calculation of Transient Electric Phenomena and Oscillations," and in "Electric Discharges, Waves and Impulses," more particularly in the fourth section of the former book.

173. Let, then, in a conductor having uniformly distributed leakage, or in that conductor section, in which the leakage can be considered as approximately uniformly distributed,

r = resistance per unit length of conductor (series resistance),
g = leakage conductance per unit length of conductor (shunted conductance),

and assume, at first, that no e.m.f. is induced in this conductor.

The voltage, de, consumed in any line element, dl, of this conductor, then is that consumed by the current, i, in the series resistance of the line element, rdl, thus:

$$de = irdl. \qquad (1)$$

The current, di, consumed in any line element, dl, that is, the difference of current between the two ends of this line element, then, is the current which leaks from the conductor in this line element, through the leakage conductance, gdl, thus:

$$di = egdl. \qquad (2)$$

Differentiating (2) and substituting into (1) gives

$$\frac{d^2i}{dl^2} = rgi. \qquad (3)$$

This equation is integrated by (see "Engineering Mathematics," Chapter II)

$$i = A\epsilon^{-al}. \qquad (4)$$

Substituting (4) into (3) gives

$$a^2A\epsilon^{-al} = rgA\epsilon^{-al}$$

hence,

$$a = \pm\sqrt{rg},$$

and thus, the *current*,

$$i = A_1\epsilon^{-\sqrt{rg}l} + A_2\epsilon^{+\sqrt{rg}l} \qquad (5)$$

where A_1 and A_2 are determined by the terminal conditions, as integration constants.

Substituting (5) into (2) gives as the *voltage*,

$$e = \sqrt{\frac{r}{g}} \left\{ A_1 \epsilon^{-\sqrt{rg}l} - A_2 \epsilon^{+\sqrt{rg}l} \right\} \qquad (6)$$

174. (*a*) If the conductor is of infinite length, that is, of such great length, that the current which reaches the end is negligible compared with the current entering the conductor, it is

$$i = 0 \text{ for } l = \infty.$$

This gives

$$A_2 = 0,$$

hence,

$$\left. \begin{aligned} i &= A\epsilon^{-\sqrt{rg}l} \\ e &= \sqrt{\frac{r}{g}} A\epsilon^{-\sqrt{rg}l} \\ &= \sqrt{\frac{r}{g}} i \end{aligned} \right\} \qquad (7)$$

That is:

A leaky conductor of infinite length, that is, of such great length that practically no current penetrates to its end, of series resistance, r, and shunted conductance, g, per unit length, has an effective resistance,

$$r_0 = \sqrt{\frac{r}{g}} \qquad (8)$$

It is interesting to note, that a change of r or g changes the effective resistance, r_0, and thus the current flowing into the conductor at constant impressed voltage, or the voltage consumed at constant-current input, much less than the change of r or g.

(*b*) If the conductor is open at the end $l = l_0$, it is

$$i = 0 \text{ for } l = l_0,$$

hence, substituted into (5)

$$0 = A_1 \epsilon^{-\sqrt{rg}l_0} + A_2 \epsilon^{+\sqrt{rg}l_0}$$

and, putting

$$A = A_1 \epsilon^{-\sqrt{rg}l_0} = -A_2 \epsilon^{+\sqrt{rg}l_0},$$

it is

$$\left. \begin{aligned} i &= A \left\{ \epsilon^{+\sqrt{rg}\,(l_0-l)} - \epsilon^{-\sqrt{rg}\,(l_0-l)} \right\} \\ e &= \sqrt{\frac{r}{g}} A \left\{ \epsilon^{+\sqrt{rg}\,(l_0-l)} + \epsilon^{-\sqrt{rg}\,(l_0-l)} \right\} \end{aligned} \right\} \qquad (9)$$

(c) If the conductor is grounded at the end $l = l_0$, it is

$$e = 0 \text{ for } l = l_0,$$

hence, substituted into (6),

$$0 = A_1 \epsilon^{-\sqrt{rg}\, l_0} - A_2 \epsilon^{+\sqrt{rg}\, l_0},$$

and, putting

$$A = A_1 \epsilon^{-\sqrt{rg}\, l_0} = A_2 \epsilon^{+\sqrt{rg}\, l_0}$$

it is

$$
\left.
\begin{aligned}
i &= A \{ \epsilon^{+\sqrt{rg}(l_0 - l)} + \epsilon^{-\sqrt{rg}(l_0 - l)} \} \\
e &= \sqrt{\frac{r}{g}} \, A \{ \epsilon^{+\sqrt{rg}(l_0 - l)} - \epsilon^{-\sqrt{rg}(l_0 - l)} \}
\end{aligned}
\right\}
\quad (10)
$$

(d) If the circuit, at $l = l_0$, is closed by a resistance, R, it is

$$\frac{e}{i} = R \text{ for } l = l_0,$$

hence, substituting (5) and (6), gives

$$\frac{A_1 \epsilon^{-\sqrt{rg}\, l_0} - A_2 \epsilon^{+\sqrt{rg}\, l_0}}{A_1 \epsilon^{-\sqrt{rg}\, l_0} + A_2 \epsilon^{+\sqrt{rg}\, l_0}} = \frac{R}{\sqrt{\dfrac{r}{g}}},$$

hence,

$$A_2 = A_1 \epsilon^{-2\sqrt{rg}\, l_0} \frac{\sqrt{\dfrac{r}{g}} - R}{\sqrt{\dfrac{r}{g}} + R}.$$

Thus,

$$
\left.
\begin{aligned}
i &= A \{ \epsilon^{-\sqrt{rg}\, l} - \frac{R - \sqrt{\dfrac{r}{g}}}{R + \sqrt{\dfrac{r}{g}}} \epsilon^{-(2 l_0 - l)\sqrt{rg}} \} \\[2em]
e &= \sqrt{\frac{r}{g}} \, A \{ \epsilon^{-\sqrt{rg}\, l} + \frac{R - \sqrt{\dfrac{r}{g}}}{R + \sqrt{\dfrac{r}{g}}} \epsilon^{-(2 l_0 - l)\sqrt{rg}} \}
\end{aligned}
\right\}
\quad (11)
$$

175. Substituting,

$$r_0 = \sqrt{\frac{r}{g}} \qquad (8)$$

as the "effective resistance of the leaky conductor of infinite length,"

and

$$a = \sqrt{rg} \qquad (12)$$

as the "attenuation constant" of the leaky conductor, it is

$$
\left.\begin{aligned}
i &= A\{\epsilon^{-al} - \frac{R - r_0}{R + r_0}\epsilon^{-a(2\,l_0-l)}\} \\[2mm]
e &= r_0 A\{\epsilon^{-al} + \frac{R - r_0}{R + r_0}\epsilon^{-a(2\,l_0-l)}\}
\end{aligned}\right\} \qquad (13)
$$

These equations (13) can be written in various different forms. They are interesting in showing in a direct-current circuit features which usually are considered as characteristic of wave transmission, that is, of alternating-current circuits with distributed capacity.

The first term of equations (13) may be considered as the outflowing components of current and voltage respectively, the second terms as the reflected components, and at the end of the circuit of distributed leakage, reflection would be considered as occurring at the resistance, R.

If $R > r_0$, the second term is positive, that is, partial reflection of current occurs, while the return voltage adds itself to the incoming voltage. If $R = \infty$, the reflection of current is complete.

If $R < r_0$, the second term is negative, that is, partial reflection of voltage occurs, while the return current adds itself to the incoming current. If $R = 0$, the reflection of voltage is complete.

If $R = r_0$, the second term vanishes, and equations (13) become those of (7), of an infinitely long conductor. That is:

A resistance, R, equal to the effective resistance, $r_0 = \sqrt{\dfrac{r}{g}}$, of the infinitely long conductor of distributed resistance and shunted conductance, as terminal of a finite conductor of this character passes current and voltage without reflection. A higher resistance partially reflects the current and increases the voltage, and a lower resistance partially reflects the voltage and increases the current. Infinite resistance gives complete reflection of current and doubles the voltage, while zero resistance gives complete reflection of voltage and doubles the current.

The term, $r_0 = \sqrt{\dfrac{r}{g}}$, thus takes in direct-current circuits the same position as the "surge impedance" $\sqrt{\dfrac{L}{C}}$ or $\sqrt{\dfrac{Z}{Y}}$ in alternating-current circuits.

176. Consider an instance: it has been proposed, for the purpose of effectively grounding the overhead ground wire used for protection of transmission lines, to run a bare underground conductor, a few feet below the ground surface, and to connect the overhead ground wire to the underground wire at every pole. Assuming the underground conductor to be a bare copper wire having 0.41 cm. diameter, the overhead ground wire a steel cable equivalent in conductivity to a copper wire of 0.52 cm. diameter. What is the effective ground resistance of the underground wire alone, what that of the underground and overhead wire together? Assuming the leakage resistance of the underground wire to be 3×10^{-3} mhos per meter?

The resistance of the underground wire is 1.3×10^{-3} ohms, that of the overhead ground wire is 0.82×10^{-3} ohms per meter.

The effective resistance of one underground wire then is

$$r_0 = \sqrt{\frac{r}{g}}$$

$r = 1.3 \times 10^{-3}$; $g = 3 \times 10^{-3}$, hence,

$$r_0 = 0.66 \text{ ohm}$$

thus, two underground wires in multiple, in the two different directions, give an effective ground resistance of

$$0.33 \text{ ohm,}$$

including the overhead ground wire, the resistance is

$$r = \cfrac{1}{\cfrac{1}{1.3 \times 10^{-3}} + \cfrac{1}{0.82 \times 10^{-3}}} = 0.5 \times 10^{-3}; g = 3 \times 10^{-3},$$

hence,

$$r_0 = 0.41 \text{ ohm,}$$

thus, the two underground and two overhead wires together give an effective resistance of

$$0.205 \text{ ohm.}$$

This is a very much lower ground resistance than most local grounds possess.

Assuming that i_0 is the current which enters this ground wire at one point, $l = 0$, then the equation of current distribution, by (7), is

$$r = 0.25 \times 10^{-3}; \quad g = 6 \times 10^{-3} \text{ (two in multiple, in}$$

the two opposite directions)

hence

$$r_0 = \sqrt{\frac{r}{g}} = 0.205$$

thus,

$$a = \sqrt{rg} = 1.225 \times 10^{-3}$$

$$i = i_0 \epsilon^{-1.225l}$$

$$e = 0.205\, i_0 \epsilon^{-1.225l}$$

where l is given in kilometers.

At various distances from the starting point, the current in the conductors thus is:

distance: 0. .	0.5	1.0	1.5	2.0	2.5	3.0	4.0	5 km.
$i \times i_0 \times 1.$.	0.54	0.294	0.155	0.086	0.046	0.025	0.0074	0.0021

As seen, beyond 2 km. distance, the current in the conductor is practically nothing.

177. If the current, i, is an alternating current, and the condition such that inductance and capacity are negligible, the equations (7), (9), (10), (11) and (13) remain the same, except that i, e and A are vector quantities, or general numbers: $\dot{I}$, $\dot{E}$, $\dot{A}$.

Considering thus the more general case, where a voltage is induced in the leaky conductor. Such for instance is the case in the lead armor of a single-conductor alternating-current cable. Let, then,

r = resistance per unit length,

g = shunted conductance per unit length,

$\dot{E}_0$ = voltage induced in the conductor, per unit length.

It is, then, in a line element, dl,

$$\frac{d\dot{E}}{dl} = r\dot{I} - \dot{E}_0 \tag{14}$$

$$\frac{d\dot{I}}{dl} = g\dot{E} \tag{15}$$

Differentiating (15) and substituting into (14) gives

$$\frac{d^2\dot{I}}{dl^2} = rg\left(\dot{I} - \frac{\dot{E}_0}{r}\right) \tag{16}$$

This is integrated by

$$\dot{I} - \frac{\dot{E}_0}{r} = \dot{A}\epsilon^{-al} \tag{17}$$

and by substituting (17) into (16), we get

$$a^2 = rg \tag{18}$$

hence, the *current*,

$$I = A_1\epsilon^{-al} + A_2\epsilon^{+al} + \frac{\dot{E}_0}{r} \tag{19}$$

where

$$a = +\sqrt{rg}$$

and A_1 and A_2 are complex imaginary integration constants.

Substituting (18) into (15) gives the *voltage*,

$$E = r_0\{A_1\epsilon^{-al} - A_2\epsilon^{+al}\} \tag{20}$$

where

$$r_0 = \sqrt{\frac{r}{g}} \tag{21}$$

178. Suppose now no voltage is impressed upon the conductor, but the only existing voltage is that induced in the conductor, as for instance the cable armor.

(a) Suppose the conductor is open at both ends: $l = +l_0$ and $l = -l_0$, having the length $2\ l_0$.

It then is

$$I = 0 \text{ for } l = \pm l_0$$

Substituting this in (19) gives

$$A_1\epsilon^{-al_0} + A_2\epsilon^{+al_0} + \frac{\dot{E}_0}{r} = 0$$

$$A_1\epsilon^{+al_0} + A_2\epsilon^{-al_0} + \frac{\dot{E}_0}{r} = 0$$

hence,

$$A_1 = A_2 = \frac{-\dot{E}_0}{r(\epsilon^{+al_0} + \epsilon^{-al_0})}$$

and

$$\left.\begin{array}{l} I = \frac{\dot{E}_0}{r}\left\{1 - \frac{\epsilon^{+al} + \epsilon^{-al}}{\epsilon^{-al_0} + \epsilon^{-al_0}}\right\} \\[3mm] E = \frac{r_0}{r}\dot{E}_0\left\{\frac{\epsilon^{+al} - \epsilon^{-al}}{\epsilon^{+al_0} + \epsilon^{-al_0}}\right\} \end{array}\right\} \tag{22}$$

in the center of the conductor, for $l = 0$, it is

$$I = \frac{\dot{E}_0}{r}\left\{1 - \frac{2}{\epsilon^{+al_0} + \epsilon^{-al}}\right\}$$

$$E = 0$$

at the ends of the conductor, for $l = \pm l_0$, it is

$$I = 0$$

$$E = \pm \frac{r_0}{r} E_0 \left\{ \frac{\epsilon^{+al_0} - \epsilon^{-al_0}}{\epsilon^{+al_0} + \epsilon^{-al_0}} \right\}$$

hence, if the conductor is long, so that ϵ^{-al_0} is negligible compared with ϵ^{+al_0}, it is

$$E = \pm \frac{r_0}{r} E_0 = \pm \frac{\dot{E}_0}{\sqrt{rg}}.$$

For an infinitely long conductor, $l_0 = \infty$, equations (22) become

$$\left. \begin{aligned} I &= \frac{\dot{E}_0}{r} \\ E &= 0 \end{aligned} \right\} \tag{23}$$

as was to be expected.

(b) Suppose the conductor is grounded at one end, $l = 0$, and open at the other end, $l = l_0$. It is, then,

$$E = 0 \text{ for } l = 0$$
$$I = 0 \text{ for } l = l_0,$$

hence, the equations are the same as (22). That is, a conductor grounded at one end and open at the other is the same as a conductor of twice the length, open at both ends.

A conductor grounded at both ends gives the same equation as an infinitely long conductor (23).

Suppose al_0 is large, so that ϵ^{-al_0} is negligible compared with ϵ^{+al_0}, in equation (22). Then for all values of l, except those very close to l_0, E and the exponential term of I are negligible.

That is, for the entire length of the leaky conductor, except very close to the ends, it is, approximately,

$$\left. \begin{aligned} I &= \frac{\dot{E}_0}{r} \\ E &= 0 \end{aligned} \right\} \tag{24}$$

Near the ends of the conductor, where l is near to l_0, ϵ^{-al} is negligible compared with ϵ^{+al}, and equations (22) thus assume the form,

$$\left. \begin{aligned} I &= \frac{\dot{E}_0}{r} \left\{ 1 - \epsilon^{-a(l_0 - l)} \right\} \\ E &= \frac{r_0}{r} \dot{E}_0 \epsilon^{-a(l_0 - l)} \end{aligned} \right\} \tag{25}$$

179. As an instance, consider the lead armor of a single-conductor cable, 10 km. long, carrying an alternating current such that it induces 60 volts per kilometer. The armor is open at either end, and of internal diameter of 4.2 cm., external diameter of 4.6 cm. The leakage conductance from the cable armor to ground is 1 mho per kilometer. What is the voltage and current distribution in the cable? What is it with 10 mhos. what with 0.1 mho per kilometer leakage conductance?

A lead section of the armor of $(2.3^2 - 2.1^2) \pi = 2.7$ cm.^{2}, at the specific resistance of lead $\rho = 19 \times 10^{-6}$, gives: $r = 0.7$ ohm per kilometer.

It is, then,

$$r = 0.7$$
$$g = 1$$
$$l_0 = 5$$
$$E_0 = 60$$

thus,

$$a = 0.84$$
$$r_0 = 0.84$$

hence,

$$\left. \begin{array}{l} I = 86\{1 - 0.015\,(\epsilon^{+0.84l} + \epsilon^{-0.84l})\} \\ E = 1.08\{\epsilon^{+0.84l} - \epsilon^{-0.84l}\} \end{array} \right\} \tag{26}$$

Thus the maximum current in the cable armor is, $l = 0: I = 83.4$ amp., and this current decreases very slowly, and is still, for $l = 2: I = 79$ amp.

The maximum voltage between cable armor and ground is, for $l = \pm 5: E = 72$ volts, and decreases fairly rapidly, being, for $l = \pm 4: E = 31.1$ volts.

If the cable is laid in very well-conducting soil,

$$g = 10$$

it is

$$a = 2.65$$
$$r_0 = 0.265.$$

$$\left. \begin{array}{l} I = 86\{1 - 1.75 \times 10^{-6}\,(\epsilon^{+2.65l} + \epsilon^{-2.65l})\} \\ E = 40 \times 10^{-6}\{\epsilon^{+2.65l} - \epsilon^{-2.65l}\} \end{array} \right\} \tag{27}$$

in this case, the current is practically constant, $I = 86$, and the voltage zero over the entire cable armor except very near the ends, where it rises to $E = 22.7$ volts for $l = 5$. Within 1 km. from the ends, or for $l = 4$, it is still: $I = 80; E = 1.6$. That is, over most of the length, the cable armor already acts as an infinitely long conductor.

Hence, for values of l near the end of the conductor, I and E are more conveniently expressed by the equation (25),

$$I = 86\{1 - \epsilon^{-2.65(5-l)}\}$$
$$E = 22.7\,\epsilon^{-2.65(5-l)} \qquad\qquad (28)$$

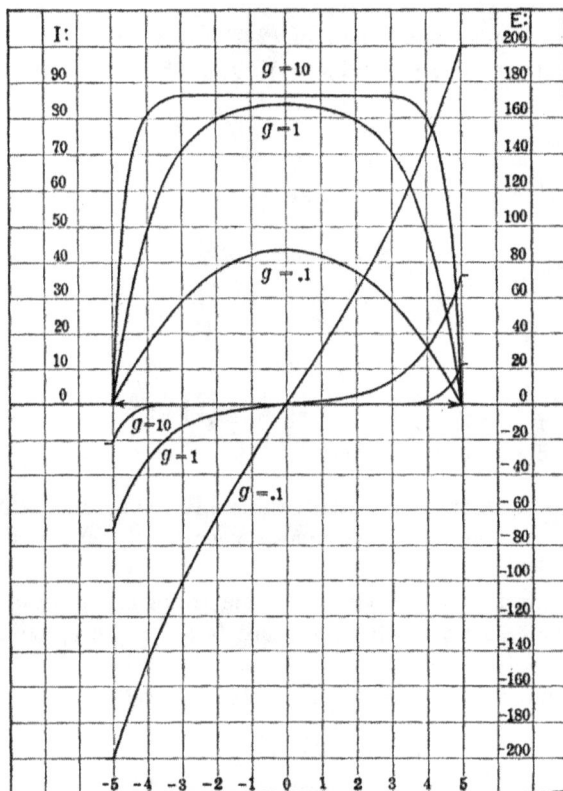

Fig. 128.

Inversely, if the cable is laid in ducts, which are fairly dry, and the leakage conductance thus is only

$$g = 0.1$$

it is

$$a = 0.265$$
$$r_0 = 2.65$$

hence,

$$I = 86 \left\{ 1 - 0.25 \left(\epsilon^{+0.265l} + \epsilon^{-0.265l} \right) \right\}$$
$$E = 57 \left\{ \epsilon^{+0.265l} - \epsilon^{-0.265l} \right\} \tag{29}$$

In this case, the maximum voltage between cable armor and ground is, at $l = \pm 5 : E = 200$.

As illustrations are shown, in Fig. 128, with l as abscissæ, the curves of I and E, calculated from equations (26), (28) and (29).

180. Considering now the case of a conductor, which is not connected to a source of voltage, nor has any voltage induced in it, but is laid in a ground in which a potential difference exists, due to stray currents passing through the ground. Such, for instance, may be a water pipe laid in the ground parallel with a poorly bonded railway circuit.

Assuming the potential difference, e_0, exists in the ground, per unit length of conductor. The conditions obviously are the same, as if the ground were at constant potential, and the potential difference, $-e_0$, existed in the conductor per unit length. Thus we get the same equations as (22) and (23). If the potential difference is continuous, as when due to a direct-current railway circuit, obviously the quantities I, E, A_1 and A_2 are not alternating vector quantities, but scalar numbers: i, e, etc. That is,

$$i = \frac{e_0}{r} \left\{ 1 - \frac{\epsilon^{+al} + \epsilon^{-al}}{\epsilon^{+al_0} + \epsilon^{-al_0}} \right\}$$
$$e = \frac{r_0}{r} e_0 \frac{\epsilon^{+al} - \epsilon^{-al}}{\epsilon^{+al_0} + \epsilon^{-al_0}} \tag{30}$$

Assuming thus as an instance a water pipe of 5 km. length: $l_0 = 5$, extending through a territory having 50 volts potential difference, or : $e_0 = 10$. Assuming that it is connected with the return circuit so that there is no potential difference at one end:

$$e = 0 \text{ for } l = 0.$$

Let the resistance of the water pipe be $r = 0.01$ ohm per kilometer, and the leakage conductance be $g = 10$ mhos per kilometer.

It is, then,

$$r = 0.01$$
$$g = 10$$
$$l_0 = 5$$
$$e_0 = 10$$

thus,

$$a = 0.316$$
$$r_0 = 0.0316$$

hence,

$$i = 1000 \{ 1 - 0.2 (\epsilon^{+0.316l} + \epsilon^{-0.316l}) \}$$
$$e = 6.2 \{ \epsilon^{+.316l} - \epsilon^{-0.316l} \} \tag{31}$$

hence, the maximum current, for $l = 0 : i = 600$ amp.
the maximum voltage, for $l = 5 : e = 28.8$ volts.

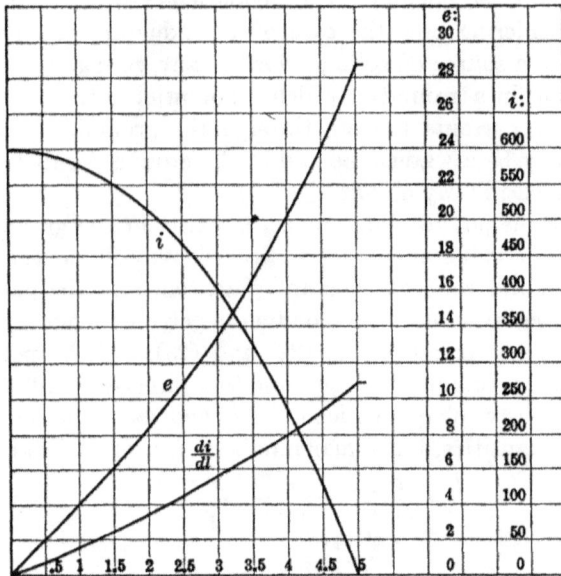

Fig. 129.

As seen, a very considerable current may flow under these conditions.

Fig. 129 shows, with l as abscissæ, the current, i, and voltage, e, and the current which enters the conductor per unit length, $\dfrac{di}{dl}$.

CHAPTER XVIII

OSCILLATING CURRENTS

Introduction

181. An electric current varying periodically between constant maximum and minimum values—that is, in equal time intervals repeating the same values—is called an alternating current if the arithmetic mean value equals zero; and is called a pulsating current if the arithmetic mean value differs from zero.

Assuming the wave as a sine curve, or replacing it by the equivalent sine wave, the alternating current is characterized by the period or the time of one complete cyclic change, and the amplitude or the maximum value of the current. Period and amplitude are constant in the alternating current.

A very important class are the currents of constant period, but geometrically varying amplitude; that is, currents in which the amplitude of each following wave bears to that of the preceding wave a constant ratio. Such currents consist of a series of waves of constant length, decreasing in amplitude, that is, in strength, in constant proportion. They are called oscillating currents in analogy with mechanical oscillations—for instance of the pendulum—in which the amplitude of the vibration decreases in constant proportion.

Since the amplitude of the oscillating current varies, constantly decreasing, the oscillating current differs from the alternating current in so far that it starts at a definite time and gradually dies out, reaching zero value theoretically at infinite time, practically in a very short time, short usually even in comparison with the time of one alternating half-wave. Characteristic constants of the oscillating current are the period, T, or frequency, $f = \dfrac{1}{T}$, the first amplitude and the ratio of any two successive amplitudes, the latter being called the decrement of the wave. The oscillating current will thus be represented by the product of a periodic function, and a function decreasing in geometric proportion with the time. The latter is the exponential function, $A^{f-\varrho t}$.

343

182. Thus, the general expression of the oscillating current is

$$I = A^{f-gt} \cos (2 \pi ft - \theta).$$

Since
$$A^{f-gt} = A^f A^{-gt} = i\epsilon^{-bt},$$

where ϵ = basis of natural logarithms, the current may be expressed,

$$I = i\epsilon^{-bt} \cos (2 \pi ft - \theta) = i\epsilon^{-a\phi} \cos (\phi - \theta),$$

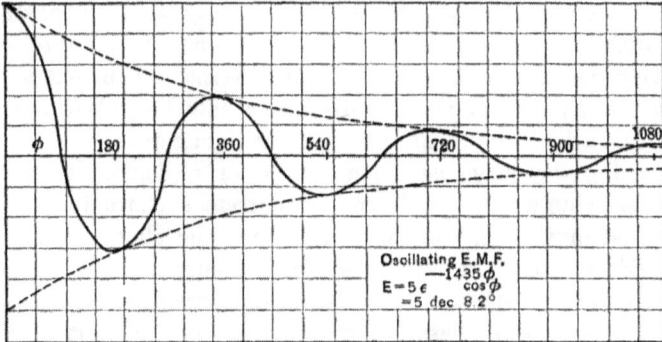

Fig. 130.

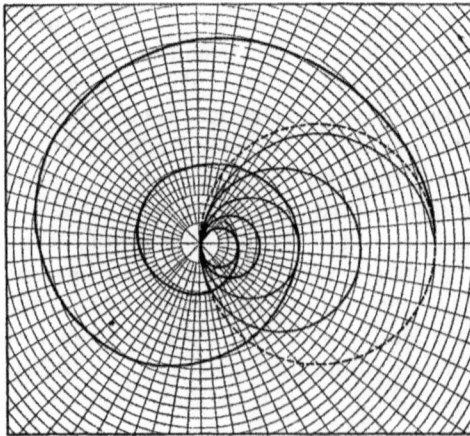

Fig. 131.

where $\phi = 2 \pi ft$; that is, the period is represented by a complete revolution.

In the same way an oscillating e.m.f. will be represented by

$$E = e\epsilon^{-a\phi} \cos (\phi - \theta).$$

Such an oscillating e.m.f. for the values,

$$e = 5, \ a = 0.1435 \text{ or } \epsilon^{-2\pi a} = 0.4, \ \theta = 0,$$

is represented in rectangular coördinates in Fig. 130, and in polar coördinates in Fig. 131. As seen from Fig. 130 the oscillating wave in rectangular coördinates is tangent to the two exponential curves,

$$y = \pm e\epsilon^{-a\phi}$$

In polar coördinates, the oscillating wave is represented in Fig. 131 by a spiral curve passing the zero point twice per period, and tangent to the exponential spiral,

$$y = \pm e\epsilon^{-a\phi}.$$

The latter are called the envelopes of a system of oscillating waves. One of them is shown separately, with the same constants as Figs. 130 and 131, in Fig. 132. Its characteristic feature is: The angle which any concentric circle makes with the curve, $y = e\epsilon^{-a\phi}$, is

$$\tan \alpha = \frac{dy}{y d\phi} = -a,$$

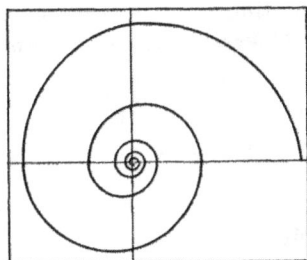 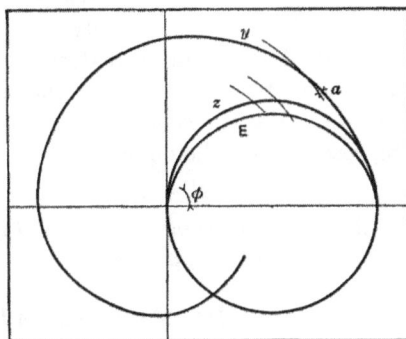

Fig. 132.   Fig. 133.

which is, therefore, constant; or, in other words: "The envelope of the oscillating current is the exponential spiral, which is characterized by a constant angle of intersection with all concentric circles or all radii vectores." The oscillating current wave is the product of the sine wave and the exponential or loxodromic spiral.

183. In Fig. 133 let $y = e\epsilon^{-a\phi}$ represent the exponential spiral;

let $\qquad\qquad z = e \cos (\phi - \theta)$

represent the sine wave;

and let $\qquad\qquad E = e\epsilon^{-a\phi} \cos (\phi - \theta)$

represent the oscillating wave.

We have then

$$\tan \beta = \frac{dE}{Ed\phi}$$

$$= \frac{-\sin(\phi - \theta) - a\cos(\phi - \theta)}{\cos(\phi - \theta)}$$

$$= -\{\tan(\phi - \theta) + a\};$$

that is, while the slope of the sine wave, $z = e\cos(\phi - \theta)$, is represented by

$$\tan \gamma = -\tan(\phi - \theta),$$

the slope of the exponential spiral, $y = e\epsilon^{-a\phi}$, is

$$\tan \alpha = -a = \text{constant},$$

that of the oscillating wave, $E = e\epsilon^{-a\phi}\cos(\phi - \theta)$, is

$$\tan \beta = -\{\tan(\phi - \theta) + a\}.$$

Hence, it is increased over that of the alternating sine wave by the constant, a.

The ratio of the amplitudes of two consequent periods is

$$A = \frac{E_{2\pi}}{E_0} = \epsilon^{-2\pi a}.$$

A is called the numerical decrement of the oscillating wave, a the exponential decrement of the oscillating wave, α the angular decrement of the oscillating wave. The oscillating wave can be represented by the equation,

$$E = e\epsilon^{-\phi \tan \alpha}\cos(\phi - \theta).$$

In the example represented by Figs. 130 and 131, we have $A = 0.4$, $a = 0.1435$, $\alpha = 8.2°$.

Impedance and Admittance

184. In complex imaginary quantities, the alternating wave,

$$z = e\cos(\phi - \theta).$$

is represented by the symbol,

$$E = e(\cos \theta - j\sin \theta) = e_1 - je_2.$$

By an extension of the meaning of this symbolic expression, the oscillating wave, $E = e\epsilon^{-a\phi}\cos(\phi - \theta)$, can be expressed by the symbol,

$$E = e(\cos \theta - j\sin \theta)\dec \alpha = (e_1 - je_2)\dec \alpha,$$

where $a = \tan \alpha$ is the exponential decrement, α the angular decrement, $\epsilon^{-2\pi a}$ the numerical decrement.

Inductance

185. Let r = resistance, L = inductance, and $x = 2\pi fL$ = reactance,
in a circuit excited by the oscillating current,

$$I = i\epsilon^{-a\phi}\cos(\phi - \theta) = i(\cos\theta + j\sin\theta)\,\text{dec }\alpha$$
$$= (i_1 + ji_2)\,\text{dec }\alpha,$$

where $i_1 = i\cos\theta$, $i_2 = i\sin\theta$, $a = \tan\alpha$.

We have then,
the e.m.f. consumed by the resistance, r, of the circuit,

$$E_r = rI\,\text{dec }\alpha.$$

The e.m.f. consumed due to the inductance, L, of the circuit,

$$E_x = L\frac{dI}{dt} = 2\pi fL\frac{dI}{d\phi} = x\frac{dI}{d\phi}.$$

Hence $E_x = -\,xi\epsilon^{-a\phi}\{\sin(\phi - \theta) + a\cos(\phi - \theta)\}$

$$= -\,\frac{xi\epsilon^{-a\phi}}{\cos\alpha}\sin(\phi - \theta + \alpha).$$

Thus, in symbolic expression,

$$E_x = -\,\frac{xi}{\cos\alpha}\{-\sin(\theta - \alpha) - j\cos(\theta - \alpha)\}\,\text{dec }\alpha$$
$$= -\,xi(a - j)\,(\cos\theta - j\sin\theta)\,\text{dec }\alpha;$$

that is, $E_x = -\,xI\,(a - j)\,\text{dec }\alpha.$

Hence the apparent reactance of the oscillating-current circuit is, in symbolic expression,

$$X = x(a - j)\,\text{dec }\alpha.$$

Hence it contains a power component, ax, and the impedance is

$$Z = (r - X)\,\text{dec }\alpha = \{r - x(a - j)\}\,\text{dec }\alpha = (r - ax + jx)\,\text{dec }\alpha.$$

Capacity

186. Let r = resistance, C = capacity, and $x_c = \dfrac{1}{2\pi fC}$ = condensive reactance. In a circuit excited by the oscillating current, I, the e.m.f. consumed due to the capacity, C, is

$$E_{x_c} = \frac{1}{C}\int I\,dt = \frac{1}{2\pi fC}\int I\,d\phi = k\int I\,d\phi;$$

or, by substitution,

$$E_{x_c} = x \int i\epsilon^{-a\phi} \cos(\phi - \theta) \, d\phi$$

$$= \frac{x}{1 + a^2} i\epsilon^{-a\phi} \{\sin(\phi - \omega) - \alpha \cos(\phi - \theta)\}$$

$$= \frac{xi\epsilon^{-a\phi}}{(1 + a^2) \cos a} \sin(\phi - \theta - \alpha);$$

hence, in symbolic expression,

$$E_{x_c} = \frac{xi}{(1 + a^2) \cos \alpha} \{-\sin(\theta + \alpha) - j \cos(\theta + \alpha)\} \text{ dec } \alpha$$

$$= \frac{xi}{1 + a^2} (a - j)(\cos \theta - j \sin \theta) \text{ dec } \alpha;$$

hence,

$$E_{x_c} = \frac{x}{1 + a^2} (-a - j) I \text{ dec } \alpha;$$

that is, the apparent capacity reactance of the oscillating circuit is, in symbolic expression,

$$X_c = \frac{x_c}{1 + a^2} (-a - j) \text{ dec } \alpha.$$

187. We have then:

in an oscillating-current circuit of resistance, r, inductive reactance, x, and condensive reactance, x_c, with an exponential decrement a, the apparent impedance, in symbolic expression, is,

$$Z = \left\{ r - x(a - j) + \frac{x_c}{1 + a^2} (-a - j) \right\} \text{ dec } \alpha$$

$$= \left\{ r - a\left(x + \frac{x_c}{1 + a^2}\right) + j\left(x - \frac{x_c}{1 + a^2}\right) \right\} \text{ dec } \alpha$$

$$= r_a + jx_a;$$

and, absolute,

$$z_a = \sqrt{r_a{}^2 + x_a{}^2}$$

$$= \sqrt{\left[r - a\left(x + \frac{x_c}{1 + a^2}\right)\right]^2 + \left[x - \frac{x_c}{1 + a^2}\right]^2}.$$

Admittance

188. Let

$$I = i\epsilon^{-a\phi} \cos(\phi - \theta) = \text{current}.$$

Then from the preceding discussion, the e.m.f. consumed by resistance, r, inductive reactance, x, and condensive reactance, x_c, is

$$E = i\epsilon^{-a\phi} \left\{ \cos(\phi - \theta)\left[r - ax - \frac{a}{1 + a^2}x_c\right] - \sin(\phi - \theta) \left[x - \frac{x_c}{1 + a^2}\right] \right\}$$

$$= iz_a\epsilon^{-a\phi} \cos(\phi - \theta + \delta),$$

where
$$\tan \delta = \frac{x - \dfrac{x_c}{1 + a^2}}{r - ax - \dfrac{a}{1 + a^2} x_c},$$

$$z_a = \sqrt{\left(x - \frac{x_c}{1 + a^2}\right)^2 + \left(r - ax - \frac{a}{1 + a^2} x_c\right)^2};$$

substituting $\theta + \delta$ for θ, and $e = iz_a$ we have
$$E = e\epsilon^{-a\phi} \cos(\phi - \theta),$$

$$I = \frac{e}{z_a} \epsilon^{-a\phi} \cos(\phi - \theta - \delta)$$

$$= e\epsilon^{-a\phi} \left\{ \frac{\cos \delta}{z_a} \cos(\phi - \theta) + \frac{\sin \delta}{z_a} \sin(\phi - \theta) \right\};$$

hence in complex quantities,
$$E = e(\cos \theta - j \sin \theta) \operatorname{dec} \alpha,$$

$$I = E \left\{ \frac{\cos \delta}{z_a} - j \frac{\sin \delta}{z_a} \right\} \operatorname{dec} \alpha;$$

or, substituting,

$$I = E \left\{ \frac{r - ax - \dfrac{a}{1 + a^2} x_c}{\left(x - \dfrac{x_c}{1 + a^2}\right)^2 + \left(r - ax - \dfrac{a}{1 + a^2} x_c\right)^2} \right.$$

$$\left. - j \frac{x - \dfrac{x_c}{1 + a^2}}{\left(x - \dfrac{x_c}{1 + a^2}\right)^2 + \left(r - ax - \dfrac{a}{1 + a^2} x_c\right)^2} \right\} \operatorname{dec} \alpha.$$

189. Thus in complex quantities, for oscillating currents, we have: conductance,

$$g = \frac{r - ax - \dfrac{a}{1 + a^2} x_c}{\left(x - \dfrac{x_c}{1 + a^2}\right)^2 + \left(r - ax - \dfrac{a}{1 + a^2} x_c\right)^2};$$

susceptance,

$$b = \frac{x - \dfrac{x_c}{1 + a^2}}{\left(x - \dfrac{x_c}{1 + a^2}\right)^2 + \left(r - ax - \dfrac{a}{1 + a^2} x_c\right)^2};$$

admittance, in absolute values,

$$y = \sqrt{g^2 + b^2} = \frac{1}{\sqrt{\left(x - \dfrac{x_c}{1 + a^2}\right)^2 + \left(r - ax - \dfrac{a}{1 + a^2} x_c\right)}^2};$$

in symbolic expression,

$$Y = g - jb = \frac{\left(r - ax - \dfrac{a}{1 + a^2} x_c\right) - j\left(x - \dfrac{x_c}{1 + a^2}\right)}{\left(x - \dfrac{x_c}{1 + a^2}\right)^2 + \left(r - ax - \dfrac{a}{1 + a^2} x_c\right)^2}.$$

Since the impedance is

$$Z = \left(r - ax - \frac{a}{1 + a^2} x_c\right) + j\left(x - \frac{x_c}{1 + a^2}\right) = r_a + jx_a,$$

we have

$$Y = \frac{1}{Z}; \ y = \frac{1}{z_a}; \ g = \frac{r_a}{z_a^2}; \ b = \frac{x_a}{z_a^2};$$

that is, the same relations as in the complex quantities in alternating-current circuits, except that in the present case all the constants, r_a, x_a, z_a, g, z, y, depend upon the decrement, a.

It is interesting to note that with oscillating currents, resistance as well as conductance have a negative term added, which depends on the decrement a. Such a negative resistance represents energy production, and its meaning in the present case is, that with the decrease of the oscillating current and voltage, their stored magnetic and dielectric energy become available.

Circuits of Zero Impedance

190. In an oscillating-current circuit of decrement, a, of resistance, r, inductive reactance, x, and condensive reactance, x_c, the impedance was represented in symbolic expression by

$$Z = r_a + jx_a = \left(r - ax - \frac{a}{1 + a^2} x_c\right) + j\left(x - \frac{x_c}{1 + a^2}\right),$$

or numerically by

$$z = \sqrt{r_a^2 + x_a^2} = \sqrt{\left(r - ax - \frac{a}{1 + a^2} x_c\right)^2 + \left(x - \frac{x_c}{1 + a^2}\right)^2}.$$

Thus the inductive reactance, x, as well as the condensive reactance, x_c, do not represent wattless e.m.fs. as in an alternating-current circuit, but introduce power components of negative sign,

$$- ax - \frac{a}{1 + a^2} x_c;$$

that means, in an oscillating-current circuit, the counter e.m.fs. of self-induction is not in quadrature behind the current, but lags less than 90°, or a quarter period, and the charging current of a condenser is less than 90°, or a quarter period, ahead of the impressed e.m.f.

191. In consequence of the existence of negative power components of reactance in an oscillating-current circuit, a phenomenon can exist which has no analogy in an alternating-current circuit; that is, under certain conditions the total impedance of the oscillating-current circuit can equal zero:

$$Z = 0.$$

In this case we have

$$r - ax - \frac{a}{1 + a^2} x_c = 0; \quad x - \frac{x_c}{1 + a^2} = 0,$$

substituting in this equation,

$$x = 2 \pi f L; \quad x_c = \frac{1}{2 \pi f C};$$

and expanding, we have

$$a = \frac{1}{\sqrt{\dfrac{4L}{r^2 C} - 1}}$$

$$2 \pi f = \frac{r}{2L} \sqrt{\frac{4L}{r^2 C} - 1} = \frac{r}{2 aL}.$$

That is, if in an oscillating-current circuit, the decrement,

$$a = \frac{1}{\sqrt{\dfrac{4L}{r^2 C} - 1}},$$

and the frequency $f = \frac{r}{4 \pi aL}$, the total impedance of the circuit is zero; that is, the oscillating current, when started once, will continue without external energy being impressed upon the circuit.

192. The physical meaning of this is: If upon an electric circuit a certain amount of energy is impressed and then the circuit left to itself, the current in the circuit will become oscillating, and the oscillations assume the frequency, $f = \frac{r}{4 \pi aL}$, and the decrement,

$$a = \frac{1}{\sqrt{\dfrac{4L}{r^2 C} - 1}}.$$

That is, the oscillating currents are the phenomenon by which an electric circuit of disturbed equilibrium returns to equilibrium.

This feature shows the origin of the oscillating currents, and the means of producing such currents by disturbing the equilibrium of the electric circuit; for instance, by the discharge of a condenser, by make-and-break of the circuit, by sudden electro-static charge, as lightning, etc. Obviously, the most important oscillating currents are those in a circuit of zero impedance, representing oscillating discharges of the circuit. Lightning strokes frequently belong to this class.

Oscillating Discharges

193. The condition of an oscillating discharge is $Z = 0$, that is,

$$a = \frac{1}{\sqrt{\frac{4L}{r^2C} - 1}}, \quad 2\pi f = \frac{r}{2\,aL} = \frac{r}{2L}\sqrt{\frac{4L}{r^2C} - 1}.$$

If $r = 0$, that is, in a circuit without resistance, we have $a = 0$, $f = \dfrac{1}{2\pi\sqrt{LC}}$; that is, the currents are alternating with no decre-ment, and the frequency is that of resonance.

If $\dfrac{4L}{r^2C - 1} < 0$, that is, $r > 2\sqrt{\dfrac{L}{C}}$, a and f become imaginary; that is, the discharge ceases to be oscillatory. An electrical discharge assumes an oscillating nature only, if $r < 2\sqrt{\dfrac{L}{C}}$. In the case $r = 2\sqrt{\dfrac{L}{C}}$ we have $a = \infty$, $f = 0$; that is, the current dies out without oscillation.

From the foregoing we have seen that oscillating discharges —as for instance the phenomena taking place if a condenser charged to a given potential is discharged through a given circuit, or if lightning strikes the line circuit—are defined by the equation, $Z = 0$ dec α.

Since

$$I = (i_1 - ji_2)\, \text{dec } \alpha, \qquad E_r = Ir\, \text{dec } \alpha,$$

$$E_x = -xI(a - j)\, \text{dec } \alpha, \qquad E_{x_c} = \frac{x_c}{1 + a^2}\, I(-a - j)\, \text{dec } \alpha,$$

we have

$$r - ax - \frac{a}{1 + a^2}\, x_c = 0,$$

$$-x + \frac{x_c}{1 + a^2} = 0;$$

hence, by substitution,

$$E_{x_c} = xI\,(-a - j)\, \text{dec } \alpha.$$

The two constants, i_1 and i_2, of the discharge, are determined by the initial conditions—that is, the e.m.f. and the current at the time, $t = 0$.

194. Let a condenser of capacity, C, be discharged through a circuit of resistance, r, and inductance, L. Let $e =$ e.m.f. at the condenser in the moment of closing the circuit—that is, at the time $t = 0$ or $\phi = 0$. At this moment the current is zero—that is,

$$I = ji_2, \quad i_1 = 0.$$

Since $\qquad E_{x_c} = xI\,(-a-j)\,\text{dec }\alpha = e$ at $\phi = 0$,

we have $\quad xi_2\sqrt{1+a^2} = e$ or $i_2 = \dfrac{e}{x\sqrt{1+a^2}}.$

Substituting this, we have,

$$I = -j\frac{e}{x\sqrt{1+a^2}}\,\text{dec }\alpha, \qquad E_r = -je\frac{r}{x\sqrt{1+a^2}}\,\text{dec }\alpha,$$

$$E_x = \frac{e}{\sqrt{1+a^2}}\,(1+ja)\,\text{dec }\alpha, \quad E_{x_c} = -\frac{e}{\sqrt{1+a^2}}\,(1-ja)\,\text{dec }\alpha,$$

the equations of the oscillating discharge of a condenser of initial voltage, e.

Since

$$x = 2\pi f L,$$

$$a = \frac{1}{\sqrt{\dfrac{4L}{r^2C}-1}},$$

$$2\pi f = \frac{r}{2aL},$$

we have

$$x = \frac{r}{2a} = \frac{r}{2}\sqrt{\frac{4L}{r^2C}-1};$$

hence, by substitution,

$$I = -je\sqrt{\frac{C}{L}}\,\text{dec }\alpha, \quad E_r = -jer\sqrt{\frac{C}{L}}\,\text{dec }\alpha,$$

$$E_x = \frac{er}{2}\sqrt{\frac{C}{L}}\left(\sqrt{\frac{4L}{r^2C}-1}+j\right)\text{dec }\alpha,$$

$$E_{x_c} = -\frac{er}{2}\sqrt{\frac{C}{L}}\left(\sqrt{\frac{4L}{r^2C}-1}-j\right)\text{dec }\alpha,$$

$$a = \frac{1}{\sqrt{\dfrac{4L}{r^2C}-1}}, \quad f = \frac{r\sqrt{\dfrac{4L}{r^2C}-1}}{4\pi L},$$

the final equations of the oscillating discharge, in symbolic expression.

23

INDEX

A

Admittance, with oscillating currents, 348
Air gap in magnetic circuit reducing wave distortion, 145
Alloys, resistance, 2
Alternating component of power of general system, 317
current electromagnet, 95
magnetic characteristic, 51
Alternations by capacity inductance shunt to arc, 187
Aluminum cell as condenser, 10
Amorphous carbon resistance, 23
Annealing, magnetic effect, 78
Anode, 6
Anthracite, resistance, 23
Apparatus economy of constant potential, constant current transformation, 281
of monocyclic square, 276
of T connection, 265
Arc as alternating current power generator, 187
characteristics, 34
condition of stability on constant current, 173
on constant voltage, 169
conduction, 28, 31, 42
constants, 36
effective negative resistance, 191
equations, 35
as oscillator, 189
parallel operation on constant current, 175
shunted by capacity, 178, 184
and inductance, 184
by resistance on constant current, 172
singing and rasping, 188, 189
tending to unstability, 164
transient characteristic, 192
as unstable conductor, 167

A

Arcing ground on transmission lines, 199
Area of BH relation, 53
Armature flux of alternator, 233
reactance flux of alternator, 232
reaction of alternator, 236
Attenuation constant, leaky conductor, 334
of synchronous machine oscillation, 213

B

Balance of quarterphase system on singlephase load, 322
of singlephase load, 319
of threephase system on singlephase load, 325
of unbalanced power of system, 319
Bends in magnetic reluctivity curve, 49
Bismuth, diamagnetism, 77
Bridged gap in magnetic circuit, wave distortion, 148

C

Cable armor as circuit, 330
equation of induced current, 336
Capacity, 1
and inductance shunting circuit, 181
inductance shunt to arc producing alternations, 187
with oscillating current, 347
and reactance as wave screen, 154
in series regulating for constant current, 247
shunt to arc, 178, 184
to circuit, 178
Carbon, resistance, 21
Cathode, 6
Cell, 7

355

www.ingramcontent.com/pod-product-compliance
Lightning Source LLC
Chambersburg PA
CBHW022109210326
41521CB00028B/171

9 781603 863179